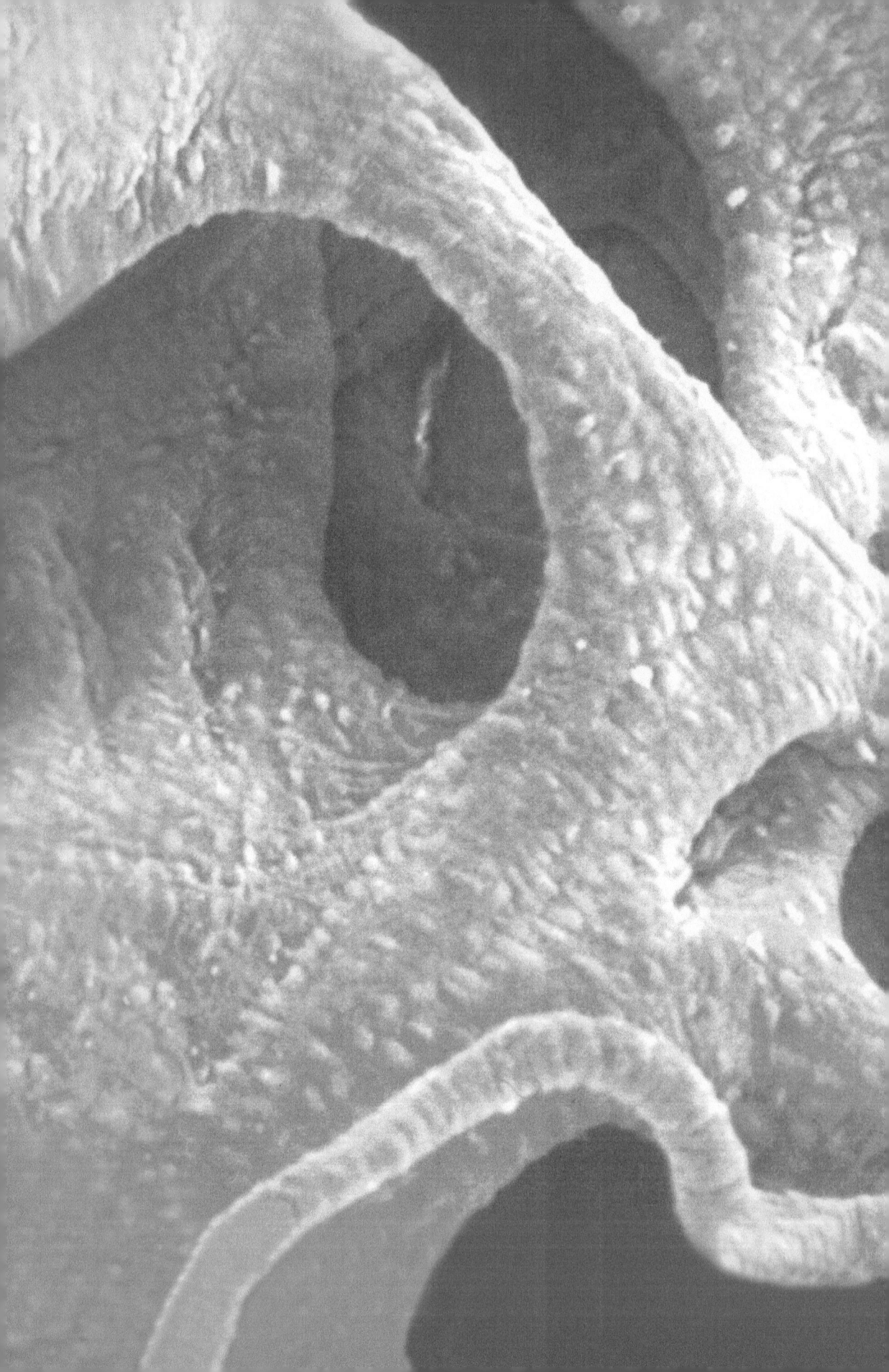

MEDICAL
INTELLIGENCE
UNIT

MOLECULAR BIOLOGY OF CARDIAC DEVELOPMENT AND GROWTH

Paul J.R. Barton, Ph.D.
Kenneth R. Boheler, Ph.D.
Nigel J. Brand, Ph.D.
Penny S. Thomas, D.Phil.

Molecular Biology Group
Department of Cardiothoracic Surgery
National Heart and Lung Institute
London, United Kingdom

Springer-Verlag Berlin Heidelberg GmbH

R.G. LANDES COMPANY
AUSTIN

MEDICAL INTELLIGENCE UNIT

MOLECULAR BIOLOGY OF CARDIAC DEVELOPMENT AND GROWTH

R.G. LANDES COMPANY
Austin, Texas, U.S.A.

Submitted: July 1995
Published: September 1995

Please address all inquiries to the Publisher:
R.G. Landes Company, 909 Pine Street, Georgetown, Texas, U.S.A. 78626
or
P.O. Box 4858, Austin, Texas, U.S.A. 78765
Phone: 512/ 863 7762; FAX: 512/ 863 0081

U.S. and Canada

International Copyright © 1995 Springer-Verlag Berlin Heidelberg
Ursprünglich erschienen bei Springer-Verlag Berlin Heidelberg New York 1995
Softcover reprint of the hardcover 1st edition 1995

ISBN 978-3-662-22194-5   ISBN 978-3-662-22192-1 (eBook)
DOI 10.1007/978-3-662-22192-1

While the authors, editors and publisher believe that drug selection and dosage and the specifications and usage of equipment and devices, as set forth in this book, are in accord with current recommendations and practice at the time of publication, they make no warranty, expressed or implied, with respect to material described in this book. In view of the ongoing research, equipment development, changes in governmental regulations and the rapid accumulation of information relating to the biomedical sciences, the reader is urged to carefully review and evaluate the information provided herein.

Library of Congress Cataloging-in-Publication Data

Barton, Paul, Ph.D.
    Molecular biology of cardiac development and growth / Paul Barton,
        p. cm.
    Includes bibliographical references and index.

    1. Heart—Growth—Molecular aspects. 2. Heart—Diseases—Molecular aspects. I. Title.
        [DNLM: 1. Heart—growth & development. 2. Heart Diseases—genetics. 3.
    Contractile Proteins—genetics. 4. Myocardial Contraction—physiology. WG 202
    B293m 1995]
    QP114.M65B37 1995
    612.1'7—dc20
    DNLM/DLC
    for Library of Congress   95-31277
                                                                                    CIP

Publisher's Note

R.G. Landes Company publishes five book series: *Medical Intelligence Unit, Molecular Biology Intelligence Unit, Neuroscience Intelligence Unit, Tissue Engineering Intelligence Unit* and *Biotechnology Intelligence Unit.* The authors of our books are acknowledged leaders in their fields and the topics are unique. Almost without exception, no other similar books exist on these topics.

Our goal is to publish books in important and rapidly changing areas of medicine for sophisticated researchers and clinicians. To achieve this goal, we have accelerated our publishing program to conform to the fast pace in which information grows in biomedical science. Most of our books are published within 90 to 120 days of receipt of the manuscript. We would like to thank our readers for their continuing interest and welcome any comments or suggestions they may have for future books.

Deborah Muir Molsberry
Publications Director
R.G. Landes Company

CONTENTS

FOREWORD

The molecular biology of cardiac development and growth is a rapidly expanding and exciting area of research in which to work. It is also a complex one. This book attempts to summarize some of the current research and excitement in this field and is the product of a joint effort by the four co-authors over a period of time ending in June 1995. Our aim has been to incorporate a wide range of topics within the field of molecular biology as it relates to cardiac development and growth. Inevitably we concentrated individually on particular areas - PJRB/NJB, basic molecular biology (Chapter 1); PST, cardiac development (Chapter 2); PJRB, contractile protein genes (Chapter 3); NJB, regulation of gene transcription (Chapter 4); KRB, cardiac hypertrophy (Chapter 5); NJB/KRB, genetic diseases of the heart (Chapter 6) while contributing to other sections. We hope that the book will help to provide a basis for those wishing to enter the field as well as offering a review of current thinking for those with an established interest in cardiac development and growth.

Paul J.R. Barton, Ph.D.
Kenneth R. Boheler, Ph.D.
Nigel J. Brand, Ph.D.
Penny S. Thomas, D.Phil.

Molecular Biology Group
Department of Cardiothoracic Surgery
National Heart and Lung Institute
London
June 1995

ACKNOWLEDGMENTS

We would like to take this opportunity to thank Professor Sir Madgi Yacoub for his continued enthusiastic support for our research in the molecular biology of the heart. We thank the British Heart Foundation for financial support for our research and Mrs. Joanna Harwood for her sympathetic and patient attitude to our demands for secretarial assistance during the preparation of our manuscript.

CHAPTER 1

BASIC CONCEPTS IN MOLECULAR BIOLOGY

OVERVIEW

The term molecular biology is generally used to refer to basic and clinical research which involves the techniques of molecular cloning and genetic engineering. Within this rather broad range of interests three overlapping and complementary themes can be identified. Firstly, molecular biology can be thought of as a basic science in its own right. In this respect it refers to research into the molecular mechanisms which are involved in gene expression and its regulation. This area of research is of fundamental importance to the understanding of embryonic and fetal development which is characterized by precise temporal and spatial patterns of gene expression leading, ultimately, to the formation of specialized tissue types.[1] Secondly, molecular biology has provided the technological basis for rapid advances in the field of medical genetics.[2] The ability to identify and isolate individual genes has led to the characterization of gene defects involved in inherited disorders including cystic fibrosis,[3] Duchenne muscular dystrophy,[4] Marfan syndrome[5] and familial cases of hypertrophic cardiomyopathy[6,7] (see chapter 6). Thirdly, molecular biology provides powerful and versatile tools for examining and manipulating biological processes at the molecular level. DNA can be manipulated in vitro and reintroduced into cells in culture or into whole animals (using for example transgenic techniques), allowing the functional role of cellular components to be examined in vivo.

Genetic information is contained in the nucleus of the cell in the form of deoxyribonucleic acid (DNA).[8,9] The DNA is packaged into individual chromosomes in association with a variety of proteins called histones. This complex of DNA and proteins is called chromatin. DNA is a double-stranded linear polymer composed of four bases linked to

Molecular Biology of Cardiac Development and Growth, by Paul J.R. Barton, Kenneth R. Boheler, Nigel J. Brand, Penny S. Thomas. © 1995 R.G. Landes Company.

a deoxyribose sugar-phosphate backbone. The four bases are adenine (A), guanine (G), thymine (T) and cytosine (C). The two opposing strands of DNA are held together by non-covalent bonds between bases such that A always pairs with T and G always pairs with C (Fig. 1.1). This pattern of base-pairing ensures that one strand of DNA is a complementary copy of the other and underpins two important events in a cell: the expression of the genetic information contained in the DNA sequence, and the replication of DNA to ensure that during cell division each daughter cell receives identical copies of the parental cell's DNA. The sequence of nucleotides along the DNA molecule constitutes the genetic information of the organism and is organized into functional units or genes (Fig. 1.2). Most genes encode information for the production of a protein constituent of the cell, and the process by which this information gives rise to the corresponding gene product is referred to as gene expression.

THE PROCESS OF GENE EXPRESSION

A central question in molecular biology is how the use of genetic information is regulated. Cells differ from one another because they express different sets of genes. For example, terminal differentiation of skeletal muscle involves the expression of various muscle-specific genes including those encoding proteins of the contractile apparatus. Yet, with few exceptions, all cells of an organism contain exactly the same number of chromosomes and the same number of genes. This selec-

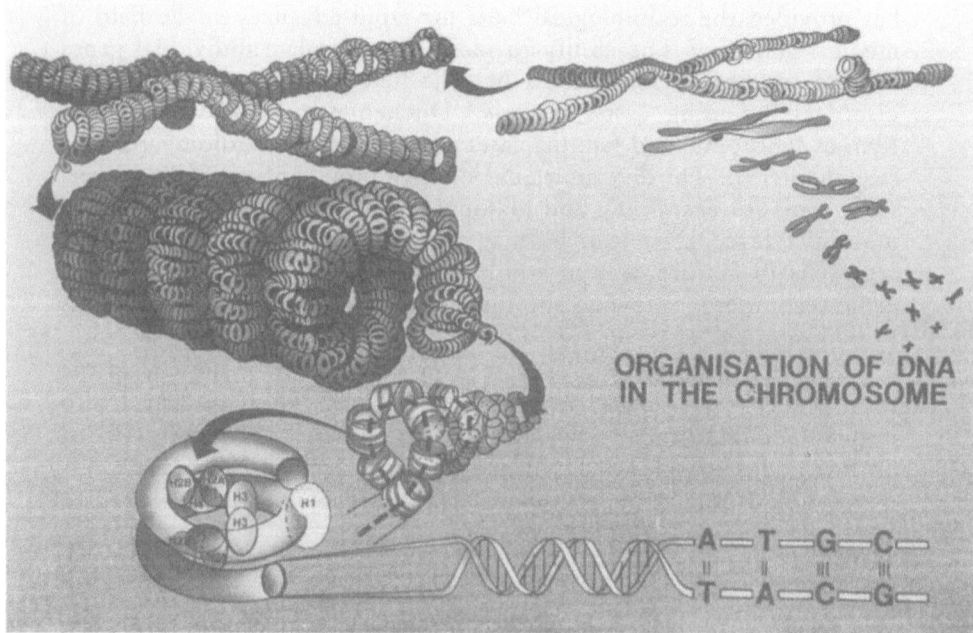

ORGANISATION OF DNA
IN THE CHROMOSOME

A—T—G—C—
T—A—C—G—

Fig. 1.1. Organization of DNA in the human chromosome. (From Barton, Annual Review of Cardiac Surgery 1990-91:3-12 with permission).

tive expression of genes is central to the process of cell differentiation whereby cells establish their identity during development and to the maintenance of the mature cell phenotype.

In order to address the question of how gene expression may be regulated, it is necessary to examine the molecular and biochemical events involved in the overall process of gene expression. Genetic information is encoded by the DNA sequence in the form of genes within

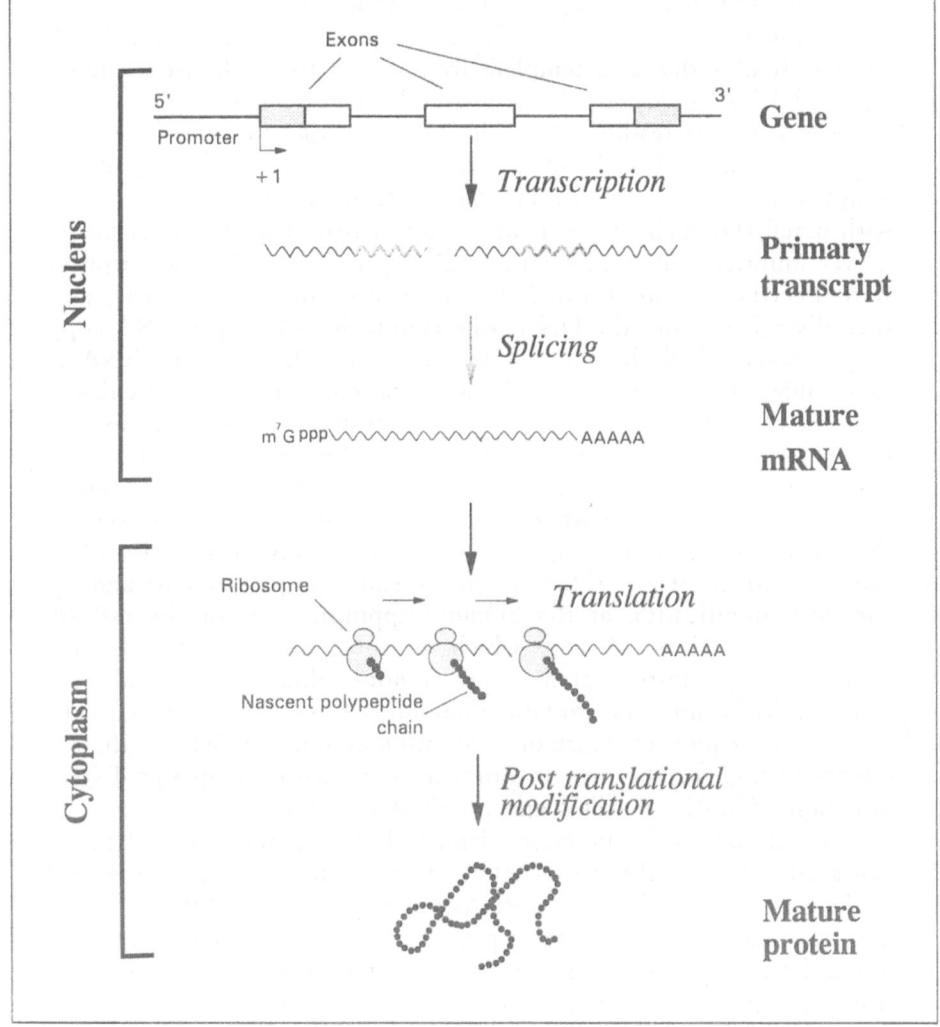

Fig. 1.2. The process of gene expression. From the top: structure of a simple gene; exons are shown as rectangles with stippled regions representing 5' and 3' UTR sequences, non-stippled regions representing protein coding sequences and lines between exons represent intronic sequences. Intron-derived sequences in the primary RNA transcript (thick wavy line) are removed by splicing. m⁷G^PPP is the chemical modification (cap) of the 5' end of the mature mRNA and AAAAA represents the 3' poly-A tail. Translation of the mRNA results in production of a nascent polypeptide (filled circles representing amino acid residues).

the nucleus of the cell. In order for this information to be expressed as protein products it must be transmitted to the cytoplasm of the cell where the machinery required for protein synthesis is located (Fig. 1.2). This requires copying the nucleotide sequence of the gene into an RNA molecule in the form of a messenger RNA (mRNA). Firstly, the linear sequence of nucleotides of the gene is converted into a direct RNA copy by the process of transcription. This is carried out by an enzyme called RNA polymerase II (RNA pol II). The enzyme passes along the DNA double helix, starting at a specific point (the transcription start-site or nucleotide +1) and "reads" the sequence of one of the DNA strands. It uses this as a template from which to synthesize a single-stranded RNA molecule which is complementary in sequence to the DNA strand it is reading. As with DNA, the specificity of base-pairing ensures that the RNA chain synthesized is a faithful copy of the template strand. Note, however, that in RNA thymine (T) is replaced with uracil (U) such that A pairs U and G pairs with C. RNA pol II moves unidirectionally along the entire gene, from the transcription start-site (the 5' end) through to the 3' extremity of the gene, and then dissociates from the DNA, whereupon the full-length RNA copy of the gene, called the primary transcript, is released. This RNA is then subject to various enzymatic modifications. The genes of eukaryotes are often referred to as split genes as the majority are divided into blocks (exons) separated by intervening sequence (introns). Exons contain the sequence information which is represented in the mature mRNA. Intron-derived parts of the primary transcript are removed by the process of RNA splicing. Further modifications include the addition of a string of A residues to the 3'-end (the poly-A tail) and by chemical modification at the 5'-end (capping). Parts of the mRNA located at the 5' and 3' ends which do not code for protein are referred to as the untranslated (UTR or non-coding) regions. The resulting mRNA molecule contains therefore a 5'-UTR, the AUG (ATG in the gene sequence) translation initiation codon (see below), the sequence coding for the protein product, a translation stop signal (termination codon), a 3'-UTR and a poly-A tail (Fig 1.2).

In addition to the processes described above, transcripts of many genes exhibit the ability to use alternative splicing pathways, resulting in the generation of different mRNAs put together from different exon-derived sequences from the same primary transcript.[10,11] This is an economical way for the cell to produce different isoforms of a protein from one gene. Alternative splicing can be achieved in a variety of ways (Fig. 1.3) giving rise to mRNAs which are identical except for the alternatively spliced region(s) and which encode protein isoforms which, therefore, differ only in the corresponding region(s). Alternative splicing pathways can be developmentally regulated and exhibit tissue-specificity, allowing the restriction of particular protein isoforms to a specific cell-type or a particular time-point during development.

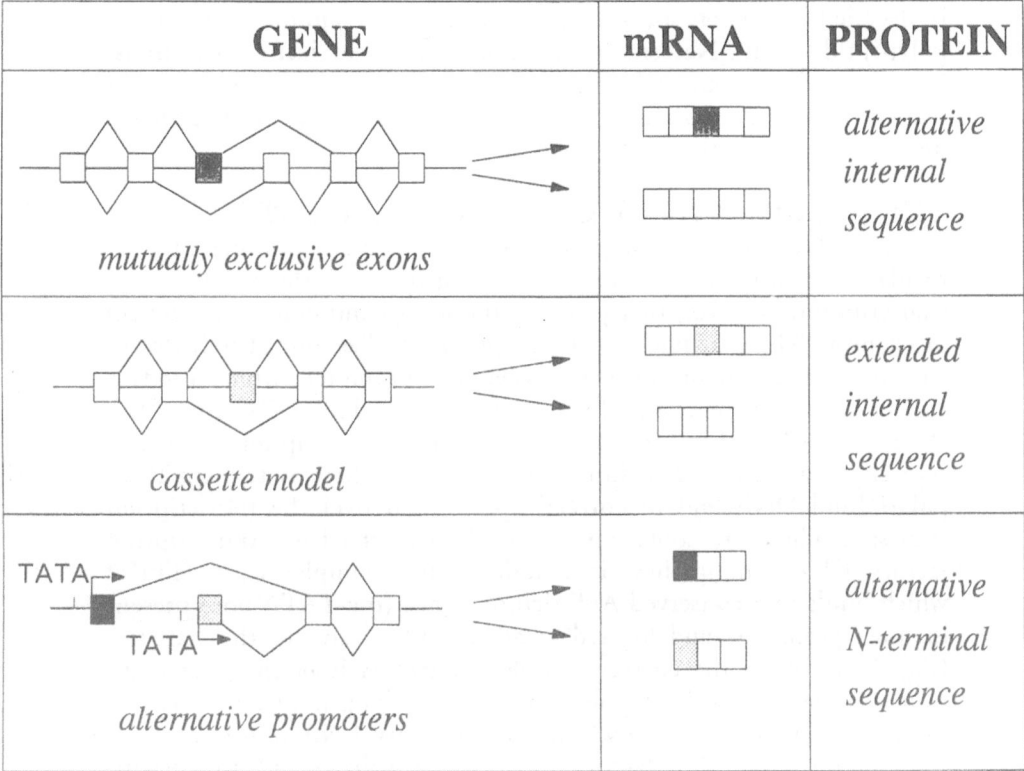

GENE	mRNA	PROTEIN
mutually exclusive exons		*alternative internal sequence*
cassette model		*extended internal sequence*
alternative promoters		*alternative N-terminal sequence*

Fig. 1.3. Alternative splicing of RNA. (From Barton, Molecular Biology of the Heart, Techniques and Applications. In: Julian et al. Disease of The Heart. 2nd Ed. in press.)

The mature mRNA is exported from the nucleus to the cytoplasm where it serves as template for the synthesis of the encoded protein product, a process called translation (Fig. 1.2). The mRNA is bound by a complex of proteins called the ribosome. The ribosome moves from the 5' end to the 3' end of the mRNA and synthesis of the polypeptide proceeds by deciphering the triplet code of the mRNA through the successive binding of adaptor RNA molecules called transfer RNAs (tRNA). These bring the correct amino acid to the ribosome such that during translation each triplet of bases of the coding part of the mRNA is recognized by a specific tRNA to which is attached the corresponding amino acid. Successive binding of tRNA molecules as each triplet is read leads to the orderly translation of the polypeptide encoded by the transcript, by the successive addition of amino acids to the growing polypeptide chain. The site of initiation of translation in the mRNA is signaled by the triplet AUG (ATG in the gene) encoding the initiator methionine. The end of translation is signaled by one of three different termination codons UAA, UAG or UGA (TAA, TAG and TGA respectively in the gene). The linear sequence of bases is decoded ultimately into a linear sequence of amino acids in the polypeptide chain. Nascent polypeptides may be subject to post-translational modification, such as the removal of a leader peptide, attachment of carbohydrate resides (glycosylation) or assembly into multimeric units, before becoming fully functional.

THE REGULATION OF GENE TRANSCRIPTION

Any of the steps leading from gene to protein product may be regulated in order to control gene expression. It is the regulation of transcription, however, that provides the most commonly used control mechanism. The start-point of transcription (+1) lies immediately downstream of a region of the gene called the promoter, which binds a variety of proteins necessary for transcription, including RNA pol II.[12-14] Proteins which bind to genes and regulate transcription are called transcription factors. The function of some of these is to help RNA pol II bind DNA and to correctly position it over the transcription start-site. These are sometimes referred to as the basal transcription factors. Chief among these is a multi-protein complex called TFIID which binds to a conserved A/T rich sequence (the TATA box) present in most promoters and located at about -30 relative to the start-site (Fig. 1.4). When the correct complex of factors is bound, transcription is initiated and RNA polymerase moves along the gene transcribing a linear sequence of DNA into the primary RNA transcript.

Other transcription factors are more specialized and are responsible in the main for stimulating high levels of transcription, often in specific cell types or at a particular point during development. There are distinct families of transcription factors, as described in the following section, all of which recognize specific DNA sequence and are encoded by related genes. The sequences they bind are often referred to as enhancers (Fig. 1.4) and will be described in more detail in chapter 4. Enhancers are usually present within or upstream of the promoter, but occasionally occur elsewhere such as within an intron. Once bound to DNA, such transcription factors interact with RNA polymerase II and other factors as shown in Figure 1.4, boosting, but sometimes suppressing, the rate at which transcription occurs. Transcription factors thus act as powerful molecular switches within the nucleus which serve to raise or lower the basal level at which a particular gene is transcribed. The flexibility of DNA means that the double-helix can loop back on itself, bringing factors bound to the promoter in close proximity to proteins bound to enhancers present upstream or elsewhere in the gene (see Fig. 1.4).

The nucleus of a cell contains many different transcription factors. The particular subset which interact on any particular gene will be determined by the DNA sequence elements (required for binding the factors) present in that gene. The combination of proteins which then bind to that gene and the effect of the interactions they make as

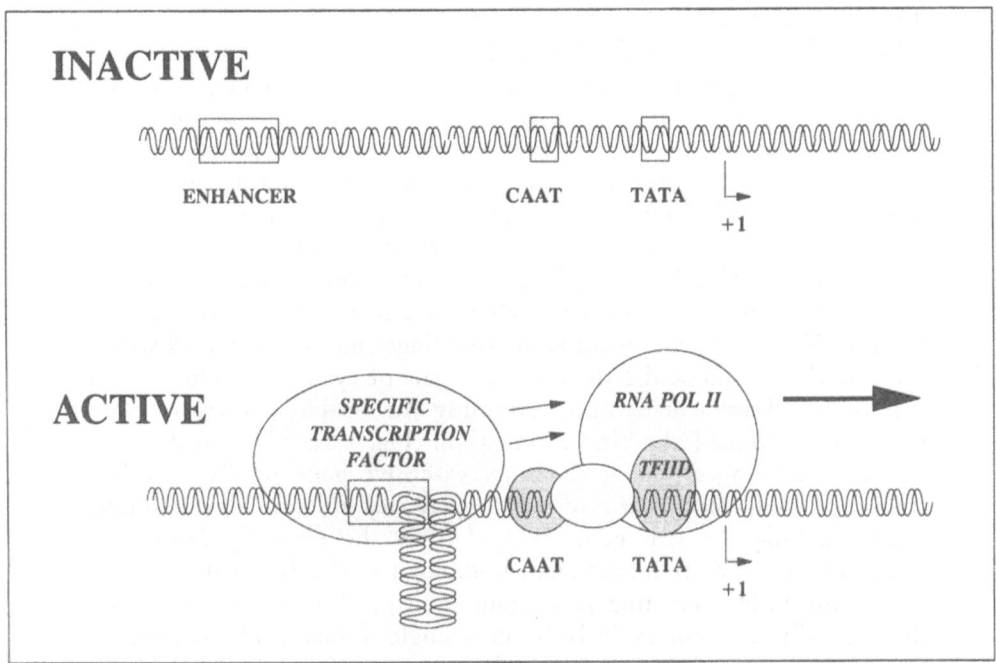

INACTIVE

ENHANCER   CAAT   TATA
                                                      +1

ACTIVE   SPECIFIC   RNA POL II
              TRANSCRIPTION
              FACTOR   TFIID

                              CAAT   TATA
                                                 +1

Fig. 1.4. Transcription proteins binding the promoter. The promoter contains binding sites, such as TATA and CAAT sequences, for transcription factors; in its inactive form (top) it does not bind factors. The active promoter (bottom) binds RNA pol II and a variety of proteins, including the TATA binding complex TFIID. High levels of transcription are driven through the binding of specific transcription factors to discreet DNA sequences called enhancers, which may be located at some distance from the promoter, even within the gene (see text). Small arrows, protein-protein interactions between specific factors and RNA pol II. +1, start-site of transcription.

a complex will determine whether transcription of that gene is essentially switched on or off. The genes encoding transcription factors are themselves usually expressed in a tissue-specific manner or at particular times in development and so gene expression can be regulated very accurately. The responsiveness of certain genes to hormones or retinoids can often be explained in this way.[15-17] For example, expression of the cardiac myosin heavy chain (MHC) genes is influenced by levels of circulating thyroid hormone.[18] This effect is mediated through the thyroid hormone receptor, itself a transcription factor. In the absence of hormone the thyroid hormone receptor is unable to bind to its cognate DNA sequence element, the thyroid hormone response element (TRE), located in the promoter of the MHC and other thyroid hormone responsive genes. In the presence of thyroid hormone, the receptor becomes activated and is able to bind tightly to the TRE. The bound receptor then interacts with RNA polymerase and basal transcription factors and has a direct influence on the transcriptional activity of these genes.

TRANSCRIPTION FACTORS BELONG
TO DISTINCT GENE FAMILIES

The majority of transcription factors identified to date may be classed into a few families based on their sequence similarity (homology) at the DNA and amino acid levels. Sequence similarity usually reflects domains within proteins with the same function, such as the formation of a DNA-binding cleft or a protein dimerization interface. This is rather like the conservation of key aspartate, histidine and serine residues which play key roles in catalysis in serine proteases. Some of the major families are shown in Table 1.1 and Figure 1.5. A common form of DNA-binding domain is the zinc finger motif. Conserved within the primary amino acid sequence are a pair of cysteine residues and a pair of histidines which, upon assuming the tertiary structure, come together to tetrahedrally bind a zinc atom (Fig. 1.5). This motif, called a C_2H_2 zinc finger, characterizes an extensive gene family encoding transcription factors, many of which play key roles in development. These include the gap gene *krüppel* from the fruit fly *Drosophila melanogaster*[19] and its mammalian homologues, the *krox* genes. Genes containing C_2H_2 zinc fingers contain multiple fingers linked in tandem, usually, but not exclusively, in a single domain. The number of fingers can range from as few as three to over 20. Recently a C_2H_2

Table 1.1. Major transcription factor gene families

Family	Key Domain	Domain Function	Examples
zinc finger[a]	C_2H_2 motif	DNA-binding	*krox* genes (mammals) *krüppel* (fly), TFIIIA (frog)
zinc twist[b]	C_4 motif	DNA-binding	steroid & thyroid hormone, retinoid receptors
homeoprotein	helix-turn-helix (HTH) homeodomain	DNA-binding	*Antennapedia* (fly) *HOX/Hox*[c] genes (mammals)
POU	POU-specific domain adjacent to a homeo-domain	DNA-binding	*Pit-1, Oct-1* (mammals) *unc-86* (nematode)
bHLH	basic region linked to helix-loop-helix (HLH)	basic region binds DNA, HLH mediates dimerization	MyoD, myogenin
bZIP	basic region linked to leucine zipper	basic region binds DNA, zipper mediates dimerization	*c-fos, c-myc* proto-oncogenes

[a] The zinc finger is a repeated DNA-binding motif in which a pair each of cysteines and histidines (C_2H_2) coordinate a zinc ion so that the amino acids between these two pairs loop out to form a "finger" that can interact with target DNA sequences (see ref. 4).

[b] Also known as a C_4 finger but structurally distinct from the (C_2H_2) motif, the zinc twist binds DNA via two copies of the C_4 motif (see refs. 21,22).

[c] Upper case denotes human and lower case mouse genes.

zinc finger factor with three fingers has been identified by molecular cloning which is expressed at high levels in cardiac muscle and regulates expression of the rat myosin light chain (MLC) 2V gene by binding to a specific sequence in the promoter.[20]

The thyroid hormone receptors and other members of the steroid hormone receptor family, which includes receptors for retinoids and vitamin D,[15-17] contain zinc-binding structures distinct from those belonging to the C_2H_2 class. The DNA-binding domain contains only two "fingers", each of which co-ordinates a zinc atom via four cysteine residues. Although often referred to as C_4 zinc fingers, the entire DNA-binding domain is more correctly termed a zinc "twist".[21] Structural studies on purified protein has revealed that the two fingers constitute a single protein domain, whereas each finger of a C_2H_2 protein is a separate DNA-binding entity.[21,22] All members of the steroid hormone

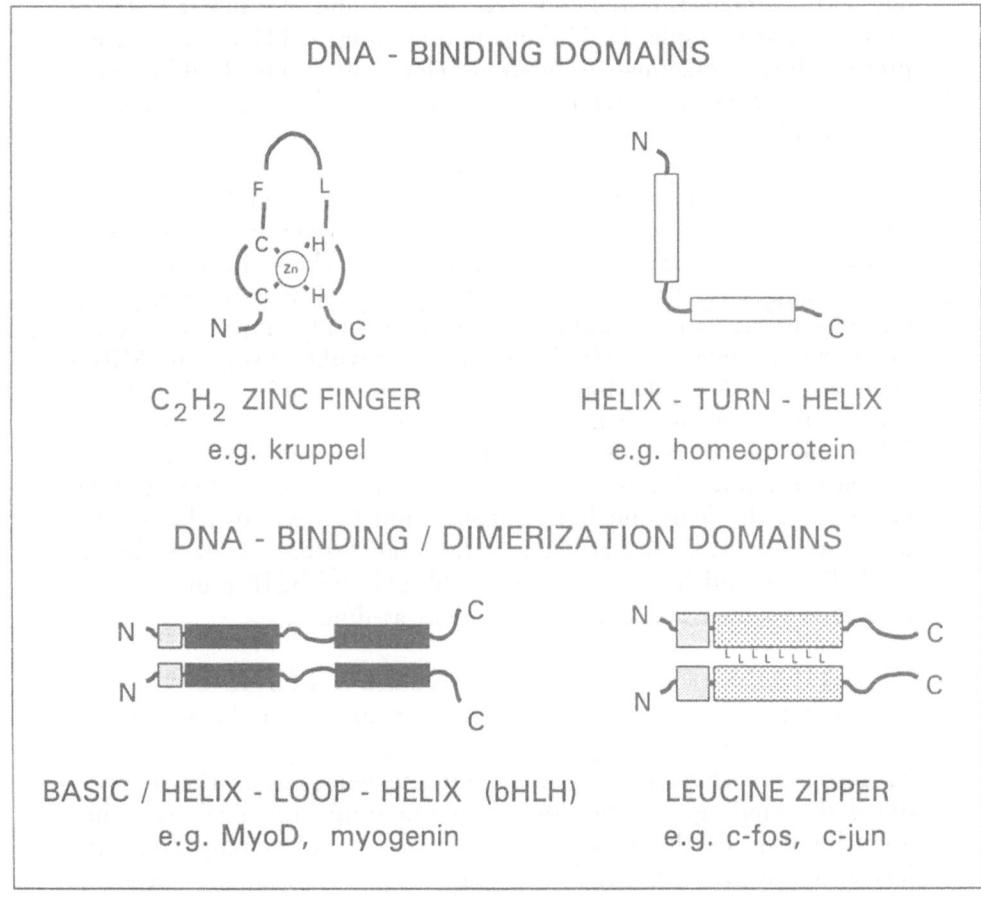

Fig. 1.5. Transcription factor DNA-binding domains.

receptor family are ligand-activated transcription factors, as described earlier for the regulation of cardiac MHC gene expression by thyroid hormone. This introduces another level of control to the process of gene expression: the receptor only binds DNA if first induced to do so by the presence of its cognate ligand within the cell.

Two other common DNA binding motifs are the helix-turn-helix (HTH) domain and the POU-domain. Two regions of α-helix, separated by a short turn of three amino acids placing one helix at right angles to the other, form the HTH domain (see Fig. 1.5). This ancient motif, present in lower eukaryotes such as yeast, is found in homeoproteins. These play key roles in the establishment of the three germ layers and the development of the body plan in *Drosophila*.[1] From comparison of cloned DNA sequences it is clear that some of these exist in vertebrates including man. Their pattern of expression suggests that they play major roles in the development of specific mammalian tissues including the central nervous system and skeleton.[23,24] The POU-domain is a well-conserved motif found in a subset of HTH proteins. For example, POU-domain containing HTH factors are expressed during mammalian brain development.[25] The POU-domain appears to be necessary for tight, sequence-specific binding of the adjacent HTH domain.

Other transcription factor families are characterized by domains which participate in protein dimerization as well as binding to DNA. Members of the basic/helix-loop-helix (bHLH) family bind DNA as dimers via an α-helical region rich in positively charged (basic) amino acid residues, and dimerize through the helix-loop-helix domain. This motif is present in a number of developmentally important factors including the myogenic regulatory factors MyoD, myogenin, MRF4 and *myf-5*.[26-32] A similar but distinct domain is seen in bZIP proteins such as the proto-oncogenes *c-fos* and *c-jun*. The basic region binds DNA and the adjacent leucine zipper, in which leucine residues occurs at every seventh position in an α-helix and are thus aligned along one face of the helix, mediates dimerization between two bZIP proteins (see Fig. 1.5). Indeed, it is a common feature of many factors (including steroid hormone receptors, bHLH and bZIP proteins) that they bind to their cognate DNA element as dimers.

IDENTIFYING PROTEIN-DNA INTERACTIONS

An important part of studying gene regulation is being able to identify the DNA sequences to which transcription factors bind. As we describe in chapter 4, this is essential to demonstrate the specificity of the binding and the effect that mutating individual bases has upon binding. Furthermore, it is a valuable way of gaining direct information on the nature and tissue distribution of a factor. The main techniques are DNaseI footprinting, methylation interference assay and band-shift assay.[33]

DNaseI Footprinting

Footprinting identifies regions of DNA that transcription factors bind to within a gene promoter or other sequence. A radiolabeled fragment of a particular gene sequence (e.g., a promoter fragment) is mixed with transcription factors (either purified or as a crude nuclear protein extract) and protein-DNA complexes allowed to form. These are then treated with the enzyme DNaseI, under conditions which result in one strand of DNA being nicked randomly. DNaseI cannot get access to DNA sequences which have proteins bound to them, due to steric hindrance, so those sequences will not be nicked. The digested fragments are separated by electrophoresis on a sequencing gel and run alongside a sample of naked DNA (i.e., no protein complexed to DNA). The naked DNA will contain molecules that have been nicked at every position along the DNA; the nucleotides these bands represent and their position in the overall DNA sequence may be identified by running a DNA sequencing ladder alongside. DNA that has been mixed with nuclear proteins prior to DNaseI treatment will have characteristic gaps in the ladder of bands which reflect sequences which were inaccessible to DNaseI because they had proteins bound to them (Fig. 1.6). This technique is a useful first step in identifying binding

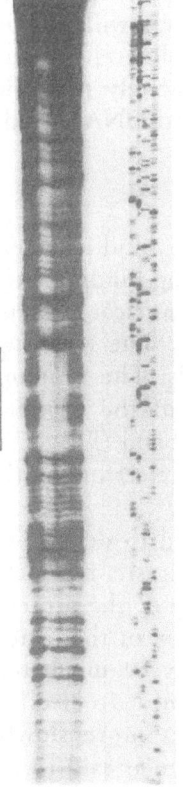

Fig. 1.6. DNaseI footprinting of a binding site for glucocorticoid receptor (GR) in the MMTV-LTR. The DNaseI footprint is shown on the left: open circles, footprints made with protein complexed to DNA, solid circles, no protein. The line indicates the region of the GR footprint in the lanes containing protein. A dideoxy sequencing ladder (GACT) is shown on the right. (Photograph courtesy of Dr. Ian Adcock).

●○○●   G A C T

sites for transcription factors within a gene. By comparing footprints obtained for one gene promoter using extracts from different cells or tissues, it is possible to determine which footprints are due to ubiquitously expressed factors (such as TFIID or other basal transcription factors) and which are due to tissue-specific factors.

INTERFERENCE ASSAYS

In contrast to footprinting this technique assays the interference which methylation of DNA exerts on the binding of protein.[33] The target DNA sequence is first end-labeled with ^{32}P and modified chemically by partial methylation of guanine residues such that on average one guanine per DNA molecule is methylated. This is then mixed with protein. DNA molecules which have bound protein are separated from those which have not (using electrophoresis) and the DNA recovered. The two samples are then subjected to chemical degradation with piperidine, which cleaves DNA at methylated G residues only. The resulting samples are run on a denaturing gel against a DNA sequence ladder for orientation. After autoradiography, bands will be apparent at all positions in the 'unbound' DNA sample. In the 'bound' DNA sample bands will be absent from those regions where protein binds (the absence of band(s) in a particular region results from the fact that protein will only bind to DNA molecules where methylation had not occurred in the protein-binding region; as the DNA will not be methylated at these sites it will not be cleaved). By labeling either one or the other of the DNA strands it is possible to separately map the sites of contact between bound protein on each of the two DNA strands (see Fig. 4.6).

BAND-SHIFT ASSAYS

Transcription factors bind tightly to their cognate binding sites. Radiolabeled double-stranded oligonucleotides containing binding sites are mixed with purified transcription factors or crude nuclear protein extracts and protein-DNA complexes allowed to form. The resulting samples are run on non-denaturing polyacrylamide gels. The protein-DNA complex migrates more slowly due to the size of the protein, resulting in a retardation or shift in the position of the band (Fig. 1.7). The assay is also referred to as a gel retardation assay or electrophoretic mobility shift assay (EMSA).[33]

Band-shift assays have many uses. They can be used to determine whether the protein which recognizes a particular binding-site is widely distributed or tissue-restricted, even though the identity of the protein or proteins binding the DNA is unknown. If the identity of the factor is known or suspected, the protein-DNA complex may be incubated with an antibody specific for that transcription factor prior to electrophoresis. The antibody-factor-DNA complex migrates more slowly through the gel than the factor-DNA complex alone and is said to be "super-shifted" as a consequence. Because the interaction of a transcription factor with its cognate binding site is highly specific, the radiolabeled binding site can be competed away from the protein-DNA complex if binding is allowed to occur in the presence of a large excess of unlabeled binding site (i.e., using the same oligonucleotide). Specificity of binding and the contribution made by individual base pairs can be determined by competing with unrelated sequence or with mutated sequence respectively.

MOLECULAR CLONING TECHNIQUES

Molecular cloning techniques allow individual mRNA and genes to be identified and isolated.[33] Both approaches make use of the fact that DNA can be propagated in bacteria in the form of self replicating units called vectors. In order for this to be achieved DNA is inserted into a vector molecule which carries all the genetic information

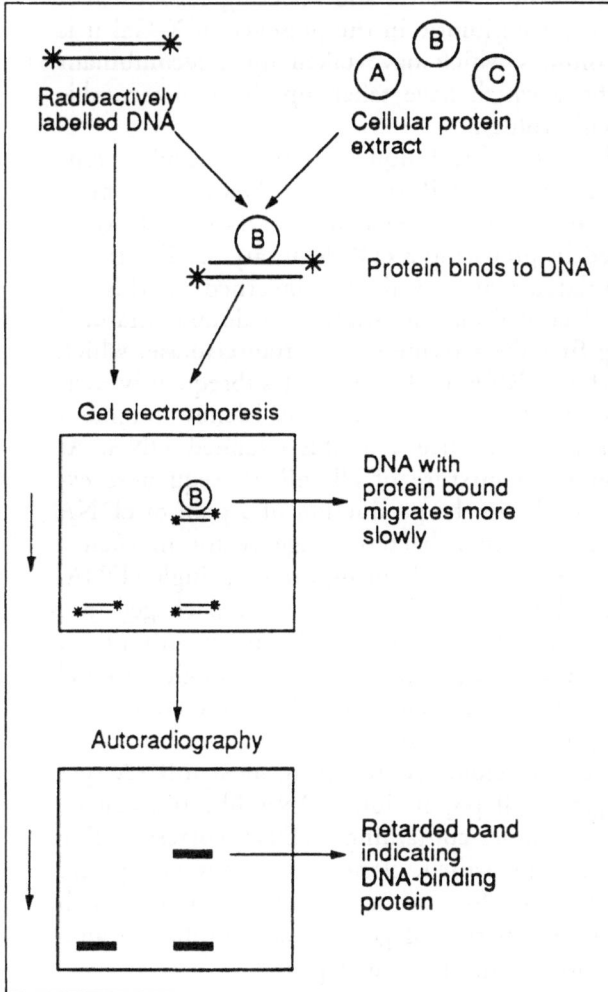

Radioactively labelled DNA

Cellular protein extract

Protein binds to DNA

Gel electrophoresis

DNA with protein bound migrates more slowly

Autoradiography

Retarded band indicating DNA-binding protein

Fig. 1.7. Band-shift assay. Top, a cellular (nuclear) protein extract containing three different transcription factors A, B and C is mixed with a double-stranded DNA cassette (asterisk denotes end-labeled with ^{32}P) bearing a binding site for factor B. Only B binds to the DNA. Middle, the protein-DNA complex is separated from free DNA on the basis of molecular mass by electrophoresis through a non-denaturing polyacrylamide gel. As a control, DNA that was not mixed with protein extract is run on the left. Bottom, the gel is autoradiographed to visualize the retarded band corresponding to the specific binding of factor B to the DNA in the sample that was mixed with protein. (From Dent and Latchman, Transcription Factors: a Practical Approach; 1993: 1-26. Reprinted by permission of Oxford University Press).

necessary for its stable replication in *Escherichia coli*. This is known as a recombinant DNA molecule because a fragment of foreign DNA has been recombined with a bacterial vector molecule to give an artificial hybrid molecule distinct from the parent vector. Most vectors also carry genetic traits which make them both useful and easy to handle. These include antibiotic resistance genes, which allow the selection of bacteria containing the vector on media containing the antibiotic. Most vectors contain a multiple cloning site (MCS) consisting of several unique sites recognized by restriction enzymes, allowing the insertion of DNA fragments prepared with the same enzyme. An enzymatic marker gene called *lacZ*, which encodes the bacterial enzyme β-galactosidase, is found in many vectors with the MCS placed within it. The enzyme is functional when the MCS is not disrupted by the insertion of foreign DNA fragment and enzyme activity can be detected using a colorless substrate such as X-Gal, which becomes blue upon cleavage by the enzyme. The enzyme is non-functional in a recombinant molecule. By transforming a DNA mixture into bacteria and selecting on solid media containing the appropriate antibiotic in the presence of X-Gal it is possible to distinguish colonies which have taken up a recombinant molecule (white) from those which have taken up the non-recombinant parent vector molecule (blue).

There are now a wide range of techniques for the molecular cloning of mRNA (see, for example, PCR techniques below) but most commonly this involves the use of complementary DNA (cDNA) libraries. The steps involved in preparing a cDNA library are illustrated in Figure 1.8. RNA is isolated from the tissue concerned, in this example cardiac muscle, and copied enzymatically into double stranded DNA. This is done using first the enzyme reverse transcriptase, which produces a DNA copy of the RNA molecule, and subsequently synthesizing the complementary DNA strand using the Klenow fragment of *E. coli* DNA polymerase to produce a double-stranded DNA. As the RNA source will contain a mixture of all mRNA sequences expressed in that tissue this results in the production of a pool of cDNA molecules. These are inserted into a bacteriophage vector in such a way as to ensure that each vector molecule incorporates a single cDNA. The resulting pool of recombinant cDNA molecules are packaged into bacteriophage particles, mixed with *E. coli* cells and subsequently plated on solid media such that each plaque results from a single original cDNA molecule (i.e., a true clone). The pool of cDNA clones obtained in this way are collectively referred to as a cDNA library. A good cDNA library will contain clones representing most mRNA species expressed in the original cell population. cDNA libraries can be screened in order to identify clones containing a cDNA corresponding to the gene of interest. Once identified, clones can be grown up and large quantities of DNA purified for use in subsequent analyses such as DNA sequencing or the production of gene-specific probes to analyze the pattern of expression of the particular gene.

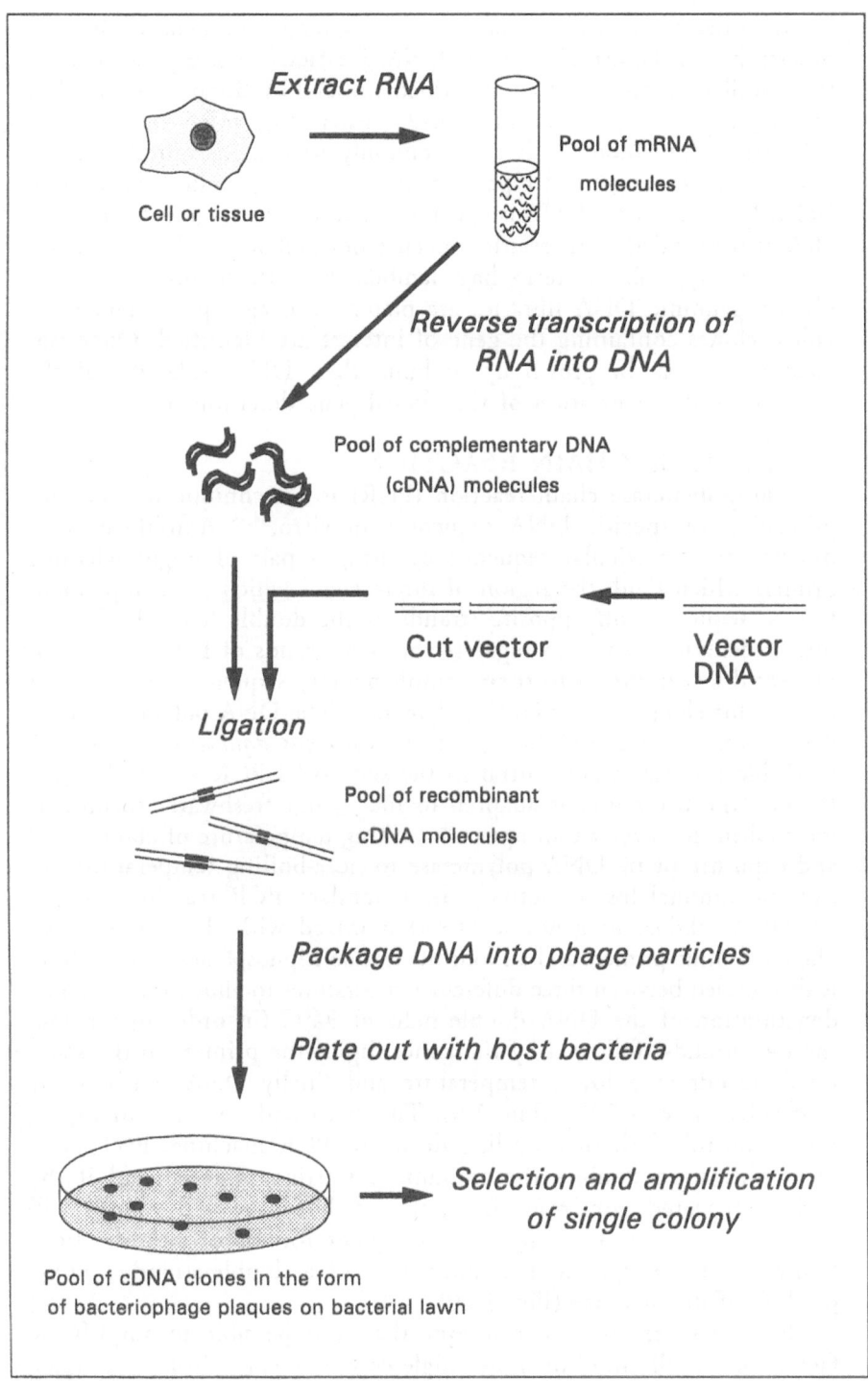

Fig. 1.8. Preparation of a cDNA library. (From Barton, Molecular Biology of the Heart, Techniques and Applications. In: Julian et al. Disease of The Heart. 2nd Ed. in press.)

In order to identify, isolate and characterize the gene itself, it is necessary to make use of genomic DNA libraries. These are constructed in a similar manner to that described for cDNA libraries above, but use fragments of chromosomal DNA rather than mRNA as starting material. White blood cells are commonly used as a source of chromosomal DNA as all cell types contain the same genetic information. Isolated chromosomal DNA is cut into suitably sized fragments (15,000-50,000 base pairs) with restriction enzymes and inserted into a cloning vector, typically bacteriophage lambda. As with the use of a cDNA library, genomic DNA libraries are plated onto agar plates and individual clones containing the gene of interest are identified. Once isolated clones can be grown up in bulk, their DNA isolated and the sequence and organization of the cloned gene determined.

POLYMERASE CHAIN REACTION

The polymerase chain reaction (PCR) is a technique for the amplification of specific DNA sequences in vitro.[34,35] Amplification is directed to a particular sequence by using a pair of oligonucleotide primers which flank the region of interest and which are complementary to sequences on opposite strands of the double helix. If the duplex is denatured so as to separate the two strands of DNA the oligonucleotides will anneal to their complementary sequences and serve as primers for elongation by DNA polymerase. The DNA polymerase used comes from a thermophilic bacterium *Thermus aquaticus* and its remarkable properties are central to the success of PCR as a technique. Because the bacterium is adapted to life in hot freshwater springs its metabolism has evolved an optimal working temperature of about 72°C and exposure of its DNA polymerase to near-boiling temperatures results in minimal loss of activity. In a standard PCR reaction the target DNA (cDNA or genomic DNA) is mixed with the enzyme, two oligonucleotide primers and all four nucleotide triphosphates. The mixture is then cycled between three different temperatures to allow, first, thermal denaturation of the DNA double helix at 94°C (in order to separate the two strands) followed by the annealing of the primers to the separated strands at a lower temperature and finally DNA synthesis by *Taq* polymerase at 72°C (Fig. 1.9). This is carried out in a microprocessor-controlled thermal cycling device or PCR machine. Each cycle of amplification doubles the amount of starting material and if the cycle is repeated numerous times, the newly-synthesized strands will themselves serve as templates in subsequent rounds of amplification, leading to the exponential accumulation of a double-stranded DNA product of defined size (Fig. 1.10).

PCR is so sensitive a technique that it is possible to amplify by factors of a millionfold or more single-copy sequences from a complex background of many unrelated sequences. This makes it ideal for cloning DNA from vanishingly small amounts of starting material. This has been shown, for example, by amplifying gene sequences from the genomic DNA extracted from a single hair root.[36] It is ideal for studying gene expression during development because it is possible to amplify specific gene sequences from small amounts of fetal or embryonic tissues.[37,38] The efficiency and reproducibility of the reaction make it suitable to many procedures in the clinical or pathology laboratory.[39] PCR allows rapid identification of genetic variation and is a major tool in forensic medicine ("DNA fingerprinting"). An early use of PCR was to improve methods of prenatal diagnosis of sickle-cell anaemia.[40] A variety of other genetic disorders can be studied by PCR, including β-thalassaemia, cystic fibrosis and Duchenne muscular dystrophy (DMD), all characterized by numerous, often single-base, changes in the gene sequence.[41] Tissue typing by PCR at the highly polymorphic HLA class II locus on chromosome 6 is used to match donor to recipient in heart transplantation. It provides a faster, more accurate analysis than conventional serotyping or mixed-lymphocyte culture in determining the best match and may contribute significantly to graft survival in the long term.[42] Chromosomal rearrangements or point mutations in proto-oncogenes are associated with many human cancers and PCR can be used to identify altered DNA sequences in such cases.[43,44]

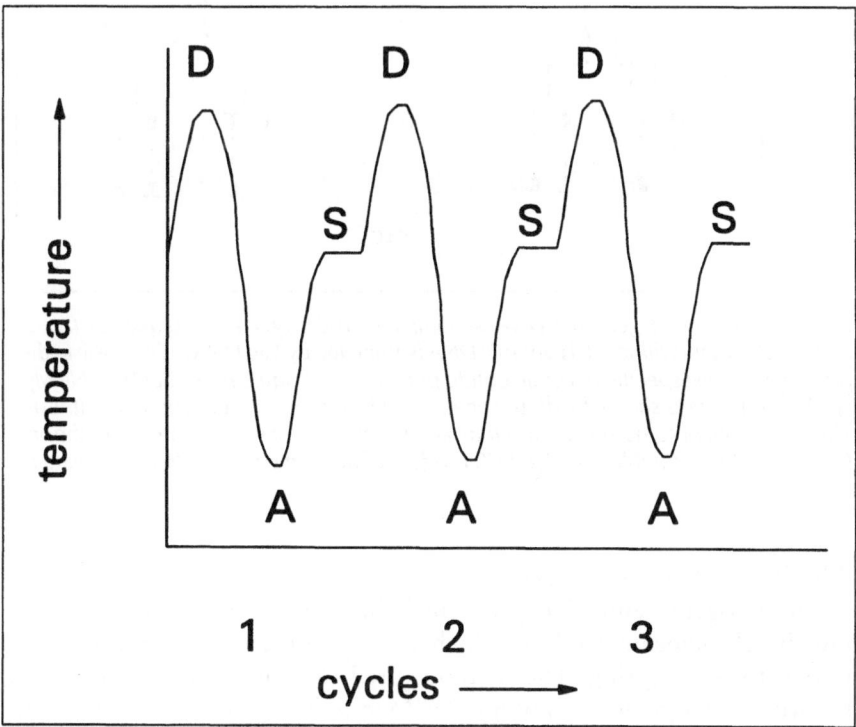

Fig. 1.9. The PCR cycle. Each PCR cycle has three steps: denaturation (D) at 94°C, followed by primer annealing (A) at a lower temperature, then DNA synthesis (S) at 72°C.

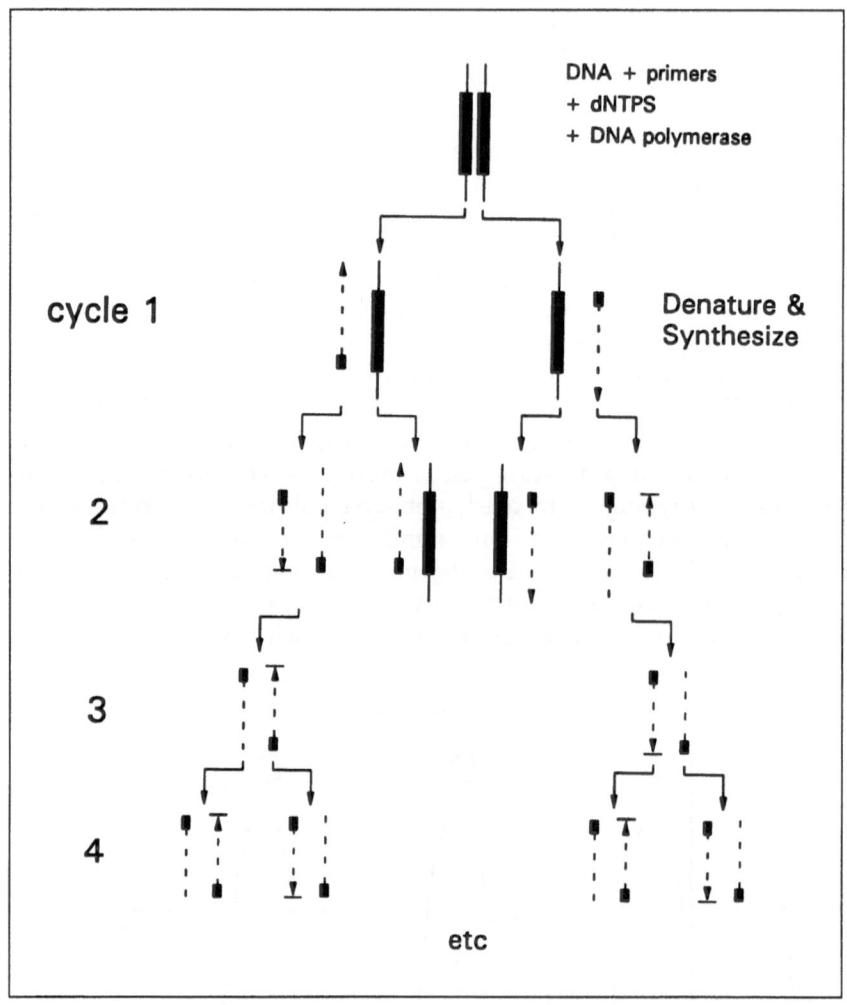

Fig. 1.10. PCR using genomic DNA as template. The target DNA sequence (black rectangle) present within total genomic DNA is amplified by Taq DNA polymerase in the presence of gene-specific oligonucleotide primers (black squares) and dNTPs. Newly-synthesized DNA is shown by dotted lines, arrowheads indicate direction of synthesis. The bar in front of some arrowheads indicates termination of synthesis from the shorter PCR templates. A double-stranded PCR product of defined size starts to accumulate in cycle 3.

TRANSGENIC ANIMALS

A transgenic animal is one which has foreign DNA incorporated into its chromosomal DNA, which is, as a consequence, stably transmitted to its offspring. The methods used to generate these animals is known as transgenic technology.[45,46] Manipulating the genetic makeup of animals by the introduction of foreign DNA into the developing embryo offers a powerful method for studying molecular and cellular processes during development and has the potential for generating animal models which mimic genetically inherited disorders. This approach is increasingly being applied to the study of the cardiovascular system.[47-49]

The laboratory mouse is the most commonly used animal for studies involving transgenic techniques, as knowledge of the genetics of mice is considerably more advanced than for other animals.[46] Similar techniques can, however, be applied to larger animals more suited to physiological studies on heart function and with contractile characteristics more similar to those of the human heart. Transgenic mice have most often been produced by micro-injection of DNA into the pronucleus of individual fertilized eggs although a variety of other approaches can be employed (see Fig 1.11). DNA becomes incorporated into the chromosomal DNA of the recipient embryo and will replicate along with the chromosomal DNA. All cells of the resulting animal (including the germ-line cells) will, therefore, contain identical copies of the foreign DNA. Breeding these animals allows the generation of a transgenic strain carrying the genetic characteristics of the introduced foreign DNA. For example, the foreign DNA may contain a marker gene under the control of a tissue-specific gene promoter which is normally active only in certain cells.

An important alternative approach to that of direct injection into fertilized eggs arose from the observation that cells derived from the inner cell mass of the early blastula embryo can be maintained in cell culture. These so-called embryonic stem (ES) cells can be re-introduced into foster embryos to produce chimeric animals (i.e., animals which are derived from two different cell populations). Foreign DNA can be introduced into embryonic stem cells before their introduction into foster embryos. If the embryonic stem cells contribute to the germ-line tissue of the resulting chimera, the introduced foreign DNA will be passed to the resulting progeny thereby creating a transgenic strain of animals. A major advantage of this approach is that the exact genetic makeup of the manipulated ES cells can be determined before they are transferred into the foster embryo, thereby allowing selection of cells where the DNA has integrated into the genome in a defined way. This has been developed to establish techniques which enhance the likelihood that the DNA will integrate at a defined site and which allow such events to be positively selected. In this way the site of integration of the foreign DNA can be directed to a region encoding a particular gene, resulting in disrupted function of that gene. Such gene targeting or gene "knock-out" experiments allow assessment of the role of the targeted gene and can provide models of genetic disease.

Transgenic techniques are finding a wide variety of applications in cardiovascular research. These include analyzing the role of regulatory genes in development (see chapter 2) and the analysis of regulatory regions of cardiac genes (see chapter 4). In other experiments the aim is to examine the pathophysiological effects of over-expression of the

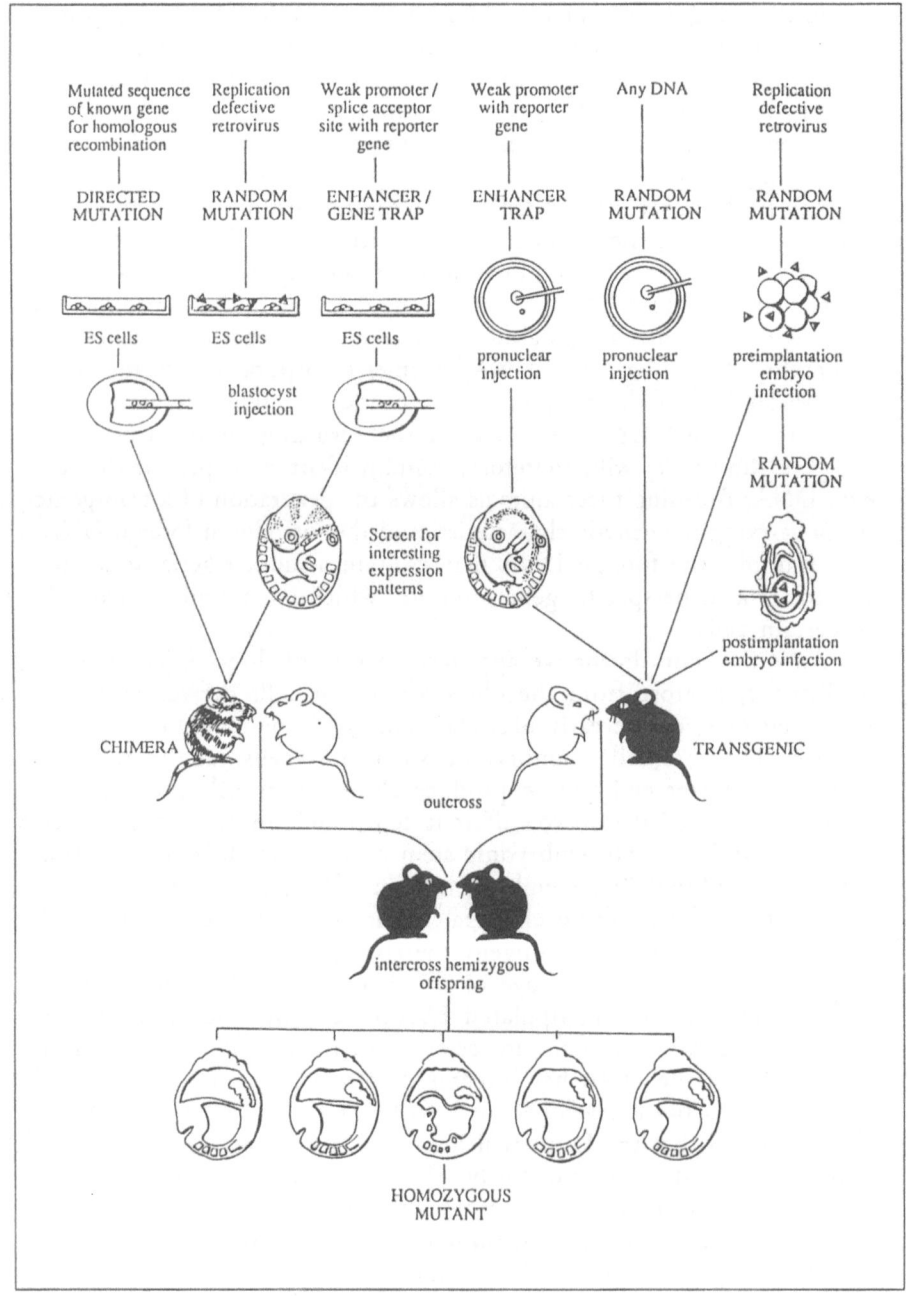

Fig. 1.11. Approaches for the preparation of transgenic animals. Numerous complementary approaches have been devised for making transgenic animals, some of which are illustrated in this figure. Top left; methods based on embryonic stem cell (ES) techniques in which DNA is transfected into ES cells and selected for integration at specific gene loci (homologous recombination) or at random (leading to random mutation and identification of new genes, known as gene trap). Top right; injection or infection of fertilized eggs or embryos. (From Beddington RSP, Transgenic Strategies in mouse embryology and develoment. In: Transgenic Animals. Grosveld F and Kollias eds London, Academic Press: 47-78.)

introduced gene. Strains of transgenic rats have been produced which carry the mouse renin gene and which, as a consequence of its over-expression, exhibit hypertension.[50] The developmental role of proto-oncogenes has also been addressed using transgenic techniques. In the case of the proto-oncogene *c-myc*, which plays a role in the control of cell growth and division in many cells, transgenic mice with constitutive *c-myc* expression in the heart have a significantly increased cardiac mass and cardiocyte cell number.[51] Transgenic mice expressing the *v-fps* proto-oncogene which encodes a protein tyrosine kinase, show cardio-vascular abnormalities including progressive atrophy and necrosis resulting in congestive heart failure[52] and may provide useful models for investigating molecular mechanisms underlying myocardial damage. Transgenic animals have also provided examples of myocardial tumors[53] which offer the possibility of generating cell lines derived from these transformed myocytes.[54,55]

REFERENCES

1. Gilbert SF. Developmental Biology, 3rd ed, Sunderland:Sinauer Associates, Inc. 1991.
2. Weatherall DJ. The New Genetics and Clinical Practice, 3rd ed, Oxford:Oxford University Press, 1991, 1-376.
3. Rommens JM, Iannuzzi MC, Kerem BS et al. Identification of the cystic fibrosis gene: chromosome walking and jumping. Science 1989; 245:1059-65.
4. Koenig M, Hoffman EP, Bertelson LJ et al. Complete cloning of the Duchenne muscular dystrophy (DMD) cDNA and preliminary genomic organization of the DMD gene in normal and affected individuals. Cell 1987; 50:509-17.
5. McKusick VA. The defect in Marfan syndrome. Nature 1991; 352:279-81.
6. Geisterfer-Lowrance AA, Kass S, Tanigawa G et al. A molecular basis for familial hypertrophic cardiomyopathy: a beta cardiac myosin heavy chain gene missense mutation. Cell 1990; 62:999-1006.
7. Thierfelder L, Watkins H, MacRae C et al. Alpha-tropomyosin and cardiac troponin T mutations cause familial hypertrophic cardiomyopathy: a disease of the sarcomere. Cell 1994; 77:701-12.
8. Lewin B. Genes V, Oxford:Oxford University Press, 1994.
9. Alberts B, Bray D, Lewis J, Raff M, Roberts K, Watson JD. Molecular Biology of the Cell, 2nd ed. New York:Garland Publishing Inc. 1989.
10. Breitbart RE, Andreadis, Nadal-Ginard B. Alternative splicing: A ubiquitous mechanism for the generation of multiple protein isoforms from a single gene. Annu Rev Biochem 1987; 56:467-95.
11. Smith CWJ, Patton JG, Nadal-Ginard B. Alternative splicing in the control of gene expression. Annu Rev Genet 1989; 23:527-77.
12. Mitchell PJ, Tjian R. Transcriptional regulation in mammalian cells by sequence-specific DNA binding proteins. Science 1989; 245:371-8.
13. Conaway RC, Conaway JW. General initiation factors for RNA polymerase II. Annu Rev Biochem 1993; 62:161-90.

14. Latchman DS. Eukaryotic Transcription Factors, London:Academic Press, 1992.
15. Evans RM. Steroid and thyroid hormone receptors as transcriptional regulators of development and physiology. Science 1988; 240:889-95.
16. Beato M. Gene regulation by steroid hormones. Cell 1989; 56:335-44.
17. Leid M, Kastner P, Chambon P. Multiplicity generates diversity in the retinoic acid signalling pathways. Trends Biochem Sciences 1992; 17:427-33.
18. Izumo S, Mahdavi V, Nadal-Ginard B. All members of the MHC multigene family respond to thyroid hormone in a highly tissue-specific manner. Science 1986; 231:231-597.
19. Gaul U, Jäckle H. How to fill a gap in the *Drosophila* embryo. Trends Genet 1987; 3:127-31.
20. Zhu H, Nguyen VT, Brown AB et al. A novel, tissue-restricted zinc finger protein (HF-1b) binds to the cardiac regulatory element (HF-1b/MEF-2) in the rat myosin light-chain 2 gene. Mol Cell Biol 1993; 13:4432-44.
21. Vallee BL, Coleman JE, Auld DS. Zinc fingers, zinc clusters and zinc twists in DNA-binding protein domains. Proc Natl Acad Sci U S A 1991; 88:999-1003.
22. Schwabe JWR, Rhodes D. Beyond zinc fingers: steroid hormone receptors have a novel structural motif for DNA recognition. Trends Biochem Sciences 1991; 16:291-6.
23. Kenyon C. If birds can fly, why can't we? Homeotic genes and evolution. Cell 1994; 78:175-80.
24. Krumlauf R. Hox genes in vertebrate development. Cell 1994; 78:191-201.
25. He X, Treacy MN, Simmons DM et al. Expression of a large family of POU-domain regulatory genes in mammalian brain development. Nature 1989; 340:35-42.
26. Davis RL, Weintraub H, Lassar HB. Expression of single transfected cDNA converts fibroblasts to myoblasts. Cell 1987; 51:987-1000.
27. Edmondson DG, Olson EN. A gene with homology to the *myc* similarity region of MyoD1 is expressed during myogenesis and is sufficient to activate the muscle differentiation program. Genes Devel 1989; 3:628-40.
28. Wright WE, Sassoon DA, Lin VK. Myogenin, a factor regulating myogenesis, has a domain homologous to MyoD. Cell 1989; 56:607-17.
29. Rhodes SJ, Konieczny SF. Identification of MRF4: a new member of the muscle regulatory factor gene family. Genes Devel 1989; 3:2050-61.
30. Braun T, Bober E, Winter B et al. *Myf-6*, a new member of the human gene family of myogenic determination factors: evidence for a gene cluster on chromosome 12. EMBO J 1990; 9:821-31.
31. Miner JH, Wold B. Herculin, a fourth member of the *MyoD* family of myogenic regulatory genes. Proc Natl Acad Sci USA 1990; 87:1089-93.
32. Braun T, Buschhausen-Denker G, Bober E et al. A novel human muscle factor related to but distinct from MyoD1 induces myogenic conversion in 10t1/2 fibroblasts. EMBO J 1989; 8:701-9.
33. Sambrook J, Fritsch EF, Maniatis T. Molecular Cloning: A Laboratory

Manual, 2nd ed, Cold Spring Harbor:Cold Spring Harbor Laboratory Press, 1989.

34. Arnheim N, Erlich H. Polymerase chain reaction strategy. Annu Rev Biochem 1992; 61:131-56.

35. Erlich H, Arnheim N. Genetic analysis using the polymerase chain reaction. Annu Rev Genet 1992; 26:479-506.

36. Higuchi R, von Beroldingen CH, Sensabaugh SA et al. DNA typing from single hairs. Nature 1988; 332:543-6.

37. Bugawan TL, Saiki RK, Levenson CH et al. The use of non-radioactive oligonucleotide probes to analyze enzymatically amplified DNA for prenatal diagnosis and forensic HLA typing. Biotechnology 1988; 6:943-7.

38. Coutelle C, Williams C, Handyside A et al. Genetic analysis of DNA from single human oocytes: a model for preimplantation diagnosis of cystic fibrosis. British Medical Journal 1989; 299:22-4.

39. Williamson R. Molecular genetics and the transformation of clinical chemistry. Clin Chem 1989; 35:2165-8.

40. Saiki R, Scharf S, Faloona F et al. Enzymatic amplification of β-globin genomic sequences and restriction site analysis for diagnosis of sickle cell anemia. Science 1985; 230:1350-4.

41. Brand NJ. Principles and applications of the polymerase chain reaction. In: Latchman DS ed: PCR Applications in Pathology. Oxford:Oxford University Press, 1995: 1-16.

42. Opelz G, Mytilineos J, Schever S et al. Survival of DNA HLA-DR typed and matched cadaver kidney transplants. Lancet 1991; 338:461-3.

43. Lyons J. The polymerase chain reaction and cancer diagnostics. Cancer 1992; 69:1527-31.

44. van Mansfeld ADM, Bos JL. PCR-based approaches for detectin of mutated ras genes. PCR Meths Applications 1992; 1:211-6.

45. Transgenic animals, Gosveld F, Kollias G, eds. London:Academic Press, 1992.

46. Hanahan D. Transgenic mice as probes into complex systems. Science 1989; 246:1265-57.

47. Field LJ. Cardiovascular research in transgenic animals. Trends Cardiovasc Med 1991; 1:141-6.

48. Wagner J, Zeh K, Paul M. Transgenic rats in hypertension research. J Hypertens 1992; 10:601-5.

49. Farza H, Barton P. Transgenic mice: Techniques and applications. In: Yacoub M, Pepper J eds: Annual of Cardiac Surgery. London:Current Science, 1992: 9-16.

50. Mullins JJ, Peters J, Ganten D. Fulminant hypertension in transgenic rats harbouring the mouse Ren-2 gene. Nature 1991; 344:541-4.

51. Jackson T, Allard MF, Sreenan CM et al. The *c-myc* proto-oncogene regulates cardiac development in transgenic mice. Mol Cell Biol 1990; 7:3709-16.

52. Yee SS, Mock D, Maltby V et al. Cardiac and neurological abnormalities in v-fps transgenic mice. Proc Natl Acad Sci U S A 1989; 86:5873-7.

53. Field LJ. Atrial natriuretic factor-SV40 T antigen transgenes produce tumors and cardiac arrhythmias in mice. Science 1988; 239:1029-33.

54. Delcarpio JB, Lanson NA, Field LJ et al. Morphological characterization of cardiomyocytes isolated from a transplantable cardiac tumour derived from transgenic mouse atria (AT-1 cells). Circ Res 1991; 69:1591-600.

55. Lanson NA, Glembotski CC, Steinhelper ME et al. Gene expression and atrial natriuretic factor processing and secretion in cultured AT-1 cardiac myocytes. Circulation 1992; 85:1835-41.

CHAPTER 2

CARDIAC DEVELOPMENT

The adult heart functions through a combination of properties: its muscle phenotypes, morphology, and integration in the circulatory system. From a very early ('tubular') stage in its morphological development, the heart must also pump blood to fulfill the requirements of the developing embryo. This function requires both temporal and regional differences in phenotype within the heart, requiring different patterns of gene expression. How are these controlled? Are development of regional muscular and morphological phenotypes co-ordinated? Are there transcription factors that control both?

Information about the hierarchy of control of gene expression can be gained by determining the pattern (both temporal and spatial) of transcription factor expression during development and relating it to that of other genes: those which the transcription factors are proposed to regulate directly and those involved in giving the cells the properties described above. This can be accomplished using techniques such as Northern blotting[†] and in situ hybridization[†], which can detect the presence of specific RNA sequences. At present, it is not known how many transcription factors are involved in cardiac development; the search for them, their identification and individual roles are essential and exciting areas of research described in more detail in chapter 4.

Antibodies are frequently used to detect the protein products of gene expression. As they detect specific epitopes (usually composed of a region of protein or sugar moiety), it is possible for the products of different, possibly unrelated gene products to be detected using the same antibody. Detection by antibody is not as direct a measure of transcriptional pattern as probes for RNA sequences but may be expected to correlate better with phenotypes such as contractile behavior or cell morphology. Thus immunocytochemistry helps to bridge the gap between what is known about specific genes and what is known about morphological or functional characteristics. Such evidence, although indirect, can contribute to models of control mechanisms at

[†] *See glossary*

Molecular Biology of Cardiac Development and Growth, by Paul J.R. Barton, Kenneth R. Boheler, Nigel J. Brand, Penny S. Thomas. © 1995 R.G. Landes Company.

the level of transcription, even when all the specific transcription factors involved remain unidentified.

There remains a further 'layer' of information that can help address the dual questions of transcriptional control of contractile function and that of morphology. This information is the 'non-molecular' phenotype of cardiac cells, both in normal development and following perturbation experiments such as teratogenesis and the culture of cardiac cells in vitro. The results of experiments investigating such questions as: 'Is there any evidence that cells can form a tube in the absence of developing a contractile phenotype? Can cells develop the range of contractile/conduction phenotypes found arranged along the heart tube when not arranged in this way?' have implications for the way in which the developmental processes occurring in the heart are controlled at the molecular level.

As well as the potential for 'direct' effecters (such as transcription factors) interlinking contractile properties with morphology, there are those that arise as a consequence of the heart being connected to a (developing) circulatory system containing blood: cells within it develop mechanisms that respond to being stretched both in contractile behavior and in growth. Thus physical properties of the circulatory system (blood pressure) can also affect development of the heart—which is itself part of that system. Investigating the development of the heart involves not only identifying individual molecular components and their effects in single, simple situations but determining how they contribute in the context of the whole heart, with its extra layers of regulation such as feedback control, physical properties, and contributions from outside the heart (such as innervation and hormones) and how all this follows from the properties of a gastrulating embryo. This chapter will consider events up to and including septation, what evidence there is for the role of transcriptional control in the development of regional morphological and functional events in the heart, and whether or not the two affect one another. A simple general description of development of a four-chambered heart, concentrating on morphological features, is followed by consideration of a number of morphological and functional events, and a discussion of the role of transcription factors in the onset of the cardiac phenotype and regionalization within the heart.

AN OVERVIEW OF HEART DEVELOPMENT

Terms such as 'atrial' or 'left ventricular' are used to describe regions of the developing heart prior to the presence of mature compartmental structure. Although generally coinciding with their normal fate it is important to remember that the full developmental potential of these cells has rarely been established. In the short description below, the same principle has been applied: cardiac precursor regions are those that include cells that fate-map to the heart prior to the appearance of, or collectively, myocardial and endocardial precursor populations (promyocardium and proendocardium) and so on. As well as location and physical appearance, reference to specific properties of the cells is given briefly and as accurately as possible, without implication as to the developmental or molecular status of the cells, nor to the function or complete composition of a particular part of the heart relative to the mature structure. Certain events are considered in more detail in the subsequent section, including the onset of the cardiac phenotype, aspects of myocardial development, and septation in the atrioventricular and outflow tract regions. For a more detailed general review of the morphogenesis, especially of the chicken heart, see Icardo and Manasek.[1]

Following ingression through the primitive streak, the cells from which the heart normally forms migrate to a position in the anterior[†] part of the mesodermal layer, where they eventually compose the splanchnic part of the lateral mesoderm. They are distributed symmetrically as a continuous population on each side of the embryo, in a crescent shape to which cells are gradually added posteriorly. The morphogenetic process by which cells from the anterior lateral mesoderm form a heart tube is illustrated in Figure 2.1. Cells of the lateral mesoderm on the left and right of the antero-posterior axis come into apposition and fuse and the adjacent endoderm layer forms the foregut tube. The cardiac precursor cells within the lateral mesoderm become promyocardium and proendocardium; these eventually form two concentric tubes, myocardium enclosing endocardium. During this process, the heart 'tube' (although not completely tubular) lies approximately in the midline of the embryo, parallel to the antero-posterior axis. The tube is formed gradually, the atrial (most caudal) region after the ventricular. Cranially, the cells are continuous with the mesoderm of the prospective pharyngeal arch region where the aortic sac forms. Caudally, the cells are continuous with mesoderm that comes to surround the endodermal tube and the developing venous sinus with the vessels draining the embryonic and extraembryonic circulation. Before this process is complete, the heart tube starts to become bent. As a result, the mid portion (ventricular) becomes displaced, so the tube makes an 's' shape (Fig. 2.2A), which becomes a 'c' shape as the more caudal (atrioventricular and atrial) part moves cranially, to a more dorsal position; finally, the tube resembles a loop or a 'u', with the atria and outflow tract adjacent (Fig. 2.2B).

[†]*Relative location along the long axis of the embryo (i.e., parallel to the eventual neural tube) is commonly given using anterior, rostral, cephalad or cranial, and posterior or caudal, although the embryo has no beak/face, head, skull or tail at some of the stages at which these terms are used. In this chapter, cranial/caudal will be used when referring to cells composing the heart, and anterior/posterior otherwise.*

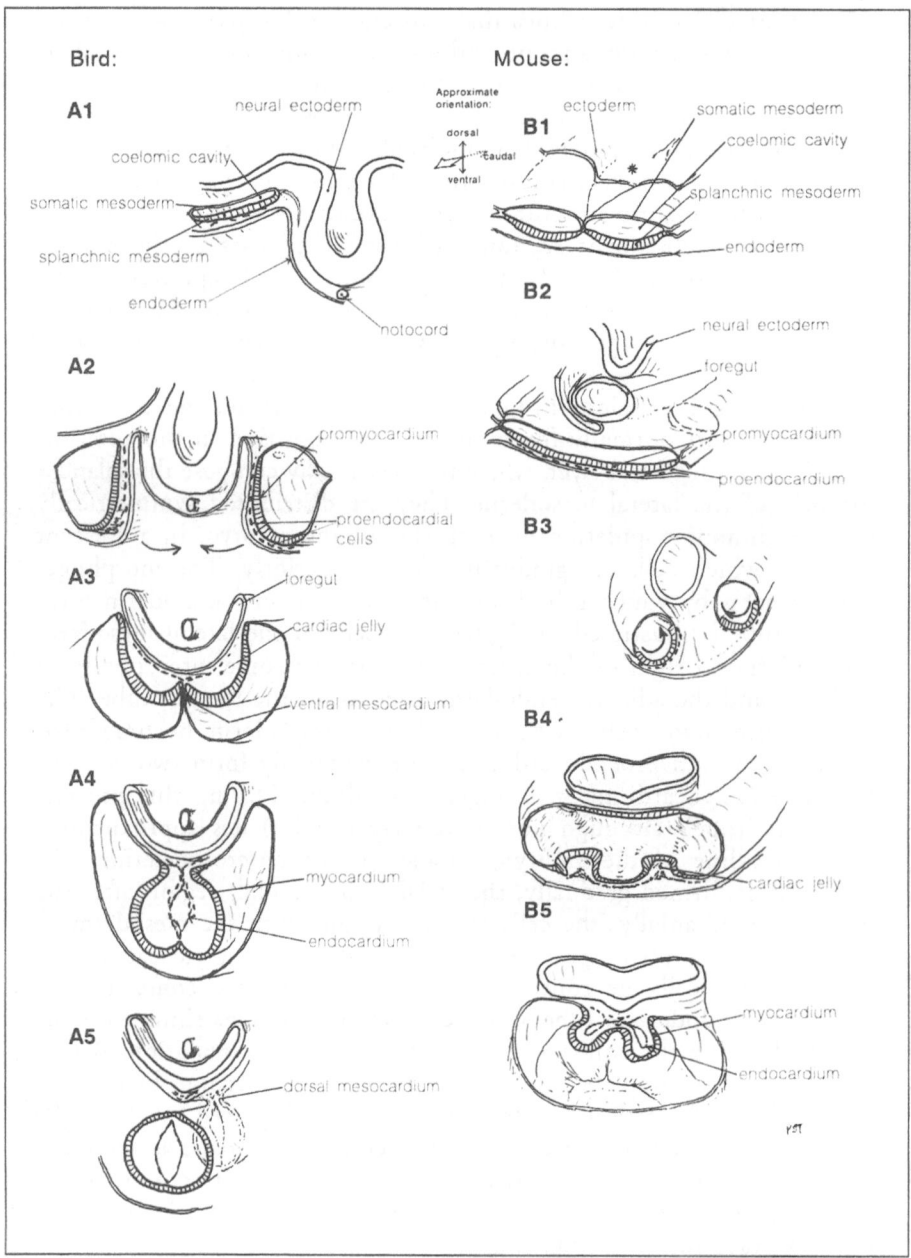

Fig. 2.1. Heart tube formation in bird and mouse. Cartoons depict embryos cut transverse to the axis of the eventual tube. Some morphogenetic events differ somewhat in timing/location, such as the initial fusion of the splanchnic mesoderm (A3 and B1), including events relative to extra-cardiac development such as foregut formation and somite number. The relative positions of 'equivalent' cells can differ (compare A3 and B4). In both bird and mouse, formation of the pharyngeal part of the foregut is accompanied by a movement of the heart-forming cells from the ventral to the dorsal side of the coelomic cavity; the resultant relative disposition of the cells is the same (A4 and B5). See main text for more details.
* Denotes the position of cells that form the buccopharyngeal membrane; those of the bird embryo would already be in front of the plane of the paper.

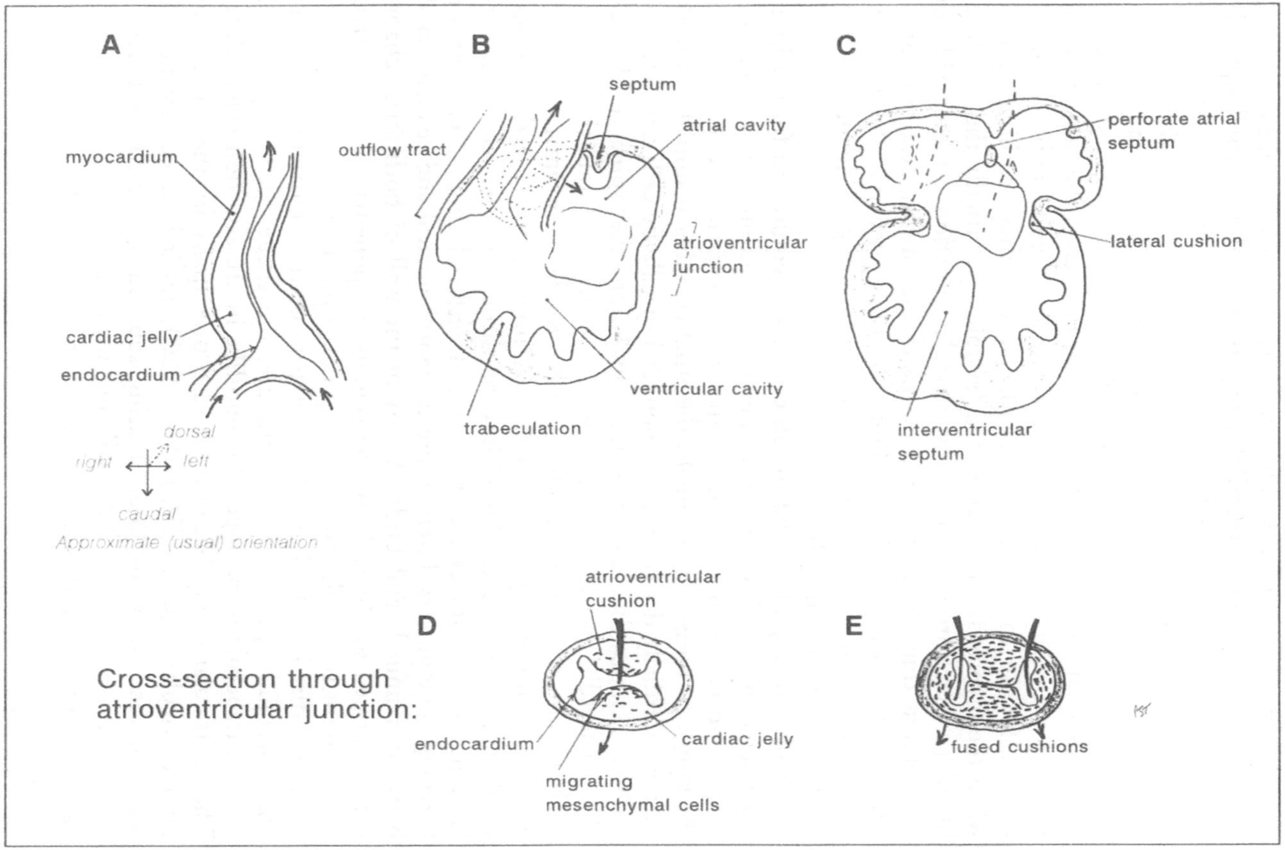

Fig. 2.2. The develop-
ment of regional mor-
phology and septation.
Cartoons depict three
(generalized) stages of
heart development as
longitudinal- or four
chamber-type views.
Relative movement of
the atrial (inflow) end of
the 'tubular' heart (A)
cranially and dorsally results in a looped heart,
(B), in which regional
phenotypes develop (in-
cluding trabeculation
and atrioventricular en-
docardial cushions), and
then (C) (the position of
the outflow tract is
marked but omitted for
clarity). At the atrioventricular junction (and
outflow tract), opposing
regions of extracellular
matrix become filled with
mesenchymal cells fol-
lowing delamination
from the endocardial cell
layer (D). Eventually
these fuse, dividing the
blood flow path in these
regions (E). See main text
for more details. Arrows
denote blood flow di-
rection.

The two layers of the tube are initially separated from one another by a prominent extracellular matrix layer along their length. In the looped heart, a thick extracellular matrix layer remains in the atrioventricular and outflow tract regions, where cells from the endocardium delaminate to form a mesenchymal population that migrates into the matrix ('cushion cells'). At the atrioventricular junction, this occurs principally on the dorsal and ventral aspects of the tube, forming dorsal (or inferior) and ventral (or superior) endocardial cushions (Fig. 2.2D). In the normal heart, the faces of these cushions appose when the myocardium is contracted (systole). Eventually the two fuse, blood passing through the atrioventricular junction on each side of this septum (Fig. 2.2E). Cellular events in the outflow tract are somewhat similar, two apposing cushions or ridges of extracellular matrix populated by mesenchymal cells eventually uniting to divide this part of the tube longitudinally.

Additional septation and fusion events are required to form a four-chambered heart. Prior to development of a pulmonary venous system, blood enters on one side of the common atrial cavity (normally the right). In birds and mammals, the atrial chamber is initially septated by a myocardially-derived septum bounded distally by cells derived from a dorsal extra-cardiac mesenchymal population, the 'spina vestibulae'. The myocardial portion of this septum becomes perforate (Fig. 2.2C), enabling blood to enter the left atrium.[2,3] As the atrioventricular cushions fuse with the spina vestibulae, an intact primary atrial septum would eventually prevent blood flowing through the left atrioventricular junction. An intact atrial septum is restored soon after birth or hatching. The process differs in precise morphological and cellular detail between mammals and birds. A part of the wall of both atrial chambers is formed by the walls of the venous components, in which atrial myocardium-type cells appear during development.[4]

More distally, the inner layer of part of the ventricular myocardium becomes sponge-like in structure, the tissue becoming divided to form trabeculations, finger- or loop-like in cross-section (Fig. 2.2B). The ventricular cavity becomes divided principally by the myocardial interventricular septum (IVS) (Fig. 2.2C), initially a coalescence of trabeculations across its floor. During subsequent growth and development, the crest of the IVS will fuse with the atrioventricular (AV) cushions, and with the cushion or ridge tissue of the outflow tract (OFT).

Critical to appropriate septation and function of the heart is the alignment of the various septal components. The position of the outflow tract ventral to the atrioventricular junction enables one of the flow paths along it to originate from the left of the interventricular septum, so that, following fusion of the IVS, AV cushions and OFT tissue, blood leaving the left ventricular cavity is unable to pass through the right. Distally, continuity of the OFT septum and the aorticopulmonary septum links each ventricular cavity exclusively to its arterial trunk.

The development of the valve leaflets at the atrioventricular junction and distal outflow tract both involve endocardial cushion tissue. Their precise cellular and morphological development differ, principally in there being a myocardial contribution to the atrioventricular valves, such that the valve leaflets are attached to the myocardial wall via tendinous cords and papillary muscles. The semilunar valves, by contrast, are principally, if not exclusively, non-muscular and are not so tethered.[5,6] Extra-cardiac epicardial and subepicardial cells cover the external surface of the heart, enabling the coronary circulatory system to develop.[7,8]

EMBRYONIC ORIGIN OF CARDIAC CELLS

Most studies investigating development of the heart have been performed using bird or rodent embryos and rely on accurate and reliable staging. The system most commonly used for bird embryos is that of Hamburger and Hamilton[9] (HH). This is based on distinctive morphological features, such as somite number, and is preferable to 'hours of incubation', especially as developmental rate can be affected by incubation temperature. An equivalent staging system exists for mice (Theiler stage[10,11]), and it is to be hoped that its use will increase, especially since its incorporation into other publications.[12] The commonly used 'days post coitum' (dpc: the morning after mating is 0.5 dpc) can be very imprecise, particularly as there is variation in degree of development within one litter. This is also the case for 'gestational/ embryonic day' (the morning after mating is 0 gd/ed) often used to stage rat embryos.

Regions of the avian embryo which contribute to the heart have been determined by using labeled (^{3}H-thymidine, DiI[†], quail in quail/ chick chimaeras) cells or groups of cells and mapping their location at subsequent stages of development.[13-15] The cells that normally give rise to the chick heart ingress through the primitive streak principally at HH stage 3a-3b.[14] These experiments indicate that, although not clearly regionally segregated from non-cardiac precursors prior to gastrulation, the relative positions of prospective cardiac cells along the antero-posterior axis whilst going through the primitive streak and once in the lateral mesoderm layer is, by and large, matched by their eventual relative cranio-caudal locations along the heart tube. This is the case for both myocardial and endocardial precursors, which are intermixed to a large degree and in register with respect to position along the axis. By HH stage 4, only a few cells, from the most anterior part of the primitive streak, contribute to the heart (cranial-most endocardial precursors).

A number of processes occur subsequently, and are summarized in Figure 2.3. From HH stage 5, the anterior lateral mesoderm becomes clearly separated into splanchnic (ventral, adjacent to the endoderm)

[†] *DiI is a lipophilic fluorescent carbocyanine dye.*

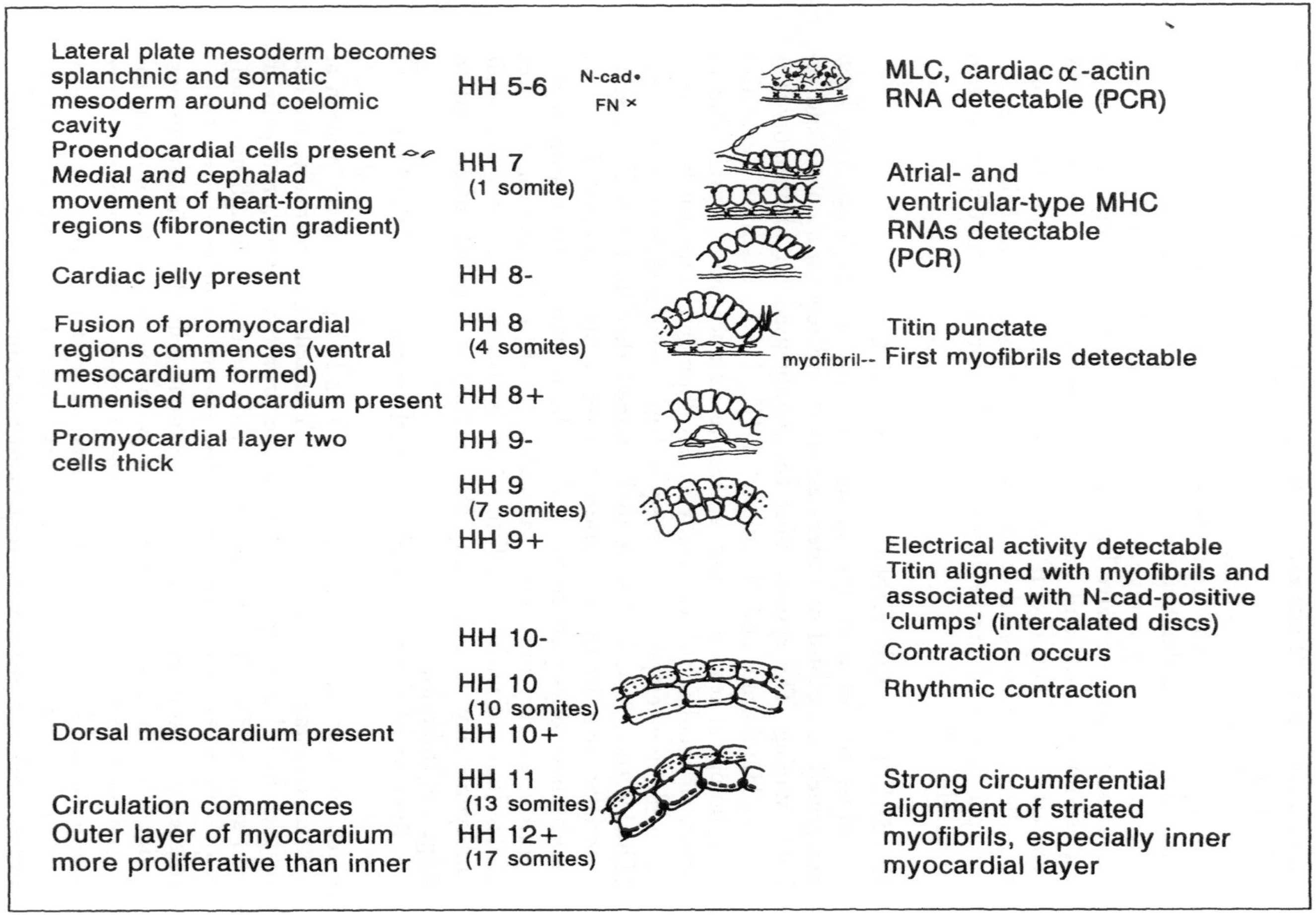

Lateral plate mesoderm becomes splanchnic and somatic mesoderm around coelomic cavity

HH 5-6   N-cad•
          FN ×

MLC, cardiac α-actin RNA detectable (PCR)

Proendocardial cells present

Medial and cephalad movement of heart-forming regions (fibronectin gradient)

HH 7 (1 somite)

Atrial- and ventricular-type MHC RNAs detectable (PCR)

Cardiac jelly present

HH 8-

Fusion of promyocardial regions commences (ventral mesocardium formed)

HH 8 (4 somites)

Titin punctate

Lumenised endocardium present

HH 8+

First myofibrils detectable

Promyocardial layer two cells thick

HH 9-

myofibril--

HH 9 (7 somites)

HH 9+

HH 10-

Electrical activity detectable

Titin aligned with myofibrils and associated with N-cad-positive 'clumps' (intercalated discs)

Contraction occurs

Dorsal mesocardium present

HH 10 (10 somites)

HH 10+

Rhythmic contraction

Circulation commences

Outer layer of myocardium more proliferative than inner

HH 11 (13 somites)

HH 12+ (17 somites)

Strong circumferential alignment of striated myofibrils, especially inner myocardial layer

Fig. 2.3. Events during development of the 'tubular' bird heart. Cartoons depict simplified tissue phenotype as the myocardium develops and the morphogenetic, molecular and functional events listed (first) occur. See main text for more details. HH denotes Hamburger-Hamilton stage. Between HH 7 and HH 14, somite number is the primary criterion for stage designation. At HH 7, one somite is present; HH 7+, two somites; HH 8-, three somites and so on. FN, fibronectin; N-cad, N-cadherin.

and somatic (dorsal, adjacent to the ectoderm) layers, the space between being the coelomic cavity (Fig. 2.1A1). This process is accompanied by a redistribution of the Ca^{2+}-dependent cell adhesion molecule N-cadherin (N-cad), present in endoderm and mesoderm cells in general. Periodic high levels are reported in groups of cells adjacent to which small cavities appear (putative forerunners of the coelomic cavity).[16] It is not clear whether these cavities correspond to the diverticulations marking the 'hinge-point' between splanchnic and somatic mesodermal layers described by DeRuiter et al.[17] Endothelial (including endocardial) precursor cells appear between the splanchnic mesoderm and endoderm, by a process proposed to resemble delamination[18] or cell sorting[19] from the splanchnic mesoderm. Endocardial precursors retain N-cad expression briefly but also express the general endothelial marker QH1[17,19] and retain expression of the more restricted marker JB3 (possibly detecting a fibrillin-related protein[20]). Endocardial cells also transiently express myosin heavy chain (HH stage 8+ to 9-),[18,21,22] a marker of myocardial differentiation. Limited retroviral infection of HH stage 4-6 embryos with a Lac-Z expression vector labels only myocardial cells, arguing against a common origin for these two cell types at these stages.[23] Although endothelial and smooth muscle cells can be infected by this method[24] the susceptibility of cells to infection[19] or their ability to express β-galactosidase might vary.

The myocardial precursor cells of the splanchnic mesoderm form a mesothelium composed of initially loosely associated cells with distinctive morphology, elongation perpendicular to the endoderm (evident by HH stage 7). N-cad expression is found on the apical side of the epithelium, integrin mostly basal and Na^+/K^+-ATPase restricted to lateral plasma membrane, in contrast to its general distribution at HH stage 5.[16] The cells of the myocardial precursor regions have a distinctive 'bulging' appearance when viewed from the coelomic cavity using scanning electron microscopy, and are separated from the diverticulations of the splanchnic/somatic junction by squamous mesothelium.[17]

As the endoderm forms tubular foregut, the anterior splanchnic mesoderm migrates relative to it (from HH stage 6) such that the endocardial precursors (by HH stage 8-) and the lateral edges of the myocardial precursor epithelia are brought into apposition (ventral to the foregut)(Fig. 2.1A2) and fusion of the ventral side of the eventual heart tube commences (HH stage 8), resulting in fusion of the coelomic cavities (Fig. 2.1A3, A4). This migration is proposed to be effected by a cranio-caudal gradient of fibronectin (FN) between the splanchnic mesoderm and endoderm, migration being increasingly slowed by a larger amount of FN more cranially. Eventually, FN also accumulates at the level of the anterior intestinal portal (AIP: the point at which the endodermal sheet becomes tubular foregut). A number of experimental approaches, including the use of anti-FN antibodies and rotation of the cardiac precursor regions, are consistent with this hypothesis.[25-27] By the time fusion commences, the myocardial precursor regions no longer lie parallel to the endoderm but bulge away from it where extracellular matrix ('cardiac jelly') separates the myocardial and endocardial precursors (Fig. 2.1A3). Only in the most medial part of the approximately bilaterally symmetrical embryo, arising from the original lateral extremes, are the endoderm, endocardial precursors and myocardial precursors in close apposition.[17,28] As the AIP becomes positioned increasingly caudally, myocardial and endocardial precursors in more caudal positions appose and fuse (Fig. 2.1A4, A5). A change in cell arrangement occurs as the myocardial precursors form a bilayer (medially at HH stage 8). The N-cad distribution differs between the two layers, being punctate on lateral walls and between the layers and forming belt-like structures around the apical surfaces of the outer layer[29] by HH stage 9 (Fig. 2.3). The Ca^{2+}-independent cell adhesion molecule N-CAM is also present in the myocardial precursor regions from at least HH stage 8.[30]

Electrical activity (conduction) becomes detectable (HH stage 9-), then rhythmic (HH stage 9+)[31,32] as the dorsal myocardial wall forms by apposition and fusion of the other boundary of the myocardial precursor regions (HH stage 10+), following the first contractions (HH stage 10- to 10). By HH stage 10, N-cad distribution includes additional, characteristically located clump-like structures at both the apical junctions of the outer- and luminal junctions of the inner layer of myocardium.[29] The endocardial precursors form a plexus, enclosed by the formation of the myocardial tube (Fig. 2.1A3, A4) and continuous with endothelial cells forming the pharyngeal arch arteries.[17] The endocardial tube becomes luminized from HH stage 8+ as part of a continuous endothelial system including the dorsal aortae (luminized by HH stage 9+). Until HH stage 12, dorsal mesocardium (Fig. 2.1A5) is continuous with mesoderm ensheathing the endoderm caudally as far as the dorsal wall of the atrial part of the heart.

The walls of the forming myocardial tube are not uniform; distinctive sulci (grooves) appear as it is formed. Labeling experiments indicate that these correlate with boundaries between regions of the heart at later stages.[13,28,33,34] The first part of the heart tube to form contributes to the ventricles. Its subsequent growth is a combination of regional expansion as well as the addition of more cells.[13,17] Labeling experiments show that the cells continue to be added downstream to the outflow tract until HH stage 15-16,[33,35] the same time that differentiation of the sinuatrial region occurs (considered by Litvin et al[36]).

Expression of contractile proteins or their RNAs starts before the formation of the tube. Cardiac α-actin and myosin light chain (MLC) 2 transcripts have been detected as early as HH stage 5 using PCR,[37] α-actin RNAs by HH stage 8 (dot blot, in situ hybridization)[38] and smooth muscle α-actin from HH stage 9- (antibody).[39] Myosin heavy chains (MHC) have been detected using the MF20 antibody,[18] and ventricular-type MHC RNA by PCR[40] from HH stage 7; atrial and ventricular-type MHCs (using separate antibodies)[22] from HH stage 8, a ventricular-type MHC RNA from HH stage 8+ and an atrial-type MHC RNA from HH stage 9- (whole mount in situ).[41] Myosin and actin distributions are diffuse until myofilaments, then striated myofibrils, appear (from HH stage 8).[18,21,29,42] Antibody localization of titin, a putative sarcomeric scaffolding protein, shows a punctate and diffuse distribution at HH stage 8 but a periodic alignment, in association with myofibrils, at HH stage 9+.[21] The ends of the myofibrils are associated with the cell membrane (including 'fasciae adherentes' or intercalated discs)[43] where N-cad 'clumps' are found.[29] That N-cad is important in myofibrillar organization is experimentally supported; anti-N-cad antibodies disrupt cell-cell contact, reduce the cytoplasmic area occupied by myofibrils and inhibit contraction of cultured cardiac myocytes.[44] Myofibrils are oriented principally circumferentially in both the outer and inner layers (although with a more branched organization in the outer layer), an organization suggested to result from the distribution of N-cad.[29] The splice variant of N-CAM containing the 'muscle-specific domain' is associated with Z-discs of the myofibril by HH stage 18[45] so could also be important in earlier myofibrillogenesis.

The morphological development of mammalian heart has not been investigated to the same degree as the more accessible avian equivalent. There are some differences between the two. Other types of study, especially molecular and genetic, have been far more extensive in mammals, especially mouse, than birds.

The coelomic cavities of the mouse embryo might also form through the coalescence of smaller cavities, but those lined by apparently uniform mesothelial cells described by Kaufman and Navaratnam[46] were not found by DeRuiter et al.[47] Generally by 7.5 days post coitum (dpc), late presomite stage,[12] the cells adjacent to the unclosed endodermal layer form a characteristic cuboidal mesothelium and comprise the myocardial precursor population. Between this and the endoderm, a mesenchymal endothelial cell population is present which forms a plexus continuous with the precursors of the dorsal aortae and eventually also the pharyngeal arch arteries. The coelomic cavities unite medially, whilst anterior to the cells forming the buccopharyngeal membrane (unlike those of the bird heart) (Fig. 2.1B1, B2). The cardiac precursors thus form a complete crescent (1-2 somite stage). As the anterior intestinal portal forms, endoderm is gradually incorporated into the foregut pocket. Whether by apparently retaining their position relative to this endoderm or by movement relative to it, the cardiac precursors move from a ventral to a dorsal position in the coelomic cavity (2-4 somite stage)(Fig. 2.1B3, B4). As extracellular matrix (cardiac jelly) appears throughout the crescent between the myocardial and endocardial precursor regions, the former bulges into the coelomic cavity. The volume of cardiac jelly is less in the medial (mid) position. Here a plexus of endocardial precursor cells connects the more lateral tubular en-docardium (Fig. 2.1B5), which is continuous with the vitelline veins caudally (5 somite stage). The left and right borders of the myocardial precursor region come together and fuse, enclosing the endocardium and forming the dorsal mesocardium.[46,47] Although a number of these morphogenetic processes resemble those in the bird, further compari-son is made hard by the difficulty of performing such studies as fate-mapping.

The roles of such molecules as N-cadherin, fibronectin and N-CAM in this process have not been investigated in as much detail in rodents as in birds. Although transgenic mice lacking fibronectin have (variably) deformed hearts, the range of other abnormalities also present[48] makes it hard to determine its role in cardiogenesis specifically. Cardia bifida (failure of fusion) is not typical of these phenotypes, nor is it when the normal interaction between fibronectin and its ligand is dis-rupted in cultured 9 gd (headfold stage) rat embryos,[49] consistent with a difference between birds and rodents in the manner in which the laterally positioned cardiac precursor regions approach the midline. The appearance of myofibrils and proteins associated with their organiza-tion in the developing mouse heart resembles that in the chick in that titin distribution is initially punctate (8.25 dpc; Theiler stage 12), eventually assuming a regular, banded organization (9 dpc; Theiler stage 14).[50] The temporal and regional patterns of expression and intracellular organization of intermediate filament and contractile proteins shows some degree of species-specificity (discussed by van der Loop et al[51]). Contraction starts before highly organized myofibrils are detectable in the rabbit[51] and might enhance myofibrillar organization: myofibrils become dissociated in quiescent adult rat myocytes, but following β-adrenergic agonist-stimulated contraction, myofibrils are reassembled from intracellular pools of cytoskeletal and contractile proteins, sug-gesting that 'mechanical forces' resulting from contraction are important in this process.[52] Although myosin heavy chain is widely distributed in the bird heart by HH stage 9-,[18] myofibril development might occur first in the more lateral regions, also the first to show electrical activity.[31,42] Collagen synthesis inhibitors also disrupt myofibrillar formation[53] and existing myofibrils, and affect contractile protein gene expression in cultured em-bryonic chick myocytes.[54] Thus, a number of proteins and processes could contribute to myofibrillogenesis at its outset.

The pattern of expression of various contractile protein genes dur-ing early mouse embryogenesis is discussed in chapter 3 and has been reviewed elsewhere.[55] α-cardiac and α-skeletal actin RNAs are present from 7.5 dpc (in situ hybridization)[56] and are still the earliest to be detected. By 8 dpc, α- and β-MHCs, and a number of MLC isoforms are amongst the other genes expressed. β-MHC and desmin are present from pre-somite stage (9 gd) in the rat (whole mount, antibody).[57] α-smooth (vascular) actin, and α- and β-MHC (antibody)[58,59] and slow skeletal troponin I (RNA)[60,61] are present in early somite stage rat hearts.

CARDIAC CELL COMMITMENT

Most experiments designed to investigate the processes required to commit cells to a cardiac fate involve creating defined combinations of embryonic cell layers and culturing the resulting tissues in defined medium (with or without potential 'signaling' molecules) for a number of days, and assessing the explants for markers of myocardium such as contractile behavior, the presence of myofibrils and expression of myosin heavy chain (MHC) or α-actin.

Only under high density (in vitro) culture conditions will a very small proportion of epiblast cells from pre-gastrulation (HH stage 3) chick embryos differentiate to form beating, MHC- and cardiac troponin I-containing cells after 48 hours;[62] this implies that the mechanism resulting in this differentiation is dependent on cell-cell interactions but not absolutely on ingression through the streak. By HH stage 4, all but the most cranial cardiac precursor population have passed through the streak.[14] By HH stage 4+, some quail mesoderm cardiac precursor region explants form beating, MHC-positive cells when cultured for 24-48 hours on fibronectin-coated plastic in a defined medium lacking potential signaling molecules such as insulin, and in the absence of ectoderm or endoderm. By HH stage 6, the survival of explants, number differentiating into beating, MHC-positive cells and the variety of culture media in which this can occur, are all much increased.[63] The sensitivity of heart formation to the teratogen bromodeoxyuridine (BrdU) in both cultured mesoderm explants and intact embryos also changes between HH stages 4 and 6. Differentiation of HH stage 4 chick mesoderm explants under specific culture conditions into MHC-positive, beating cells does not occur in the presence of 40μM BrdU, but that of HH stage 6 explants does sometimes. This effect does not seem to be the result of differential cell death or uptake of BrdU.[62] When cultured in the presence of dopamine, cell fate of chick HH stage 4 embryo aggregates can be almost exclusively cardiac myocytes.[64]

Are the cardiac myocyte-related properties of these cells affected by the presence of ectoderm or endoderm?[†] Removal of the endoderm layer, leaving ectoderm overlying the HH stage 4+ to 5+ quail cardiac precursor region of mesoderm, almost completely inhibits development of the myocyte phenotype, possibly because the cells die, and often results in floating, vesicular structures. The presence of contiguous endoderm, however, speeds up the rate of differentiation to resemble that in vivo. Explants settle stably onto a fibronectin-coated substratum and 'tubular' structures are formed.[63] By HH stage 6, however, ectoderm-mesoderm explants can settle onto fibronectin-coated plastic and some contractile protein-positive, contractile cells develop.[63,65] In the presence of the underlying endoderm, the HH stage 6 chick promyocardial region develops a (perhaps incompletely) tubular mor-

---

† *A very recent publication[260] should also be considered.*

phology.[53] Does endoderm play any role besides a 'physical' support? As endoderm and mesoderm precursors are intermixed at primitive streak stages, it is not yet possible to separate them to establish whether they, or any other part of the pre-gastrulation embryo can induce cardiac potential. Anterior endoderm which underlies the heart-forming mesoderm at HH stage 6 does not induce cardiac cells in posterior mesoderm at this stage,[65] although culture in dopamine might do so.[64] If endoderm has a role in induction or patterning during HH stages 3a-3b, it is hard to demonstrate what it is by removal or transplantation from otherwise intact embryos because endoderm regenerates and its patterning properties may 'regulate', i.e., reform within the endoderm field.[66] At HH stage 6, when cultured with underlying endoderm, whether anterior (contiguous) or posterior, or anterior ectoderm, heart-forming region mesoderm explants become multilayered, but only those with anterior endoderm differentiate into beating, α-actin-positive cells. As the two germ layers, initially not in contact, merge during the culture period, no inference with respect to the requirement of cell-cell contact can be made.[65] The resulting explants lack a tubular-type morphology, however, in contrast to the results of Antin et al[63] for which mesoderm and endoderm were not separated prior to culture, and those of Linask and Lash:[19] HH stage 5 quail heart-forming mesoderm explants cultured for 24 hours on poly-L lysine coated coverslips are described as 'tubular'. These results are consistent with a contracting phenotype not being dependent on formation of coherent tube structure, as is the onset of myofibrillar gene expression prior to tube formation in the normal avian and rodent embryos. It also implies that at least the morphogenesis to form a tube normally requires the presence of anterior endoderm, if only to supply 'physical' cues and support. The degree to which the structures arising during culture in vitro conditions are complete tubes is not reported.

In the absence of endoderm, the degree of myofibrillar organization of beating cells from endoderm-free HH stage 5 chick blastoderms seems limited.[67] Is this due to an absence of specific signals from endoderm for myofibrillogenesis or a failure to develop a coherent multicellular structure? The distribution of fibroblast growth factor (FGF) and FGF receptor 1 (FGFRI or *cek*) in embryonic chick endoderm and cardiac precursor cells and in experiments with anti-FGFR1 antibodies[68] are consistent with FGF being secreted by the endoderm from at least HH stage 6.[65] This could result in proliferation and differentiation of cardiac precursor cells, including the synthesis of FGF-2 (also known as bFGF) by the cardiac cells themselves by HH stage 10. Antisense oligonucleotide to FGF-2 inhibits proliferation and contractile behavior of cultured HH stage 6 cardiac precursor cells[69] but, although FGFR1 antibodies retard proliferation and multilayering of these cells, onset of contraction is not affected.[68] RNA for the FGFR *bek* is present in the developing mouse heart at 8 dpc.[70]

Whether FGF-2 affects both proliferation and differentiation is not clear. FGFs can regulate the expression of skeletal and cardiac α-actins.[71] Antin et al[63] noted that even small fragments of endoderm alter the morphology of adjacent mesoderm, cell contact leading to three-dimensional foci. The change from a single to multiple cell layer is presumably correlated with changes in cell-adhesion molecule distribution such as that found for N-cad.[29] The distribution and effects of other growth factors, notably TGFβ1 and TGFβ2,[72,73] are consistent with a number of such molecules regulating proliferation, morphology and differentiation during early heart development.

Are there regional differences in the properties of the cells of the heart-forming region? The results of some studies are consistent with a craniocaudal wave of differentiation of these cells. For example, in the studies using BrdU to inhibit cardiac myocyte formation, the effects on whole embryo culture are stage dependent. If cultured from HH stage 4 with BrdU, no heart tube or MHC-positive cells can be detected in embryos 24-48 hours later. If BrdU is present from HH stage 6, no heart tube is formed but two patches of MHC-positive cells are present anterior to the AIP. The cells in these patches Montgomery et al[62] propose to be the most differentiated (consistent with the increasing number of MHC-positive cells in explant cultured with BrdU from HH stages 4, 6 and 8), those that ingressed first through the primitive streak and precursor of the most cranial region of the heart tube. The mechanism by which BrdU results in these effects is unknown, although it might perturb a molecular regulatory event involving a specific transcription factor, as a BrdU-mediated effect on the expression of the transcription factor MyoD has been described.[74] The migratory behavior (length of time before ingression, and direction) of some primitive streak cells varies depending on position within the streak and may account for the origin- and stage-dependent fates of transplanted streak cells when cells normally very close to and far away from Hensen's Node (the most anterior part of the streak) are transplanted to a position very close to Hensen's Node. Those originally very close to the Node remain in the streak for longer and assume a more medial position in the mesoderm layer. Experiments in which streak cells that would normally contribute to cranial or caudal parts of the heart tube are transplanted to 'other' heart region-forming areas of the streak reveal no evidence of committed regionalization; they contribute to areas concomitant with their new, transplant, site. There is, however, a delay in the development of the caudal region when 'anterior' cardiac streak cells are transplanted to a more posterior site.[66] What could account for this delay, for example whether it results from a general property of more anterior streak cells to spend longer in the streak before migrating is not discussed. It is not possible to tell from the results of these experiments[14,66] whether time of ingression normally correlates at all with eventual craniocaudal location in the heart tube normally, or whether subsequent migration patterns result in cranio-caudal appearance of the 'differentiated' phenotype referred to by Montgomery et al.[62] However, any such 'differentiation' still allows pieces of HH stage 5-7 cardiac precursor region mesoderm plus endoderm to be transplanted from cranial to caudal (or vice versa) and assume the beat rate appropriate for their new position (caudal faster, cranial slower) and, in many cases, side of the embryo (left faster than right) after 24 hours in culture.[75] As already described, a gradient of fibronectin at the mesoderm-endoderm interface, highest cranially, also develops during this period.[26,27]

The molecular and behavioral markers used in these experiments do not define whether the cardiac myocytes that develop under these conditions resemble those of any particular part of the heart tube so it is not possible to determine to what extent development of regional contractile phenotype is dependent upon the correct cues from an intact, functional tube in an intact functional embryo.

REGULATORS OF EARLY HEART DEVELOPMENT

What molecular events accompany or regulate the changes in phenotype described above? The onset of a muscular phenotype is easiest to define molecularly by the presence of transcripts of contractile protein genes with roles only in muscular function. A number of transcription factors have been cloned that can bind to regulatory sequences and regulate contractile protein genes (see chapter 4). The patterns of expression of only a few have so far been determined in embryos during this early period of cardiac development. In the mouse, that of three members of the MEF-2 family (A, C and D) has been investigated using in situ hybridization from 7.0 dpc. MEF-2C RNA is found at 7.5 dpc in anterior splanchnic mesoderm that is α-cardiac actin-negative. It is subsequently detected throughout the myocardium of the heart tube and venous sinus, disappearing after 11.5 dpc. Expression of MEF-2A and MEF-2D RNAs is detectable by 8.5 dpc, when the heart tube is looping, and persists beyond 13.5 dpc.[76] Zinc finger transcription factor GATA-4 RNA is present between 7.0 and 7.5 dpc in anterior endoderm and mesoderm. Within the heart tube, its expression is restricted to the caudal-most end, in addition to the venous sinus and adjacent splanchnic mesoderm, until 8.5 dpc, when expression includes the whole heart tube.[77] Regionally and temporally, GATA-4 RNA precedes that of cardiac troponin T. Members of the MEF-2 and GATA families are also implicated in the onset of bird heart development. PCR detects MEF-2 RNA (probably equivalent to MEF-2A in mammals) continuously from HH stage 5 (using whole embryos). Band shift experiments indicate that another factor ('BBF-1') which can bind to a MEF-2 site is present from HH stage 5, specific to the heart at HH stage 12 and distinct from antigenically identified MEF-2.[37] Additional members of the GATA family, -5 and -6, have been cloned in bird. GATA-5 RNA is detected in a crescent over the anterior intestinal portal from HH stage 7 but shows a distinct stronger lateral and caudal distribution until HH stage 9.[78] Expression in the myocardial progenitor cells that are first to fuse has not been clearly demonstrated, and the relative boundaries of expression of these three GATAs and other markers at these early stages have not been established. As is described in chapter 4, members of the bHLH transcription factor family are important in the onset of the skeletal muscle phenotype. As yet, the sequence of a 'cardiac' equivalent has not been reported, although a protein detected by an antibody to the bHLH protein MyoD can be detected by band-shift in lateral plate mesoderm from HH stage 6 and heart from HH stage 9. The same antibody binds to myocardial cells in intact hearts and in culture at HH stage 11 but not 17.[79] As will be discussed later, some of these transcription factors could regulate the expression of other genes besides those that are markers of a muscle phenotype.

Regulation of expression of other genes apparently important in early heart development has not been investigated in detail. N-cad is encoded by two mRNAs in rat embryos[80] and three in the mouse cell line C2,[81] the relative distribution and functions of which remain to be established. The mouse N-cad gene has been cloned.[82] Its large size is contributed to greatly by that of the first and second introns in which regulatory regions could also be present. Both N-cad and N-CAM are present in the myocardial precursor mesothelium in the chick and have a similar distribution in other parts of the embryo, suggesting they could be co-ordinately regulated.[83] However, this is not the case in the apical ectodermal ridge of the limb, where N-CAM but not N-cad is expressed,[83] nor during skeletal myoblast fusion.[81] As skeletal and cardiac myofibrillogenesis differ, desmin being detectable by antibody prior to titin in the former,[84] skeletal and cardiac muscle gene regulation may differ too in this respect. N-CAM RNA expression can be directly regulated by homeobox transcription factors,[85,86] some of which are implicated in the control of regional developmental events, including patterning (reviewed by Krumlauf[87]). Two groups have independently identified a mouse homeobox gene that is strongly expressed in developing and adult myocardium (called Csx[88] or Nkx2.5[89]). Nkx2.5 RNA is present in the cuboidal myocardial precursor mesothelium from 7.5 dpc and subsequently throughout the developing myocardial tube and venous sinus. The use of probes for these transcription factors will be valuable in establishing the order of molecular events accompanying experiments designed to investigate the processes required for the onset of the 'cardiac phenotype'.

The molecular regulation of the skeletal muscle phenotype (considered in chapter 4) has led to speculation that there might be an overlap in the types of transcription factors (such as bHLH) regulating cardiac and skeletal muscle phenotypes. As the heart consists of the physical coincidence of features (striated muscle, distinctive morphology and part of the endothelial circulatory network), skeletal muscle is unlikely to provide a good model for all these aspects. The existence of a single 'cardiac specific' regulatory transcription factor is also unlikely, as the vertebrate heart has not been 'designed in isolation' but evolved from pre-existing tissues, and is functional from a very early stage in its rather complex development. As the unifying aspect is that of location, and *Nkx2.5* RNA is expressed not only in myocardial precursor cells but also in adjacent pharyngeal endoderm, developing thyroid, and some visceral mesoderm derivatives,[89] the role of Nkx-2.5 in cardiac development may have arisen from an ancestor with a role in regionalization. Other *NK* class homeobox genes are expressed in the heart and visceral mesoderm derivatives,[90,91] and expression of both an *NK* class homeodomain (*tin*) and a MEF-2 gene (*DMef*) are required for the development of the heart (a dorsal pulsatile vessel) in the fruitfly *Drosophila melanogaster* (reviewed by Bodmer[91]). The mesodermal cells from which the hearts in both vertebrates and invertebrates arise could be developmentally equivalent (as illustrated by Arendt and Nübler-Jung[92] and Bodmer[91]), so the involvement of the same classes of genes in heart development would have origin in a common ancestral organism. Such a model would predict that Amphioxus, a cephalochordate considered to resemble the 'ancestral vertebrate',[93] would also express both *NK* and MEF-2 genes in part of its circulatory system, the pulsatile regions of which include that equivalent to the venous sinus.[94,95]

DEVELOPMENT OF THE 'TUBULAR' HEART: CURVATURE, REGIONALIZATION AND SEPTATION

Is there evidence that the bending process itself is dependent on specific gene products? Culture of isolated chick heart tubes by Manning and McLachlan[96] indicates that the initial curving process can be caused by forces intrinsic to the heart at least from HH stage 9- (6 somites) although axial deviation within the embryo is present from before this stage (as discussed by Stalsberg[97] and Cooke[98]). Actin has been proposed as important in this process.[99] The same authors suggest that altering the sidedness of the heart loop also alters the side to which the head turns, a conclusion supported by Hoyle et al.[100] Experimental reversal of axial body rotation of rat embryos does not affect heart loop sidedness.[101] There is genetic evidence consistent with the hypothesis that sidedness of looping, along with other laterally asymmetrical aspects of body development, is under the control of a specific gene product, although, as yet, it has not been identified (for model and review, see Brown et al[102]). Phosphorylation of a gap-junction protein (connexin 43) has been implicated in the regulation of sidedness in humans.[103] However, abnormal laterality is not reported in knock-out mice lacking connexin 43.[104]

An intrinsic change that influences looping direction occurs in normal chick cardiac precursor cells between HH stage 5 and 6,[100] in a time at which a consistent left-right asymmetry is present elsewhere in the embryo.[97,98] The results of experiments in which caudal or cranial pieces of quail HH stage 5-7 cardiac precursor regions were transplanted homo- or heterotopically into chick cardiac precursor regions, have led Easton et al[105] to propose that a morphogen secreted by cells in the caudal end contributes to cessation of looping. Regional differences in cell division rate have also been proposed to account for the curvature and heart asymmetry.[97,106] Retinoic acid can cause a variety of heart abnormalities (time, location or concentration dependent), amongst which are alterations in siddedness of looping toward a retinoic acid-soaked bead. Although grafts of Hensen's Node into a host embryo up HH stage 5 can also affect looping direction,[107] and the Node is a source of retinoids in both chick and mouse hearts,[108,109] the mechanism by which retinoic acid affects the siddedness of looping is still unknown and it might not have a direct role.[107]

Looping of the heart tube has two aspects: the formation of a loop itself and the mechanism by which this has a siddedness. Fate mapping shows that the two myocardial precursor regions normally make asymmetric contributions to the heart tube along its cranio-caudal axis.[110] Is this lateral asymmetry reflected at the molecular level? The onset of detectable electrical activity in chick hearts, at HH stage 9-, shows an apparently inconsistent siddedness,[31] although the eventual site of the pacemaker region does show a regular siddedness, and, if fusion of the cardiac precursor regions is prevented, a predictable siddedness (left) to the faster beat rate between the resulting tubes is present.[75] The pattern of expression of genes that enable this activity to occur has yet to be established. Han et al[18] also noted that the first detectable myosin heavy chain, at HH stage 7, is sometimes, apparently randomly, sided. Hiruma and Hirakow[42] noted an apparent predominance or precedence of degree of myofibrillar organization on the right side at HH stage 10- and 10. Molecular markers of differentiation do not necessarily behave in a laterally determined way.

The siddedness of tube looping and the control of 'right-left lateral identity' of regions of the heart at subsequent stages are difficult to correlate without markers for the latter. The transition from unseptated to septated tube presents some problems as to how to categorize cells initially defined according to their position along the caudo-cranial axis that eventually contribute to structures at an 'equivalent' position along this axis to those initially at a different cranio-caudal location. Some molecular markers show an asymmetry of ventricular expression prior to completion of septation (onset of muscle creatine kinase expression in rodents[111,112] and MLC3F and a transgene including its regulatory regions in mouse[113]). In both these cases, expression initially or eventually includes other heart regions. The signals that lead to these patterns are clearly important to determine.

Cells of the heart tube cannot regulate (adjust) to produce normal morphology following gross changes in body flexion following experiments such as those performed on chick embryos by Männer et al.[114] When the cranial and cervical flexures of the neural tube are prevented from forming, the simple c-shaped bend of the heart tube is achieved but the subsequent process, by which the inflow part assumes a position more dorsal and cranial relative to the rest of the tube, fails to occur. The abnormalities that develop during subsequent incubation of these embryos suggest that a failure to achieve normal looping affects processes required for septation. It has been proposed[115] that the curvature of the tube results in asymmetric deformation of the walls; this differential deformation results in differential growth of endocardial and sub-endocardial cells. These interesting models have not been the subject of molecular study nor have the same experiments been performed on mammalian hearts. In the absence of a molecular marker demarcating the precise regions in which distinctive aspects of curvature occur, it remains possible that the physical properties of the tube itself, (see Manasek et al[116] for a model and Icardo and Manasek[1] for a review) its dimensions and stiffness, are critical in determining the final loop morphology of the heart tube, rather than the properties of a few localized 'special' cells. In vitro culture of 0-3 somite stage rat embryos in the presence of hyaluronidase for up to 72 hours prevents neither heart tube formation nor looping.[117] Hyaluronate is a major component of cardiac jelly[118] and, if tube formation and looping resulted from its physical presence, a lack of hyaluronate would be expected to result in failure of both these processes under these experimental conditions. In the presence of hyaluronidase, no cardiac jelly separating myocardium and endocardium is observed at any point during heart development (S. Baldwin, personal communication); myocardium and endocardium are in direct apposition after 24 hours culture, and remain so.[117]

Although there are no molecular markers that correlate well with sidedness of looping, those involved in muscle function show non-uniform and developmentally regulated distributions along the heart tube. By taking account of the likely effects on rate of myocardial contraction and relaxation of the properties of these molecules (as described in chapter 3), this 'molecular anatomy' enables unidirectional flow of the heart tube prior to septation and the development of valves, to be explained (reviewed by Moorman and Lamers[119,120]). In the early tubular heart, in which cardiac jelly is present throughout its length, this requires that the spontaneous depolarization and contraction rate forms a gradient (highest at the caudal end) and that, during the peristaltic wave of contraction, slow relaxation and cardiac jelly occluding the lumen of the tube prevent backflow. In the looped heart, caudocranial gradients or differences in distribution of a number of proteins have developed. For example, α- or atrial MHC is predominant in the caudal (atrial) end and β- or ventricular MHC in the ventricular and outflow tract regions. These different isoforms effect differences in contraction rate (as described in chapter 3), further reinforcing the initiation of each contraction at the caudal end of the heart tube. The peristaltic nature of the contraction wave becomes replaced by the sequential contraction of atrial and ventricular compartments, the delay between the two being effected by the slow conductance of the atrioventricular junctional myocardium. Backflow from the ventricular compartment and arteries is prevented by the nature of contraction of the atrioventricular and outflow tract myocardium in combination with the continued presence of cardiac jelly in these regions. Regional distribution of RNAs encoding, Ca^{2+}-handling proteins sarcoplasmic reticulum Ca^{2+}-ATPase (SERCA) and phospholamban, which regulate relaxation rate, is also consistent with this model.[121]

DEVELOPMENT OF REGIONAL PHENOTYPES

What regulates the development of regional variation of molecular distribution? Is the formation of a tubular heart structure necessary? Cells from the mouse embryonic stem cell line D3 can be cultured to form aggregates under conditions in which beating cardiac myocytes subsequently differentiate. Patch-clamp techniques have been used to determine the types of ion channel and action potentials, and PCR used to detect expression of α- and β-myosin heavy chains in individual cells enzymatically isolated from the aggregates. Over time, the types of cardiac myocyte defined by these properties increase in a way resembling that in normal heart development, so that eventually there are three types of cardiac myocyte in the aggregates with electrical properties resembling those found in atria, ventricles and sinus node of normal hearts.[122,123] Although the cells have a typical but variable cardiac myocyte morphology, no data presented indicates whether the distribution of these different types of cell is non-uniform in the aggregate nor how much the tissue morphology resembles that of a heart tube. The morphology of individual cells is not related to functional phenotype and there is a cell population size effect on the proportion of cells that differentiate.[122] Not only is the mechanism that allows some cardiac specification to occur not dependent on position such as would be available in an intact embryo but the development of some regional and temporal cardiac myocyte phenotypes is probably not either. However, the close link between the expression of contractile marker and particular channel phenotype implies coordination of these two aspects of cardiac myocyte function. The expression of myosin light chain (MLC)-2A, MLC2V, MLC1V, and MLC1A and atrial natriuretic factor (ANF) RNAs, and staining of cells with antibodies to MLC2V and ANF in appropriately cultured embryonic stem cells has also been reported.[124,125] MLC2A RNA expression is normally lost from the ventricle (Kubalak et al[125]) and MLC2V RNA is not normally expressed in atrium and becomes restricted to the ventricular part of the heart tube;[126] ANF usually has a restricted distribution (principally atrial and trabecular[127]). Expression of these region-specific muscle function protein markers is not dependent on normal, tubular morphology, the presence of sulci, or normal function.[124] Other experimental systems in which normal morphology is absent[64,128] present further opportunities for investigating the regulation of gene expression, phenotype and developmental potential.

Studies of regulation of contractile protein gene expression in skeletal muscle and neonatal and adult cardiac myocytes show that numerous transcription factors can affect transcription, reflecting regional and physiological aspects of the overall expression pattern, as described in chapter 4. The regional expression patterns of few of these transcription factors have been determined during embryonic heart development. Mouse MEF-2 RNAs are present during the regionalization process, one report indicating that MEF-2A RNA expression is atrium-specific at 8.5 dpc,[55] another not,[76] but no other region-specific pattern of expression has been demonstrated. MEF-2 gene products can be variably spliced (summarized by Martin et al[129] and shown in Fig. 4.5), the functional significance of which is not yet known, and the distribution of the proteins themselves in the developing heart is also unknown so they might still effect or maintain regional expression of contractile proteins. MEF-2 genes are also expressed in non-muscle tissues, so might regulate expression of other types of gene too. As described in chapter 4, they contain the 'MADS' domain, also found in factors regulating a number of regional or developmental events.

HF1b, a zinc-finger transcription factor that binds to a DNA sequence also bound by MEF-2 factors, is implicated in regulating the expression of rodent MLC2V,[130,131] which is ventricle-specific by 11 dpc.[126] The distribution of HF1b has not been determined during these stages and the restricted expression of the MLC2V gene is attributed to both positive and negative regulation occurring through several regulatory elements and factors.[132] Expression of the transcription factor GATA-4 (both RNA and protein) is initially restricted to the caudal-most end of the mouse heart. Although preceding the expression of cardiac troponin C,[77] their distribution spreads to include the whole heart tube as the regional pattern of contractile expression develops. The role of GATA-4 in the regulation of the mammalian α-MHC gene is described and discussed in chapter 4.

Fate-mapping in bird hearts[13] indicates that the cells that normally give rise to the caudal end of the heart tube are not found exclusively in the caudal-most part of the cardiac precursor regions as fusion occurs (HH stage 8+) but also in more cranial lateral-most regions. The regions expressing atrial- but not ventricular-type MHC at this and subsequent stages[22] resemble this pattern, in as much as they are the caudal and lateral regions, rather than the medial region where fusion has occurred. This also suggests that cutting unfused cardiac precursor regions transversely relative to the cranio-caudal axis would not isolate groups of cells that normally give rise to the same cranio-caudal level of the heart tube. The first detectable electrical activity (HH stage 9-) is also in a relatively 'lateral' position, away from the medial, region of fusion.[31] By this time, myosin heavy chain expression detected by MF20 antibody is more general and extends medially to the ventral fusion line (which remains noticeably less stained until at least HH stage 10-).[18] The distribution of myofibrils (determined electron microscopically) shows more fibrils in lateral rather than medial positions of the ventricle-forming region.[42] Although a 'caudal' property seems to be present by HH stage 5-6,[105] molecular or developmental evidence for precise regional determination into compartments resembling those definable after septation is poor at the pre-fusion stages. Morphologically, the sulci that fate-map to distinct locations or levels within the septating heart are present as the cardiac forming regions fuse ventrally. The molecular basis of their formation and maintenance is not known.

The results of a number of experiments show that retinoic acid affects normal regional development. Exogenous retinoic acid can inhibit fusion of HH stage 3 to 8 chick cardiac precursor regions[133] and increases the proportion of the resulting heart tubes that express an atrial-specific MHC RNA.[41] The effects are both dose- and stage- dependent, higher doses and earlier stages resulting in more severe abnormality. These effects have been ascribed to inhibition of migration toward the midline and considered similar to defects caused by the removal of the cranial-most part of the cardiac precursor region.[133] In the absence of cell death, however, the effect is also consistent with a decrease in the proportion of the tube with a cranial-most phenotype, as indicated by the molecular marker. Increasing concentrations of retinoic acid also decrease the proportion of the Zebrafish heart tube with the 'arterial'-most phenotype, defined using antibodies to atrial- and general-MHCs,[134] although this occurs by reduction in size of the heart, an effect not described by Yutzey et al[41] in retinoic acid-treated chick hearts. Modak et al[135] note that a reduction in cell population growth and caudalisation of early chick embryos by exogenous all-trans retinoic acid are at least concomitant. In the converse of this type of experiment, vitamin A-deprived quail embryos also develop cardiac abnormalities, partially rescued by exogenous retinoic acid administered between 22-28 hours incubation (HH stage 5-8).[136] The deprived embryos show a failure of the caudal end of the myocardial and endocardial precursor regions to connect with the omphalomesenteric veins from the extraembryonic circulatory system, which appears normal,[137] manifest at HH stage 9- (6-7 somites). Failure of fusion (cardia bifida) is also often found in these embryos.[137] Molecular analysis of these hearts is needed to determine if vitamin A deprivation results in a reduction or absence of caudal phenotype, such as a smaller proportion of the tube expressing atrial-type MHC, for example. The distribution of sulci is also not described in any of these experimental bird hearts, nor whether a phenotype resembling that of the atrioventricular junction develops. The severely abnormal cardiac phenotypes might preclude survival to a stage at which markers of such regions are expressed. RNA for the retinoic acid receptor RARβ2 is present in both the foregut and presumptive cardiac mesoderm at HH stage 6, becoming more restricted during subsequent development. At HH stage 12, an abrupt boundary of expression is present where the omphalo-mesenteric veins meet the venous end of the heart, the cells of which do not express RARβ2 by this stage.[138] RARβ2 expression can be induced directly by retinoic acid.[139] Although these results are consistent with retinoids, possibly retinoic acid, regulating gross caudo-cranial regionality in the cardiac precursors and early heart tube (including venous connection), as it does in other parts of the embryo (as discussed by Smith[138] and Stainier and Fishman[134]), there is still no direct evidence that retinoids or mechanisms to respond to them are present in the cardiac precursors at the appropriate stages.

Transgenic mice in which expression of various RAR or retinoid receptor (RXR) genes have been knocked out have not shown a phenotype that clearly indicates through which receptors retinoic acid affects this patterning at these early stages (see 'Ventricular myocardium'). Retinoic acid-induced differentiation of F9 and P19 cells results in the upregulation of GATA-4 and N-cad gene expression respectively.[140,141] Neither gene has been shown to be regulated directly by retinoids (containing retinoic acid response elements in their regulatory regions, for example), but the initial caudal expression of GATA-4 could result from an indirect mechanism and the distribution of transcripts from this gene altered in chick embryos treated with retinoic acid as described above. Retinoids are implicated in numerous developmental events in the vertebrate embryo, which include a number affecting cardiac development at different stages (described later).

SEPTATION OF THE TUBULAR HEART

Septation is a very obvious way in which distinct regions of the heart are generated and involves division of the atrial, atrioventricular junction, ventricular, and outflow tract regions. The process of normal septation requires both the development of distinct morphological structures (referred to here as septa) and their fusion to each other. Thus the tissue between two chambers of the septated heart can be composed of septal elements of separate origin, and 'septal defects' of the mature heart result from an abnormality of these elements themselves or the fusion between them. Septation involves myocardially and endocardially-derived, and extra-cardiac, cells and is not effected by the same cellular event at each location along the tube. Although myocardial contributions to both atrial and ventricular septation are morphologically apparent, they differ in other respects. The cellular events at the atrioventricular and outflow tract regions seem more similar: in both positions, two opposing thick extracellular matrix pads are present, which become filled with mesenchymal cells and eventually fuse.

VENTRICULAR MYOCARDIUM

Whilst the processes required for septation are occurring, the compartments eventually separated to form chambers develop distinctive morphological phenotypes. Gene expression in myocardium in atrial and ventricular regions differs from that elsewhere, and is consistent with force-generation as the primary function of myocardium in these regions ('working myocardium'[119,120,142]). In the ventricle, myocardium develops luminal trabeculations, and the interventricular septum forms. Regional differences in cell proliferation accompany these processes. In experimental studies of the bird embryo, a gradient of ^{3}H-thymidine incorporation across the ventricular myocardial wall, least on the inner (luminal) side, appears between HH stages 9+ and 12+.[143] Greater incorporation is a marker for DNA synthesis required for cell division. Although trabeculations are not yet present, intercellular myofibrillar alignment is more prevalent in the inner layer of cells. Differences in arrangement of junctional and myofibrillar components between the inner and outer myocardial layers during this period is described by Shiraishi et al[29] as previously discussed. As trabeculations appear, (from HH stage 15[144]), and cardiac jelly is lost from the ventricular region,[145] the gradient of DNA synthesis remains present. By HH stage 26, a more uniform pattern of DNA synthesis exists across the non-trabeculated part of the wall, the outermost layer of which forms a smooth outline and is covered with epicardium. By HH stage 31, the gradient of incorporation has gone, and capillaries are present in the compact zone (non-trabeculated myocardium), which increases in thickness. This 'trabecular compaction' can be affected by hemodynamics such as those resulting from gross morphological abnormalities elsewhere in the heart[146] unlike the earlier myocardial development. Tokuyasu[143] has proposed that these changes in morphology are necessary to maintain the supply of nutrients to myocardial cells prior to the development of the coronary circulation, enabling the growing heart to function efficiently. The role of trabeculation in angio-architectural development (see Rychter and Rychterová[147] for a model) and coronary circulation in general is beyond the scope of this chapter. The same pattern of DNA synthesis, low in trabeculated myocardium, has been demonstrated in bird and rat heart using BrdU incorporation[148,149] and the position of clonal descendants of ventricular myocardial cells using retroviral labeling in bird heart is also consistent with cells dividing less frequently in the more luminal and trabecular regions.[23,150] Regionalization across the myocardial wall is thus present in both degree of myofibrillar organization and rate of cell division, perhaps reflecting a limitation on the latter posed by the former (discussed by Rumyantsev[151]). Regions displaying less cell proliferation have been proposed to have an important effect on gross cardiac morphogenesis.[148,149,152] Although a number of gene products (principally those required for muscle function) are differentially expressed between trabeculated and non-trabeculated myocardium (see Lyons[55]), the distribution of transcription factors that can regulate their expression has not been determined so the basis of such a distribution remains unknown. In some cases, the compact/trabecular difference is transitory (such as muscle creatine kinase[112]) or develops after a period of time (MLC1A[153]) suggesting this distribution is unlikely to arise from a simple common mechanism.

From scanning electron microscopic observations, Steding and Seidl[115] describe trabeculation arising by excavations into the myocardium from its luminal face, the trabecular patterns arising as a consequence of regional differences in myocardial growth.[144,147] Activity of urokinase, an extra-cellular proteinase also associated with cell invasion, elevates between HH stages 14/15 and 18/19 in the quail ventricular region.[154] In culture, endocardial urokinase production is increased by the presence of a fibronectin protein fragment.[155] Staining for fibronectin increases on both myocardium and endocardium as trabeculation starts in the chick ventricle and is then lost, apparently as the cardiac jelly disappears (HH stage 17-18).[145] Both fibronectin and α_5 integrin are present in rat ventricles from at least 11 days gestation[156] and disruption of the fibronectin: $\alpha_5\beta_1$integrin interaction in early rat hearts results in multilayered endocardium apparently intermixed with myocardium.[49] The distribution of PECAM, characteristic of endothelial, but not mesenchymal, cells delaminated from endocardium[49] has not been determined in the treated rat endocardial cells. These observations could contribute to a model whereby the single cell layer-thick endocardium can come to follow the contours of the trabeculating myocardium, but not necessarily to one explaining how the trabeculation process itself is initiated, and they also apply to other, non-trabeculating regions of the heart. Another cell adhesion molecule, V-CAM, expressed in mouse cardiac precursors from at least as early as 7.5 dpc, appears restricted to the non-trabeculated regions of myocardium by 8.75 dpc, including the ventricular compact zone. The absence of V-CAM in a knock-out transgenic mouse results in a loss of normal compact zone morphology.[157] V-CAM binds to certain integrins containing α_4,[158] but α_4 is not detected in myocardium, so how it manifests its role prior to the arrival of epicardium (which does express α_4) is unknown.[157]

Are there any candidate transcription factor genes for the development of this ventricular morphology? A homeobox gene, GH6, is expressed in the ventricular compartment of the bird heart,[159] strongly at HH stage 23 then lessening until undetectable by in situ hybridization by HH stage 29. In the absence of information about its expression in the heart prior to HH stage 23 and as trabecular development is well advanced by this stage, its proposed role in ventricular trabeculation per se remains only one of many consistent with the results. It could be a rather late, transient marker of the ventricular compartment related to a stage of differentiation such as the development of the interventricular septum (see below), which is still a rather rudimentary structure at this stage, or increase in thickness of the compact zone. Expression does not appear to be restricted to the trabecular layer in the ventricle, and none outside the ventricles is reported. Two other homeobox genes, S8[160] and MHox,[161] are also expressed in myocardium but insufficient information regarding their regional distribution is available to suggest roles in regional development.[†]

Five mutant mouse genotypes involving transcription factors show abnormalities relating to the morphology of the ventricular wall. These are knock-out transgenic mice for retinoic acid receptors (RARα and RARγ), retinoid receptor RXRα, WT, TEF-1 and N-myc. In mice lacking the retinoid receptor RXRα[162,163] the ventricular myocardial wall seems morphologically normal at 10.5 dpc but the thickening of the compact zone at 12.5 dpc fails to occur, although the number of cells does not differ from that of normal hearts; the ventricular walls are composed of trabeculations, rather thinner than normal, and a very thin outer myocardial layer consisting of well differentiated cells with well organized myofibers, rather than the normal, relatively undifferentiated ones. The interventricular septum is present but thinner, less compact and often perforate. RXRα RNA expression is ubiquitous in normal mouse embryos between 10.5 and 14.5 dpc[164] and its RNA is detectable by PCR in 8.5, 12.5 and 13.5 dpc and adult mouse hearts.[165] No heart abnormalities have been reported in mice in which any single RAR is not present but a phenotype with compact zone deficiency is partially penetrant in RARα/RARγ double mutants. An apparently normal phenotype is restored by the presence of only one copy of one of the two splice variants of RARα.[166] Thin myocardium is one of the phenotypes of vitamin A deficiency in rats (discussed by Mendelsohn et al[166]). These data are consistent with the maintenance or development of the compact zone requiring retinoids, the effects of which are mediated by RXRα, and by RARα or RARγ. RARα, but not RARγ, RNA has been detected in the heart at 12.5 dpc by in situ hybridization, although whether its distribution across the myocardium is uniform is not noted.[167] Both RARα and RARγ are detectable in the atrioventricular mesenchyme, but in situ hybridization might not be sensitive enough to detect the presence of some RARs in other parts of the heart.

[†] *The recent report of the Nkx2-5 knock-out mouse cardiac phenotype[261] is of great importance.*

The phenotypes of these mice imply that the mechanism by which the trabeculating rodent heart develops compact ventricular myocardium is probably regulated by retinoic acid, particularly as vitamin A deficiency affects this feature. Although an abnormal ventricular myocardial phenotype in higher vertebrates, the absence of a thick compact zone is typical of some lower vertebrates, as discussed by Van Mierop and Kutsche,[168] so the change in phenotype following perturbation of retinoic acid/retinoid receptors could be atavistic. Other changes in morphology might reflect separate roles in other cardiac, and extracardiac, development (especially that of neural crest cells). Exogenous retinoic acid has no effect on expression of GH6,[159] but the effects of an absence of retinoic acid have not been reported. What is known of regional distribution of RAR and RXR RNAs or proteins is not consistent with regional phenotype arising from regionally restricted expression of these transcription factors. The distribution, and regulation, of retinoic acid synthesis is important to establish in this context.

The absence of several other transcription factors results in phenotypes lacking compact myocardium. Mice lacking WT-1[169] show no heart defects until 13.5 dpc, when the heart appears smaller and lacks compact zone myocardium, particularly in the free wall of the right ventricle. There are no defects in the apparently normal interventricular septum. By 14.5 dpc, the left ventricular wall is also very thin and the compartment much smaller than normal. Expression of WT-1 has been reported in the pericardium[170] but not detected in the myocardium. Mice lacking N-myc are abnormal after 10.5 dpc: their hearts are smaller, trabeculations and compact zone are both described as thinner than normal and interventricular septum apparently absent.[171] N-myc RNA is expressed in cardiac cells prior to this (9.5 dpc, correlating well with protein distribution[172]). Both N-myc and c-myc RNAs (N>c) are expressed in undifferentiated mesoderm of 7.5 dpc mouse embryos but only c-myc by 8.5 dpc, leading Downs et al[173] to propose that N-myc might be involved in the maintenance of 'primitive' mesoderm and must be silenced before mesoderm can differentiate; regions that normally express N-myc look normal at 9.5 dpc in homozygous null animals. The abnormal cardiac phenotype could result from both a lack of cell division and premature complete differentiation. In general N-myc is expressed at 9.5 dpc in places that look normal until 11.5 dpc in the null animals but, as with all knock-out models, it is important to ascertain how any resulting abnormal expression of other gene products, the effects of which may be (partially) compensatory, affect the phenotype of these mice, particularly before defining the knocked out gene product as 'redundant' prior to the appearance of an abnormal phenotype.[174] Transgenic mice carrying a constitutively expressing c-myc gene have enlarged atria and ventricles that contain more than twice as many myocytes as normal, but no details of the heart morphology are available.[175,176]

Mice with no expression of the transcription factor TEF-1 are also described as having thinner compact zone myocardium and fewer trabeculations.[177] Some evidence for subendocardial myocyte degeneration is also present. No defects in interventricular septum development are reported prior to embryonic death at 11-12 dpc. Although microscopy demonstrates no ultrastructural abnormalities in sarcomeric organization or mitochondria of the trabeculated myocardium, the relative appearance of the (remaining) compact zone myocardium is not described, so it is not clear whether a failure of the outer layer of myocardium to retain its relatively undifferentiated phenotype contributes to the abnormality. TEF-1 gene products can affect regulation of a number of contractile genes expressed in the heart (as described in chapter 4) and, whilst expression per se does not depend on the presence or absence of a single factor, the isoform composition of the sarcomeres could be affected. In another transgenic mouse model, contractile protein isoforms expressed in the heart are altered as a result of the transgene, the skeletal muscle bHLH transcription factor MyoD expressed under the control of a muscle creatine-kinase enhancer/promoter element. These mice develop a very different but grossly abnormal (and embryonic-lethal) cardiac morphology[178] which, along with its molecular anatomy, remains relatively uncharacterized.

Interventricular septal development is abnormal in some of these knock-out mouse lines. The presence of holes in its myocardium ('muscular' type defects) and the distribution of V-CAM RNA[157] could be consistent with the bulk of it being composed of compact layer myocardium, the development of which is compromised elsewhere in the ventricles. Its relative insubstantiality could also account for subsequent failure of fusion with cushion tissue, resulting in 'membranous'-type ventricular septal defects. A failure of trabeculated myocardium to compact could also result in such defects. Further careful analyses of ventricular septal development in these lines will contribute to what is currently a poorly understood process in rodents. The avian interventricular septum (IVS) is initially composed of a coalescence of apical trabeculations.[144] Ink-labeling of regions of chick HH stage 17-21 myocardium suggests that septal growth results largely from displacement of cells from the free ventricular walls[179,180] and that two myocardial proliferation centers, one in each ventricle, account for growth[179] prior to septation, rather than extra, localized, cell division or a limit to the extent of trabecular formation in the region of the ventricles where the septum develops. The development of the septum has not been specifically described following studies using retroviral clonal marking.[23,150] A relative lack of BrdU incorporation in HH stage 17-21 distinguishes the crest of the interventricular septum from other parts of the ventricular myocardium, trabeculations elsewhere becoming relatively unlabeled from HH stage 24.[148]

No molecular markers precede the appearance of the IVS as a ridge of (aligned) trabeculations at a distinct location in the ventricular cavity. However, a number of markers pick out this structure, or at least the crest of it: for example, antibodies GlN2 (human),[181] Leu 7 (rat)[182,183] and HNK-1 (chick);[184] RNA probe for homeobox gene *Msx-2* (chick)[185] and Schiff staining (mouse),[186,187] and have been associated with the development of conduction system myocytes. Labeling studies performed during chick heart tube formation suggest that the junction between cells that normally give rise to the trabeculated regions of the left and right ventricles lies perpendicular to the caudo-cranial axis[34] but continuous association of future compartment boundaries with a sulcus is species-dependent (discussed by Vuillemin and Pexieder[188]). How the position of the septum and subsequent regional phenotype of the cells of its crest are regulated is not known. Further studies, including detailed analyses of trabecular pattern,[189] are also needed to determine whether differences between left and right ventricular morphology arise through pre-existing differences in regional identity of the ventricular myocardial cells themselves.

A number of other transgenic mice exhibit abnormal cardiac morphology, often affecting more than one region of the heart. For example, NF-1 and its splice variants show complex myocardial expression patterns during mouse heart development,[190,191] and further detailed morphological analyses of the very abnormal hearts of mice lacking NF-1[192] are needed to discern its role(s). Mice lacking connexin 43, a gap junction protein associated with conduction, appear to die primarily as the result of very abnormal but little characterized myocardial morphology.[104] To take full advantage of the increasing number of lines of transgenic mice produced, recognition, and accurate and detailed description, of resulting abnormal cardiac phenotypes are essential.

ATRIOVENTRICULAR AND OUTFLOW TRACT REGIONS

The results of transgenic experiments described above suggest a number of transcription factors associated with ventricular region development. Evidence that all aspects of ventricular, or atrial, phenotype are regulated by single region-specific transcription factor is still lacking. What of the regions bordering them, the atrioventricular (AV) and outflow tract (OFT) regions? Labeling experiments suggest that both regions are little more than junctions between atrial, ventricular and extra-cardiac regions as the heart tube forms.[33,34] Later in development, there are clear differences in both myocardial and endocardial phenotype between the AV/OFT and ventricular/atrial regions, particularly the presence of cushion tissue in the former. Rather than disappearing, the extracellular matrix 'cardiac jelly' becomes filled with mesenchymal cells. Two mesenchyme-filled pads (cushions, or ridges) oppose each other across the lumen of the heart tube in each of these regions, and eventually fuse, dividing the lumen into two flow paths (Fig. 2.2D, E).

Detailed studies of molecular and cellular events underlying the endocardial delamination process have been made (reviewed by Markwald et al[193] and Eisenberg and Markwald[194]). These include the use of a culture system, in which the endocardial layer is placed on a collagen gel into which mesenchymal cells can migrate.[195] Cells from the endocardial layer delaminate in response to the presence of extracellular protein complex (ES) made by the adjacent myocardium.[196-198] Some components of this complex have been identified, including fibronectin[199] and ES130.[200] Urokinase and hyaluronidase activities are both elevated in regions of delamination, enabling cells to migrate through the extracellular matrix.[154,155,201] TGFβ isoforms are also present and required for the delamination process (in bird[202-204]) and expression by the endocardial layer in these regions is regulated by ES components.[205] TGFβ RNAs have also been detected in mouse heart in these regions during induction and delamination,[72,73] the presence of -1 and -2 being decreased by exogenous retinoic acid.[206] RNAs for the related growth factors BMP-2A and BMP-4 both become localized in the myocardium of the AV and OFT regions in overlapping but different temporal and physical patterns.[207,208] These are just some of the many growth factors, receptors and other molecules detected in these regions. By comparison of this event in the heart with other induction events elsewhere in the embryo, which share many molecular and cellular aspects, Nakajima et al[205] conclude that it is unlikely that all the cellular events that occur in the induced endocardium are controlled by a single factor, some components effecting only individual aspects of the process.

Transcription factor RNAs detected in the AV endocardium include those encoded by homeobox genes *Msx-1*[185,209-211] and *Mox-1*,[212] the HLH factor Id[213,214] and the retinoic acid receptor RARγ.[167,211] A detailed study of the distribution of *Msx-1* RNA in the bird shows a general temporal and physical correlation with delamination but as negative cells in both the endocardial layer and the mesenchyme could be found in both AV and OFT regions, it seems unlikely that continuous *Msx-1* expression is essential for the process.[185] Unlike expression in the bird endocardium, that in embryonic mouse heart is detectable prior to delamination (from at least 6 somite stage) (PST, unpublished observations;[211]), making its expression less likely to be the temporal switch for delamination. A functional study[215] suggests that Msx-1 (protein)† promotes cell proliferation and suppresses differentiation. Msx-1 can also repress transcription under experimental conditions.[216] Expression of Id, also associated (although not uniquely) with undifferentiated or proliferating cells elsewhere in the embryo, is co-expressed with *Msx-1* RNA in developing mouse hearts and at many

† *With respect to the transcription factors Msx-1, Mox-1 and Msx-2, italics will be used to denote the genes and RNAs, and upright characters denote the protein products.*

other sites.[213,214] In the bird, the cells comprising the endothelial layer lining the heart are heterogeneous with respect to their expression of an antigen detected by the antibody JB3, which may be a fibrillin-like molecule.[20] Only JB3-positive cells give rise to mesenchyme in culture, and JB3-negative cells cannot be induced to express 'JB3 antigen'. Delamination at the AV and OFT regions might arise in part by a cell-sorting process, as in the tubular heart positive and negative cells are mixed throughout its length (and perhaps in mesoderm elsewhere, as Noden[217] reported delamination from grafted quail mesoderm in the OFT of a chick host embryo). The ability of the endocardium to respond to the inductive signal appears both stage- (from HH stage 14) and region-specific and might reflect the completion of the sorting process. If endocardial *Msx-1* expression in the bird were to be restricted to JB3-positive cells, and a similar cell heterogeneity to exist in the mouse, the *Msx-1* expression pattern would imply a less mixed population of cells prior to mesenchymal induction. Endocardial cells might have diverse molecular phenotypes. Although PECAM is expressed throughout mouse endothelium, down-regulated as cells assume a mesenchymal phenotype in the heart, any regional distribution of expression of its splice variants has not yet been determined.[49] Expression of a marker gene under the control of a regulatory region of the receptor tyrosine kinase gene *tie2* shows some regional variation within the endocardium of the 8.5 dpc mouse heart, and that different regulatory elements result in expression in different populations of endothelial cells.[218] Normal *tie* expression is required for cardiac development.[219,220]

Preliminary observations suggest that *Msx-1* RNA-negative cushion mesenchyme cells at both the AV and OFT in the mouse heart express another, unrelated, homeobox gene, *Mox-1*. These cells are principally those on the myocardial side of the cushions. This adjacent/slightly overlapping expression pattern of *Msx-1* and *Mox-1* RNAs is also present in other parts of the embryo (PST, unpublished observations). *Mox-1* expression persists in the mesenchymal populations of the heart until at least 19 dpc and Candia et al[212] suggest a role for Mox-1 in specification and differentiation. Msx-1 and Mox-1 might thus mark 'undifferentiated' and 'differentiating' populations of cells in the heart. The distribution of both Id and RARγ RNAs appears to include *Msx-1* RNA-negative cells,[211,213] suggesting that the expression of these transcription factors is regulated differently, and have separate, even if overlapping, roles. It is not known what controls the expression of these genes (discussed by Lyons et al[211]), although sequence analysis of the mouse *Msx-1* gene reveals a number of putative binding sites for transcription factors, including one to which the Msx-1 homeodomain will bind in vitro.[221] The genes regulated by Msx-1, Mox-1, Id and RARγ have not been identified. Msx-1 and Mox-1 might alter expression by competing at the same DNA binding site. Cytotactin is expressed in cushions[30] and the distribution of tenascin in mouse overlaps with that of TGFβ expression.[72] The regulatory region of the cytotactin gene includes two TAATGAT elements[222] (very similar to a consensus putative Msx-1 homeodomain binding site[223]) the presence of which enables homeobox gene products to drive reporter construct expression in vitro.[224] N-CAM, downregulated in cells as they become mesenchymal,[30] can also be regulated by homeodomain proteins.[85] The distribution of *Mox-1* expression in neural crest mesenchyme is described as corresponding to loose connective tissues associated with sites of muscle assembly and attachment.[212] It is also important that the fibrous regions of the heart adhere firmly to one another (including where cushions fuse) and to the myocardium (where septa meet, and where valves abut myocardium).

Knock-out transgenic mice lacking either *Msx-1* or RARγ have not been reported to show any cardiac abnormalities.[166,225] A number of other sites at which these genes are normally expressed are also unaffected (for example, limbs). Mechanisms might exist whereby upregulation of other gene products with overlapping function occurs in these mice. Double knock-out mice with neither *Msx-1* nor RARγ might reveal such a relationship and have multiple embryonic abnormalities. The cardiac abnormalities in RXRα knock-out transgenic mice also include AV and OFT defects,[162,163] but it is unclear whether these result from a direct effect on the induction and delamination processes themselves.

In summary, the expression of homeobox genes *Msx-1* and *Mox-1*, Id and RARγ in the non-myocardial atrioventricular and outflow tract regions, whilst showing a regional distribution within the heart, seem likely to be regulating aspects of mesenchymal development rather than compartment identity per se. These and many other genes expressed in cells of the AV and OFT are also expressed in other parts of the developing embryo. One of these, the homeobox gene *Msx-2*, is expressed principally in the AV, and, to a lesser extent, OFT myocardium in the unseptated mouse heart (PST, unpublished observations). However, in the developing chick heart, strongest expression is in the myocardium of the 'primary ring', including the right AV junction and the crest of the interventricular septum, where expression remains following completion of septation, coinciding morphologically with the conduction system[185] (which forms a complete ring in the bird heart[226]). This suggests that Msx-2 is not a myocardial 'regional' regulator in a simple sense. Regions of *Msx-2* RNA expression, in both chick and mouse, overlap but do not correlate exactly with a number of cellular, morphological and developmental events: regions of low conduction rate (persisting 'primary' myocardium[119,120]), where cells eventually move apart[227] or form conduction system myocardium (reviewed by Lamers et al[142] and Virágh and Challice[228]); regions where delamination is induced in adjacent endocardium and valves form; and regions where little cell proliferation occurs (reviewed by Thompson et al[152]). The most obvious difference between bird and mammalian hearts in this context is the difference in conduction system (the primary system is less extensive and differentiated in the latter[229]). Cells in the position of the ring expressing *Msx-2* RNA in the chick heart are also molecularly distinct in the human heart before septation[181] although this is not so for all conduction system markers (e.g., Leu-7 in the rat[183]). Expression of *Msx-2* RNA is also detected in the left atrioventricular myocardium in the chick, but is weaker and lost sooner.[152,185] Although this might indicate another role for Msx-2 besides association with the conduction system, myocardium resembling that of the conduction is reported at this site in fetal human.[230] It is also possible that Msx-2 plays several roles in heart development. Outside the heart, *Msx-2* RNA is found at sites of induction (e.g., apical ectodermal ridge of the limb[231]), and although AV (and OFT) myocardium induces endocardial delamination, there is no obvious induction event in the crest of the interventricular septum in chick but not mouse heart. However, there is a difference between the eventual positions of the conduction system with respect to the interventricular septal myocardium and the fibrous component of the atrioventricular septum in birds and mammals.[232] Expression of the *Msx-2* gene in the apical ectodermal ridge, but not that in craniofacial, neural ectoderm or distal limb regions, in mouse is controlled by a highly conserved sequence element.[233] An absence of expression in the heart is not mentioned. Like Msx-1 and Mox-1, Msx-2 seems most likely to be regulating types of developmental events rather than controlling general regional phenotype.

Delamination from endocardium seems to occur by the same process in both AV and OFT, but there are differences, for example, in the amounts of extracellular matrix proteins within and between these regions in the chick heart[20,234-238] which probably have consequences on the physical properties of the region, such as elasticity and resistance to shear stresses, affecting the function of the unseptated heart. Another important difference is the position of the OFT adjacent to extra-cardiac tissues, and the contribution of cells from outside the heart to OFT septation and valve formation. Ablation of a distinct neural crest cell population in the bird heart affects these processes (reviewed by Kirby[239]), as well as coronary artery vascular smooth muscle cell gene expression.[240] Labeling studies in the chick suggest that arterial septation requires these cells[241] and that some of the cells enter the OFT septum and valves.[242,243] The range of defects that result from neural crest ablation, including those affecting the myocardium[244-246] suggests that at least some aspects of the abnormal phenotype result from indirect effects, such as hemodynamics. Abnormalities affecting the behavior of this neural crest population (knock-out transgenic mouse for the homeobox gene *Hoxa-3*[247] and the Splotch phenotype resulting from an abnormality in the *Pax-3* paired box gene[248]) might result in cardiac abnormalities secondary to the functions normally performed by these cells. That neural crest, or other extra-cardiac cells (such as cephalic mesoderm[217]) are not found in the AV region at this stage might reflect the relative inaccessibility of the region rather than an intrinsic 'non-permissive' difference. The development of the OFT valve morphology shows subtle differences between chick and mouse,[5,6] as does the precise OFT cushion or ridge disposition, thought to reflect evolutionary differences between birds and mammals[168,249] but how these differences are reflected at the molecular level is unknown. Labeling experiments in chick embryos indicate that cells from the AV cushions contribute to both the AV septum and the AV valves in the mature, septated heart.[250,251] AV valve leaflet morphology is species-specific[168,252,253] and there is no evidence that endocardial cells are pre-determined to contribute to one or the other. Morphology of both AV and OFT (arterial, semilunar) chick valves can be altered by abnormal hemodynamics, including those resulting from abnormal gross morphology,[254,255] as is the morphology of endocardial cells.[256]

DEVELOPMENT OF REGIONAL PHENOTYPES WITHIN THE HEART TUBE

Efficient function of the heart results from the development of regions with appropriate phenotypes. These phenotypes arise through the presence of a distinctive combination of protein products (controlled largely through transcription and splicing mechanisms) and correlate with morphological features (such as compartments). By at least some molecular and experimental criteria, the heart tube commences as a rather homogeneous place that develops longitudinal asymmetry in the form of (continuous) gradients of properties that become increasingly discrete. In seeking to explain how this arrangement comes about, molecular mechanisms regulating patterning in other parts of the embryo can be considered. One particularly attractive example is that of *Hox* cluster-type homeodomain protein regulation of anteroposterior identity in the neural tube and axial mesoderm (reviewed by Krumlauf[87]). In this model, the 'positional identity' of cells along the axis is reflected in the combination of *Hox* genes expressed. As these genes encode transcription factors, this molecular 'identity code' can then effect expression of the genes appropriate to positions along the antero-posterior axis to result in the appropriate phenotypes. The apparent developmental requirements of the heart tube certainly seem to resemble those of tissues in which *Hox* cluster genes are expressed. A hypothesis suggesting that Hox proteins do indeed regulate regionality in the circulatory system including the heart (illustrated by Gorski et al[257]) is not currently supported by expression data for the genes concerned (e.g., Gaunt et al[258]). Expression of *Hox* genes has not been reported in either myocardium or endocardium. Such expression is present in the neural crest cells, some of which are required for arterial septation and some cells of which can contribute to the arterial valve leaflets and distal outflow tract septum (reviewed by Kirby[239] and described earlier in this chapter). Regionality within the heart tube commences before these cells reach the heart, so any contribution to patterning would be limited.

Hox cluster genes seem unlikely to be involved in regulating heart regionality, but what of other homeobox gene families? As has been described in this chapter, a number of homeobox genes are expressed in the developing heart, including in restricted regions. None of these show the type of overlapping domains of expression typical of *Hox* genes, nor has their expression been detected at very early tube stages that might be expected if subsequent regionality were to result from such molecular patterning.

Other transcription factor families such as MEF-2 and GATA are expressed early in heart development. Both are strongly implicated in subsequent regional control of contractile protein gene expression. Expression in tissues and at times when these contractile protein genes are not expressed strongly suggests roles for MEF-2 and GATA proteins in direct or indirect regulation of expression of other types of gene. Distribution of MEF-2 RNAs later in heart development makes them candidates for maintenance of regional phenotype. Splice-variant forms of three MEF-2 gene products can exist and the distribution of the proteins they encode has not been reported. It remains possible that these (or other) proteins regulate regionality by variants detectable only at the protein level, a mechanism dependent on processes subsequent to transcriptional control. These arguments serve to shift the emphasis away from the transcriptional control of genes being regulated at the level of transcription of the transcription factors themselves, and are valid but currently scantily supported hypotheses, the question still arising: what would regulate the post-transcriptional or translational mechanisms on which regional phenotypes would then depend?

Two other aspects of the 'patterned heart tube' model require further consideration. A heart tube patterned into distinct atrial, atrioventricular, ventricular and outflow tract compartments, for example, by a mechanism resembling that of the *Hox* cluster genes elsewhere in the embryo might also be expected to show certain regional duplication/deletion phenotypes when the expression of a relevant patterning gene is defective. Genetic mutagenesis experiments in a lower vertebrate (Zebrafish, *Danio rerio*), the development of which is less dependent on heart function at early embryonic stages, have generated many abnormal heart phenotypes, none of which have been classified as being 'atrioventricular junction replacing ventricle', for example.[259] Detecting such phenotypes in higher vertebrates might be hard as such hearts are likely to be very inefficient, so that embryonic death would occur early in development before sufficient regional markers are expressed to confirm compartment identity. Although heart tube morphology can be grossly categorized, sufficient features consistent with performing this type of analysis have not yet been distinguished. This is not to say that there are no features in the forming heart tube that reflect future compartmental arrangement. The presence of sulci in the fusing myocardial precursor regions that fate-map to distinct parts of the heart presents a serious hurdle to a model whereby the heart tube can be considered homogeneous. A transcription factor regulating the expression of genes for both the muscular and morphological phenotypes of a compartment would presumably 'obey' the boundaries defined by these features, and might be expected to be expressed from the time the sulci are present, preceding what is currently considered to be homogeneous or at least continuous expression of genes that are subsequently very discretely expressed.

What alternatives are there to the 'patterned' model of heart tube regionality? Molecular and developmental evidence is consistent with a difference between myocardial precursor cells more lateral/caudal and those more medial/cranial and between left and right, as reviewed earlier in this chapter. This could result from the establishment of a more simple pattern than one in which each compartment of the septated heart is prefigured.[†] Subsequent regionalization within these grossly defined regions could result from rather local environmental cues and involve many of the same molecules. Such a mechanism would be consistent with the expression of proteins that are eventually regionally specific, but which commence as more widespread (such as MLC2A[125] and MLC3F and a transgene containing part of the MLC3F promoter[113]). Although much is known about the transcription factors that can regulate contractile gene expression in the heart, their distributions, sequences and regulation are too incompletely defined to present a coherent account of regulation of even this rather well characterized group of proteins during prenatal development.

One aspect of regional phenotype development in the heart tube that makes the modeling of its mechanism particularly challenging is that of 'longitudinal' versus 'lateral' identity, particularly pertinent to the ventricular compartment. Analysis of the development of left and right ventricular phenotypes, and the interventricular septum, requires good experimental model systems and markers, in addition to consideration of the development and structure of hearts in 'lower' vertebrates in which septation and compartmentation is less (described, for example, by Van Mierop and Kutsche[168]). Any contribution made by the presence of both left and right cardiac precursor regions is hard to determine, as descriptions of hearts derived from only one are available only at stages before the trabeculation and septum development. Purely 'transverse asymmetry' in models of regional identity suggested by the rather parallel arrangement of the mature compartments is in-

[†] *See also a recent review, ref. 262*

consistent with the apparent serial disposition of the left and right trabecular ventricle anlagen along the looped heart tube. Attempts to negate one type of model in favor of the other are currently thwarted by the lack of both reliable meaningful markers that define ventricular compartments throughout their development, and the results of experiments investigating the developmental potential of cells from different regions of the heart tube (rather than fate-mapping, or gross categorization of tissue origin in morphologically abnormal hearts). Such attempts are in some ways meaningless: that normal, septated hearts have separate connections between atrium and ventricle on the left and right sides indicates that a purely serial arrangement of left- and right- ventricularly determined cells cannot exist at earlier stages (unless some of the left ventricular cells migrate or die, for which there is no evidence).

A common molecular regulatory mechanism for the development of morphological and functional (force-generation) regionality cannot yet be proposed. Some aspects of myocardial phenotypic diversity occur in the absence of normal gross morphology, but, even so, significant experimental dissociation of morphological and molecular boundaries in the heart is hard to demonstrate. This is not strong evidence for initiation or regulation of regionality by common transcription factors, although this could be because the developmental distribution of so few has been determined. A number of aspects of heart development, such as organization of myofibrils and cell junctions, and regional differences in cell proliferation, provide links between force-generation and cellular and gross aspects of morphogenesis. Feed-back mechanisms, including those responding to hemodynamics and other physical factors, serving to maintain homeostasis are unlikely to include detectors of specific morphological abnormalities such as 'ventricular septal defect' or 'double outlet right ventricle'! The behavior of ventricular myocardial cells is not likely be determined simply by 'position relative to the interventricular septum', nor left and right atrioventricular valve morphology to require mesenchyme cells with 'left' and 'right' positional identity respectively. Whilst the morphology of the heart is undoubtedly essential to efficient function, the criteria by which heart development is regulated will be established by humans considering the 'heart cell's view' of the heart, requiring a multi-disciplinary approach and increasingly detailed and careful analyses and reports of all aspects of the developing heart.

REFERENCES

1. Icardo JM, Manasek FJ. Cardiogenesis: Development mechanisms and embryology. In:Fozzard HA, Haber E, Jennings RB, Katz AM, Morgan HE (eds): The Heart and Cardiovascular System. 2nd ed. New York:Raven Press Limited, 1992: 1563-1586.
2. Hendrix MJC, Morse DE. Atrial septation I. Scanning electron micros-

copy in the chick. Dev Biol 1977; 57:345-63.

3. Morse DE. Formation of foramina secunda in the chick. In:Pexieder T (ed): Perspectives in Cardiovascular Research: Mechanisms of Cardiac Morphogenesis and Teratogenesis. New York:Raven Press, 1981: 133-149.

4. Jones WK, Sánchez A, Robbins J. Murine pulmonary myocardium: Developmental analysis of cardiac gene expression. Dev Dynam 1994; 200:117-28.

5. Hurle JM. Scanning and light microscope studies of the development of the chick embryo semilunar heart valves. Anat Embryol 1979; 157:69-80.

6. Hurle JM, Colveé E, Blanco AM. Development of mouse semilunar valves. Anat Embryol 1980; 160:83-91.

7. Virágh S, Challice CE. The origin of the epicardium and the embryonic myocardial circulation in the mouse. Anat Rec 1981; 201:157-68.

8. Hiruma T, Hirakow R. Epicardial formation in embryonic chick heart: Computer-aided reconstruction, scanning, and transmission electron microscopic studies. Am J Anat 1989; 184:129-38.

9. Hamburger V, Hamilton HL. A series of normal stages in the development of the chick embryo. J Morphol 1951; 38:49-92.

10. Theiler K. The House Mouse:Development and Normal Stages from Fertilization to 4 Weeks of Age, Berlin: Springer-Verlag, 1972.

11. Theiler K. The House Mouse: Atlas of Embryonic Development, New York:Springer-Verlag, 1989.

12. Kaufman MH. The Atlas of Mouse Development, London:Academic Press, 1992.

13. Stalsberg H, DeHaan RL. The precardiac areas and formation of the tubular heart in the chick embryo. Dev Biol 1969; 19:128-59.

14. Garcia-Martinez V, Schoenwolf GC. Primitive-streak origin of the cardiovascular system in avian embryos. Dev Biol 1993; 159:706-19.

15. Hatada Y, Stern CD. A fate map of the epiblast of the early chick embryo. Development 1994; 120:2879-89.

16. Linask KK. N-cadherin localization in early heart development and polar expression of Na⁺, K⁺-ATPase, and integrin during pericardial coelom formation and epithelialization of the differentiating myocardium. Dev Biol 1992; 151:213-24.

17. DeRuiter MC, Poelmann RE, Mentink MMT et al. Early formation of the vascular system in quail embryos. Anat Rec 1993; 235:261-74.

18. Han Y, Dennis JE, Cohen-Gould L et al. Expression of sarcomeric myosin in the presumptive myocardium of chicken embryos occurs within six hours of myocyte commitment. Dev Dynam 1992; 193:257-65.

19. Linask KK, Lash JW. Early heart development: Dynamics of endocardial cell sorting suggests a common origin with cardiomyocytes. Dev Dynam 1993; 195:62-9.

20. Wunsch AM, Little CD, Markwald RR. Cardiac endothelial heterogeneity defines valvular development as demonstrated by the diverse expression of JB3, an antigen of the endocardial cushion tissue. Dev Biol 1994; 165:585-601.

21. Tokuyasu KT, Maher PA. Immunocytochemical studies of cardiac myofibrillogenesis in early chick embryos. I. Presence of immunofluorescent titin spots in premyofibril stages. J Cell Biol 1987; 105:2781-93.
22. De Jong F, Geerts WJC, Lamers WH et al. Isomyosin expression pattern during formation of the tubular chicken heart: A three-dimensional immunohistochemical analysis. Anat Rec 1990; 226:213-27.
23. Mikawa T, Borisov A, Brown AMC et al. Clonal analysis of cardiac morphogenesis in the chicken embryo using a replication-defective retrovirus. I. Formation of the ventricular myocardium. Dev Dynam 1992; 193:11-23.
24. Mikawa T, Fischman DA. Retroviral analysis of cardiac morphogenesis: Discontinuous formation of coronary vessels. Proc Natl Acad Sci USA 1992; 89:9504-8.
25. Linask KK, Lash JW. Precardiac cell migration: Fibronectin localization at mesoderm-endoderm interface during directional movement. Dev Biol 1986; 114:87-101.
26. Linask KK, Lash JW. A role for fibronectin in the migration of avian precardiac cells. I. Dose-dependent effects of fibronectin antibody. Dev Dynam 1988; 129:315-23.
27. Linask KK, Lash JW. A role for fibronectin in the migration of avian precardiac cells. II. Rotation of the heart-forming region during different stages and its effects. Dev Biol 1988; 129:324-9.
28. Virágh S, Szabó E, Challice CE. Formation of the primitive myo- and endocardial tubes in the chicken embryo. J Mol Cell Cardiol 1989; 21:123-37.
29. Shiraishi I, Takamatsu T, Fujita S. 3-D observation of N-cadherin expression during cardiac myofibrillogenesis of the chick embryo using a confocal laser scanning microscope. Anat Embryol 1993; 102:115-20.
30. Crossin KL, Hoffman S. Expression of adhesion molecules during the formation and differentiation of the avian endocardial cushion tissue. Dev Biol 1991; 145:277-86.
31. Hirota A, Kamino K, Komuro H et al. Mapping of early development of electrical activity in the embryonic chick heart using multiple-site optical recording. J Physiol 1987; 383:711-28.
32. Kamino K, Komuro H, Sakai T, et al. Optical assessment of functional organization of the pacemaking area in the early precontractile chick heart. In:Clark EB, Takao A (eds): Developmental Cardiology: Morphogenesis and Function. Mount Kisco, New York:Futura Publishing Company Inc, 1990: 257-271.
33. de la Cruz MV, Gómez CS, Arteaga MM et al. Experimental study of the development of the truncus and the conus in the chick embryo. J Anat 1977; 123:661-86.
34. de la Cruz MV, Sánchez-Gómez C, Palomino MA. The primitive cardiac regions in the straight tube heart (Stage 9·) and their anatomical expression in the mature heart: an experimental study in the chick embryo. J Anat 1989; 165:121-31.

35. Garcia-Peláez I, Arteaga M. Experimental study of the development of the truncus arteriosus of the chick embryo heart. I. Time of appearance. Anat Rec 1993; 237:378-84.

36. Litvin J, Montgomery M, Gonzalez-Sanchez A et al. Commitment and differentiation of cardiac myocytes. Trends Cardiovasc Med 1992; 2:27-32.

37. Goswami S, Qasba P, Ghatpande S et al. Differential expression of the myocyte enhancer factor 2 family of transcription factors in development: the cardiac factor BBF-1 is an early marker for cardiogenesis. Mol Cell Biol 1994; 14:5130-8.

38. Ruzicka DL, Schwartz RJ. Sequential activation of α-actin genes during avian cardiogenesis: Vascular smooth muscle α-actin gene transcripts mark the onset of cardiac differentiation. J Cell Biol 1988; 107:2575-86.

39. Sugi Y, Lough J. Onset of expression and regional deposition of alpha-smooth and sarcomeric actin during avian heart development. Dev Dynam 1992; 193:116-24.

40. Bisaha JG, Bader D. Identification and characterization of a ventricular-specific avian myosin heavy chain, VMHC1: Expression in differentiating cardiac and skeletal muscle. Dev Biol 1991; 148:355-64.

41. Yutzey KE, Rhee JT, Bader D. Expression of the atrial-specific myosin heavy chain AMHC1 and the establishment of anteroposterior polarity in the developing chicken heart. Development 1994; 120:871-83.

42. Hiruma T, Hirakow R. An ultrastructural topographical study on myofibrillogenesis in the heart of the chick embryo during pulsation onset period. Anat Embryol 1985; 172:325-9.

43. Tokuyasu KT. Immunocytochemical studies of cardiac myofibrillogenesis in early chick embryos. III. Generation of fasciae adherentes and costameres. J Cell Biol 1989; 108:43-53.

44. Soler AP, Knudsen KA. N-cadherin involvement in cardiac myocyte interaction and myofibrillogenesis. Dev Biol 1994; 162:9-17.

45. Byeon MK, Sugi Y, Markwald RR et al. NCAM polypeptides in heart development: Association with Z discs of forms that contain the muscle-specific domain. J Cell Biol 1995; 128:209-21.

46. Kaufman MH, Navaratnam V. Early differentiation of the heart in mouse embryos. J Anat 1981; 133:235-46.

47. DeRuiter MC, Poelmann RE, VanderPlas-de Vries I et al. The development of the myocardium and endocardium in mouse embryos. Fusion of two heart tubes?. Anat Embryol 1992; 185:461-73.

48. George EL, Georges-Labouesse EN, Patel-King RS et al. Defects in mesoderm, neural tube and vascular development in mouse embryos lacking fibronectin. Development 1993; 119:1079-91.

49. Buck CA, Baldwin HS, DeLisser H et al. Cell adhesion receptors and early mammalian heart development: an overview. C R Acad Sci Paris 1993; 316:849-59.

50. Schaart G, Viebahn C, Langmann W et al. Desmin and titin expression in early postimplantation mouse embryos. Development 1989; 107:585-96.

51. van der Loop FTL, Schaart G, Langmann W et al. Expression and organization of muscle specific proteins during the early developmental stages of the rabbit heart. Anat Embryol 1992; 185:439-50.

52. Simpson DG, Decker ML, Clark WA et al. Contractile activity and cell-cell contact regulate myofibrillar organization in cultured cardiac myocytes. J Cell Biol 1993; 123:323-36.

53. Wiens D, Sullins M, Spooner BS. Precardiac mesoderm differentiation in vitro. Actin-isotype synthetic transitions, myofibrillogenesis, initiation of heartbeat, and the possible involvement of collagen. Differentiation 1984; 28:62-72.

54. Fisher SA, Periasamy M. Collagen synthesis inhibitors disrupt embryonic cardiocyte myofibrillogenesis and alter the expression of cardiac specific genes in vitro. J Mol Cell Cardiol 1994; 26:721-31.

55. Lyons GE. In situ analysis of the cardiac muscle gene program during embryogenesis. Trends Cardiovasc Med 1994; 4:70-7.

56. Sassoon DA, Garner I, Buckingham M. Transcripts of α-cardiac and α-skeletal actins are markers for myogenesis in the mouse embryo. Development 1988; 104:155-64.

57. Baldwin HS, Jensen KL, Solursh M. Myogenic cytodifferentiation of the precardiac mesoderm in the rat. Differentiation 1991; 47:163-72.

58. Sawtell NM, Lessard JL. Cellular distribution of smooth muscle actins during mammalian embryogenesis. Expression of the α-vascular but not the α-enteric isoform in differentiating striated myocytes. J Cell Biol 1989; 109:2929-37.

59. De Groot IJM, Lamers WH, Moorman AFM. Isomyosin expression patterns during rat heart morphogenesis: An immunohistochemical study. Anat Rec 1989; 224:365-73.

60. Ausoni S, De Nardi C, Moretti P et al. Developmental expression of rat cardiac troponin I mRNA. Development 1991; 112:1041-51.

61. Schiaffino S, Gorza L, Ausoni S. Troponin isoform switching in the developing heart and its functional consequences. Trends Cardiovasc Med 1993; 3:12-7.

62. Montgomery MO, Litvin J, Gonzalez-Sanchez A et al. Staging of commitment and differentiation of avian cardiac myocytes. Dev Biol 1994; 164:63-71.

63. Antin PB, Taylor RG, Yatskievych T. Precardiac mesoderm is specified during gastrulation in quail. Dev Dynam 1994; 200:144-54.

64. Sarasa M, Climent S. Cardiac differentiation induced by dopamine in undifferentiated cells of early chick embryo. Dev Biol 1991; 148:243-8.

65. Sugi Y, Lough J. Anterior endoderm is a specific effector of terminal cardiac myocyte differentiation of cells from the embryonic heart forming region. Dev Dynam 1994; 200:155-62.

66. Inagaki T, Garcia-Martinez V, Schoenwolf GC. Regulative ability of the prospective cardiogenic and vasculogenic areas of the primitive streak during avian gastrulation. Dev Dynam 1993; 197:57-68.

67. Arias M, Garcia C, Villar JM. Ultrastructural analysis of chick embryo blastoderms explanted in vitro in absence of endoderm. Acta anat 1987; 128:27-32.
68. Sugi Y, Sasse J, Barron M et al. Developmental expression of fibroblast growth factor receptor-1 (*cek-1; flg*) during heart development. Dev Dynam 1995; 202:115-25.
69. Sugi Y, Sasse J, Lough J. Inhibition of precardiac mesoderm cell proliferation by antisense oligodeoxynucleotide complementary to fibroblast growth factor-2 (FGF-2). Dev Biol 1993; 157:28-37.
70. Orr-Urtreger A, Givol D, Yayon A et al. Developmental expression of two murine fibroblast growth factor receptors, *flg* and *bek*. Development 1991; 113:1419-34.
71. Parker TG, Chow KL, Schwartz RJ et al. Differential regulation of skeletal α-actin transcription in cardiac muscle by two fibroblast growth factors. Proc Natl Acad Sci USA 1990; 87:7066-70.
72. Akhurst RJ, Lehnert SA, Faissner A et al. TGF beta in murine morphogenetic processes: the early embryo and cardiogenesis. Development 1990; 108:645-56.
73. Dickson MC, Slager HG, Duffie E et al. RNA and protein localisations of TGFβ2 in the early mouse embryo suggest an involvement in cardiac development. Development 1993; 117:625-39.
74. Tapscott SJ, Lassar AB, Davis RL et al. 5-Bromo-2'-deoxyuridine blocks myogenesis by extinguishing expression of MyoD1. Science 1989; 245:532-6.
75. Satin J, Fujii S, DeHaan RL. Development of cardiac beat rate in early chick embryos is regulated by regional cues. Dev Biol 1988; 129:103-13.
76. Edmondson DG, Lyons GE, Martin JF et al. *Mef2* gene expression marks the cardiac and skeletal muscle lineages during mouse embryogenesis. Development 1994; 120:1251-63.
77. Heikinheimo M, Scandrett JM, Wilson DB. Localization of transcription factor GATA-4 to regions of the mouse embryo involved in cardiac development. Dev Biol 1994; 164:361-73.
78. Laverriere AC, MacNeill C, Mueller C et al. GATA-4/5/6, a subfamily of three transcription factors transcribed in developing heart and gut. J Biol Chem 1994; 269:23177-84.
79. Litvin J, Montgomery MO, Goldhamer DJ et al. Identification of DNA-binding protein(s) in the developing heart. Dev Biol 1993; 156:409-17.
80. Chen B, Blaschuk OW, Hales BF. Cadherin mRNAs during rat embryo development in vivo and in vitro. Teratology 1991; 44:581-90.
81. MacCalman CD, Bardeesy N, Holland PC et al. Noncoordinate developmental regulation of N-cadherin, N-CAM, integrin, and fibronectin mRNA levels during myoblast terminal differentiation. Dev Dynam 1992; 195:127-32.
82. Miyatani S, Copeland NG, Gilbert DJ et al. Genomic structure and chromosomal mapping of the mouse N-cadherin gene. Proc Natl Acad Sci USA 1992; 89:8443-7.

83. Hatta K, Takagi S, Fujisawa H et al. Spatial and temporal expression pattern of N-cadherin cell adhesion molecules correlated with morphogenetic processes of chicken embryos. Dev Biol 1987; 120:215-27.

84. Fürst DO, Osborn M, Weber K. Myogenesis in the mouse embryo: Differential onset of expression of myogenic proteins and the involvement of titin in myofibril assembly. J Cell Biol 1989; 109:517-27.

85. Valarché I, Tissier-Seta J-P, Hirsch M-R et al. The mouse homeodomain protein Phox2 regulates *Ncam* promoter activity in concert with Cux/CDP and is a putative determinant of neurotransmitter phenotype. Development 1993; 119:881-96.

86. Edelman GM, Jones FS. Outside and downstream of the homeobox. J Biol Chem 1993; 268:20683-6.

87. Krumlauf R. Mouse *Hox* genetic functions. Curr Op Genet Develop 1993; 3:621-5.

88. Komuro I, Izumo S. *Csx*: A murine homeobox-containing gene specifically expressed in the developing heart. Proc Natl Acad Sci USA 1993; 90:8145-9.

89. Lints TJ, Parsons LM, Hartley L et al. *Nkx-2.5*: a novel murine homeobox gene expressed in early heart progenitor cells and their myogenic descendants. Development 1993; 119:419-31.

90. Duboule D. Guidebook to the Homeobox Genes. Oxford:Sambrook & Tooze Publication (Oxford University Press), 1994: 207

91. Bodmer R. Heart development in *Drosophila* and its relationship to vertebrates. Trends Cardiovasc Med 1995; 5:21-8.

92. Arendt D, Nübler-Jung K. Inversion of dorsoventral axis? Nature 1994; 371:28

93. Gee H. Return of the amphioxus. Nature 1994; 370:504-5.

94. Rähr H. The circulatory system of Amphioxus (*Branchiostoma lanceolatum* (Pallas)). Acta zool (Stockh) 1979; 60:1-18.

95. von Skramlik E. Über den Kreislauf bei den niedersten Chordaten. Erg d Biol 1938; 15:166-308.

96. Manning A, McLachlan JC. Looping of chick embryo hearts in vitro. J Anat 1990; 168:257-63.

97. Stalsberg H. Mechanism of dextral looping of the embryonic heart. Am J Cardiol 1970; 25:265-71.

98. Cooke J. Vertebrate embryo handedness. Nature 1995; 374:681

99. Itasaki N, Nakamura H, Sumida H et al. Actin bundles on the right side in the caudal part of the heart tube play a role in dextro-looping in the embryonic chick heart. Anat Embryol 1991; 183:29-39.

100. Hoyle C, Brown NA, Wolpert L. Development of left/right handedness in the chick heart. Development 1992; 115:1071-8.

101. Fujinaga M, Baden JM. Microsurgical study on the mechanisms determining sidedness of axial rotation in rat embryos. Teratology 1993; 47:585-93.

102. Brown NA, McCarthy A, Wolpert L. Development of handed body asymmetry in mammals. In:Bock GR, Marsh J (eds): Biological Asymmetry and Handedness. Chichester:Wiley, 1991: 182-201.

103. Britz-Cunningham SH, Shah MM, Zuppan CW et al. Mutations of the *connexin43* gap-junction gene in patients with heart malformations and defects of laterality. New Eng J Med 1995; 332:1323-9.

104. Reaume AG, de Sousa PA, Kulkarni S et al. Cardiac malformation in neonatal mice lacking connexin43. Science 1995; 267:1831-4.

105. Easton H, Veini M, Bellairs R. Cardiac looping in the chick embryo: the role of the posterior precardiac mesoderm. Anat Embryol 1992; 185:249-58.

106. Stalsberg H. Regional mitotic activity in the precardiac mesoderm and differentiating heart tube in the chick embryo. Dev Biol 1969; 20:18-45.

107. Chen Y, Solursh M. Comparison of Hensen's node and retinoic acid in secondary axis induction in the early chick embryo. Dev Dynam 1992; 195:142-51.

108. Chen Y, Huang L, Russo AF et al. Retinoic acid is enriched in Hensen's node and is developmentally regulated in the early chicken embryo. Proc Natl Acad Sci USA 1992; 89:10056-9.

109. Hogan BLM, Thaller C, Eichele G. Evidence that Hensen's node is a site of retinoic acid synthesis. Nature 1992; 359:237-41.

110. Stalsberg H. The origin of heart asymmetry: Right and left contributions to the early chick embryo heart. Dev Biol 1969; 19:109-27.

111. Hasselbaink HDJ, Labruyere WT, Moorman AFM et al. Creatine kinase isozyme expression in prenatal rat heart. Anat Embryol 1990; 182:195-302.

112. Lyons GE, Mühlebach S, Moser A et al. Developmental regulation of creatine kinase gene expression by myogenic factors in embryonic mouse and chick skeletal muscle. Development 1991; 113:1017-29.

113. Kelly R, Alonso S, Tajbakhsh S, Cossu G, Buckingham M. Myosin light chain 3F regulatory sequences confer regionalised cardiac and skeletal muscle expression in transgenic mice. J Cell Biol 1995; (in press)

114. Männer J, Seidl W, Steding G. Correlation between the embryonic head flexures and cardiac development. Anat Embryol 1993; 188:269-85.

115. Steding G, Seidl W. Contribution to the development of the heart. Part I: Normal development. Thorac cardiovasc Surgeon 1980; 28:386-409.

116. Manasek FJ, Isobe Y, Shimada Y, et al. The embryonic myocardial cytoskeleton, interstitial pressure, and the control of morphogenesis. In:Nora JS, Takao A (eds): Congenital Heart Disease: Causes and Processes. New York:Futura Publishing Company, 1984: 359-376.

117. Baldwin HS, Lloyd TR, Solursh M. Hyaluronate degradation affects ventricular function of the early postlooped embryonic rat heart in situ. Circ Res 1994; 74:244-52.

118. Manasek FJ, Reid M, Vinson W et al. Glycosaminoglycan synthesis by the early embryonic chick heart. Dev Biol 1973; 35:332-48.

119. Moorman AFM, Lamers WH. Molecular anatomy of the developing heart. Trends Cardiovasc Med 1994; 4:257-64.

120. Moorman AFM, Lamers WH. A molecular approach towards the understanding of early heart development: an emerging synthesis. In:El Haj A (ed): Molecular Biology of Muscle. Cambridge:The Company of Biologists Limited, 1992:285-300.

121. Moorman AFM, Vermeulen JLM, Koban MU et al. Patterns of expression of sarcoplasmic reticulum Ca^{2+}-ATPase and phospholamban mRNAs during rat heart development. Circ Res 1995; 76:616-25.

122. Maltsev VA, Rohwedel J, Hescheler J et al. Embryonic stem cells differentiate in vitro into cardiomyocytes representing sinusnodal, atrial and ventricular cell types. Mech Develop 1993; 44:41-50.

123. Maltsev VA, Wobus AM, Rohwedel J et al. Cardiomyocytes differentiated in vitro from embryonic stem cells developmentally express cardiac-specific genes and ionic currents. Circ Res 1994; 75:233-44.

124. Miller-Hance WC, LaCorbiere M, Fuller SJ et al. In vitro chamber specification during embryonic stem cell cardiogenesis. J Biol Chem 1993; 268:25244-252.

125. Kubalak SW, Miller-Hance WC, O'Brien TX et al. Chamber specification of atrial myosin light chain-2 expression precedes septation during murine cardiogenesis. J Biol Chem 1994; 269:16961-70.

126. O'Brien TX, Lee KJ, Chien KR. Positional specification of ventricular myosin light chain 2 expression in the primitive murine heart tube. Proc Natl Acad Sci USA 1993; 90:5157-61.

127. Zeller R, Bloch KD, Williams BS et al. Localized expression of the atrial natriuretic factor gene during cardiac embryogenesis. Genes Devel 1987; 1:693-8.

128. Markwald RR. Formation and early morphogenesis of the primary heart tube. In:Clarke EB (ed): Congenital Heart Disease. Armonk, New York:Futura Publishing Company Inc, 1994: 149-156.

129. Martin JF, Miano JM, Hustad CM et al. A *Mef2* gene that generates a muscle-specific isoform via alternative mRNA splicing. Mol Cell Biol 1994; 14:1647-56.

130. Lee KJ, Ross RS, Rockman HA et al. Myosin light chain-2 luciferase transgenic mice reveal distinct regulatory programs for cardiac and skeletal muscle-specific expression of a single contractile protein gene. J Biol Chem 1992; 267:15875-85.

131. Zhu H, Nguyen VTB, Brown AB et al. A novel, tissue-restricted zinc finger protein (HF-1b) binds to the cardiac regulatory element (HF-1b/MEF-2) in the rat myosin light-chain 2 gene. Mol Cell Biol 1993; 13:4432-44.

132. Lee KJ, Hickey R, Zhu H et al. Positive regulatory elements (HF-1a and HF-1b) and a novel negative regulatory element (HF-3) mediate ventricular muscle-specific expression of myosin light-chain 2-luciferase fusion genes in transgenic mice. Mol Cell Biol 1994; 14:1220-9.

133. Osmond MK, Butler AJ, Voon FCT et al. The effects of retinoic acid on heart formation in the early chick embryo. Development 1991; 113:1405-17.

134. Stainier DYR, Fishman MC. Patterning the zebrafish heart tube: Acquisition of anteroposterior polarity. Dev Biol 1992; 153:91-101.

135. Modak SP, Ghatpande SK, Rane RK et al. Caudalization by retinoic acid is correlated with inhibition of cell population growth and expansion of chick blastoderms cultured in vitro. Int J Dev Biol 1993; 37:601-7.

136. Dersch H, Zile MH. Induction of normal cardiovascular development in the vitamin A-deprived quail embryo by natural retinoids. Dev Biol 1993; 160:424-33.

137. Heine UI, Roberts AB, Munoz EF et al. Effects of retinoid deficiency on the development of the heart and vascular system of the quail embryo. Virchows Archiv 1985; 50:135-52.

138. Smith SM. Retinoic acid receptor isoform β2 is an early marker for alimentary tract and central nervous system positional specification in the chicken. Dev Dynam 1994; 200:14-25.

139. de Thé H, del Mar Vivanco-Ruiz M, Tiollais P et al. Identification of a retinoic acid responsive element in the retinoic acid receptor β gene. Nature 1990; 343:177-80.

140. Arceci RJ, King AAJ, Simon MC et al. Mouse GATA-4: a retinoic acid-inducible GATA-binding transcription factor expressed in endodermally derived tissues and heart. Mol Cell Biol 1993; 13:2235-46.

141. Jonk LJC, de Jonge MEJ, Vervaart JMA et al. Isolation and developmental expression of retinoic-acid-induced genes. Dev Biol 1994; 161:604-14.

142. Lamers WH, De Jong F, De Groot IJM et al. The development of the avian conduction system, a review. Eur J Morphol 1991; 29:233-53.

143. Tokuyasu KT. Co-development of embryonic myocardium and myocardial circulation. In:Clark EB, Takao A (eds): Developmental Cardiology: Morphogenesis and Function. Mount Kisco, New York:Futura Publishing Company Inc, 1990: 205-218.

144. Icardo JM, Fernandez-Terán A. Morphologic study of ventricular trabeculation in the embryonic chick heart. Acta anat 1987; 130:264-74.

145. Icardo JM, Manasek FJ. Fibronectin distribution during early chick embryo heart development. Dev Biol 1983; 95:19-30.

146. Rychterová V. Principle of growth of the heart ventricular wall in the chick embryo. Folia Morphol (Praha) 1971; 19:262-72.

147. Rychter Z, Rychterová V. Angio- and myoarchitecture of the heart wall under normal and experimentally changed morphogenesis. In:Pexieder T (ed): Perspectives in Cardiovascular Research. Mechanisms of Cardiac Morphogenesis and Teratogenesis. New York:Raven Press, 1981: 431-452.

148. Thompson RP, Lindroth JR, Wong YM. Regional differences in DNA-synthetic activity in the preseptation myocardium of the chick. In:Clark EB, Takao A (eds): Developmental Cardiology: Morphogenesis and Function. Mount Kisco, New York:Futura Publishing Company Inc, 1990: 219-234.

149. Thompson RP, Lindroth JR, Alles AJ et al. Cell differentiation birthdates in the embryonic rat heart. Ann N Y Acad Sci 1990; 588:446-8.

150. Mikawa T, Cohen-Gould L, Fischman DA. Clonal analysis of cardiac morphogenesis in the chicken embryo using a replication-defective retrovirus. III: Polyclonal origin of adjacent ventricular myocytes. Dev Dynam 1992; 195:133-41.

151. Rumyantsev PP. Interrelations of the proliferation and differentiation processes during cardiac myogenesis and regeneration. Int Rev Cytol 1977; 51:187-273.

152. Thompson RP, Kanai T, Germouth PJ, et al. Organization and function of early specialized myocardium. In:Clark EB, Markwald RR, Takao RA (eds): Developmental Mechanisms of Heart Disease. Amonk New York:Futura Publishing Company Inc, 1995: 269-280.

153. Lyons GE, Schiaffino S, Sassoon D et al. Developmental regulation of myosin expression in mouse cardiac muscle. J Cell Biol 1990; 111:2427-36.

154. McGuire PG, Orkin RW. Urokinase activity in the developing avian heart: A spatial and temporal analysis. Dev Dynam 1992; 193:24-33.

155. McGuire PG, Alexander SM. Urokinase production by embryonic endocardial-derived cells: Regulation by substrate composition. Dev Biol 1993; 155:442-51.

156. Roman J, McDonald JA. Expression of fibronectin, the integrin α5, and α-smooth muscle actin in heart and lung development. Am J Respir Cell Mol Biol 1992; 6:472-80.

157. Kwee L, Baldwin HS, Shen HM et al. Defective development of the embryonic and extraembryonic circulatory systems in vascular cell adhesion molecule (VCAM-1) deficient mice. Development 1995; 121:489-503.

158. Elices MJ, Osborn L, Takada Y et al. VCAM-1 on activated endothelium interacts with the leukocyte integrin VLA-4 at a site distinct from the VLA-4/fibronectin binding site. Cell 1990; 60:577-84.

159. Stadler HS, Solursh M. Characterization of the homeobox-containing gene *GH6* identifies novel regions of homeobox gene expression in the developing chick embryo. Dev Biol 1994; 161:251-62.

160. Opstelten DE, Vogels R, Robert B et al. The mouse homeobox gene, *S8*, is expressed during embryogenesis predominantly in mesenchyme. Mech Develop 1991; 34:29-42.

161. Cserjesi P, Lilly B, Bryson L et al. MHox: a mesodermally restricted homeodomain protein that binds an essential site in the muscle creatine kinase enhancer. Development 1992; 115:1087-101.

162. Sucov HM, Dyson E, Gumeringer CL et al. RXRα mutant mice establish a genetic basis for vitamin A signaling in heart morphogenesis. Genes Devel 1994; 8:1007-18.

163. Kastner P, Grondona JM, Mark M et al. Genetic analysis of RXRα developmental function: Convergence of RXR and RAR signaling pathways in heart and eye morphogenesis. Cell 1994; 78:987-1003.

164. Dollé P, Fraulob V, Kastner P et al. Developmental expression of murine retinoid X receptor (RXR) genes. Mech Develop 1994; 45:91-104.

165. Mangelsdorf DJ, Borgmeyer U, Heyman RA et al. Characterization of three RXR genes that mediate the action of 9-*cis* retinoic acid. Genes Devel 1992; 6:329-44.

166. Mendelsohn C, Lohnes D, Décimo D et al. Function of the retinoic acid receptors (RARs) during development (II) Multiple abnormalities at various stages of organogenesis in RAR double mutants. Development 1994; 120:2749-71.

167. Dollé P, Ruberte E, Leroy P et al. Retinoic acid receptors and cellular retinoid binding proteins I. A systematic study of their differential pat-

tern of transcription during mouse organogenesis. Development 1990; 110:1133-51.

168. Van Mierop LHS, Kutsche LM. Comparative anatomy and embryology of the ventricles and arterial pole of the vertebrate heart. In:Nora JJ, Takao A (eds): Congenital Heart Disease: Causes and Processes. New York:Futura Publishing Company Inc, 1984: 459-479.

169. Kreidberg JA, Sariola H, Loring JM et al. WT-1 is required for early kidney development. Cell 1993; 74:679-91.

170. Armstrong JF, Pritchard-Jones K, Bickmore WA et al. The expression of the Wilm's tumor gene *WT-1* in the developing mammalian embryo. Mech Develop 1992; 40:85-97.

171. Sawai S, Shimono A, Wakamatsu Y et al. Defects of embryonic organogenesis resulting from targeted disruption of the N-*myc* gene in the mouse. Development 1993; 117:1445-55.

172. Kato K, Kanamori A, Kondoh H. Rapid and transient decrease of *N-myc* expression in retinoic acid-induced differentiation of OTF9 teratocarcinoma stem cells. Mol Cell Biol 1990; 10:486-91.

173. Downs KM, Martin GR, Bishop JM. Contrasting patterns of *myc* and N-*myc* expression during gastrulation of the mouse embryo. Genes Devel 1989; 3:860-9.

174. Routtenberg A. Knockout mouse fault lines. Nature 1995; 374:314-5.

175. Jackson T, Allard MF, Sreenan CM et al. The *c-myc* proto-oncogene regulates cardiac development in transgenic mice. Mol Cell Biol 1990; 7:3709-16.

176. Robbins RJ, Swain JL. C-*myc* protooncogene modulates cardiac hypertrophic growth in transgenic mice. Am J Physiol 1992; 262:H590-7.

177. Chen Z, Friedrich GA, Soriano P. Transcriptional enhancer factor 1 disruption by a retroviral gene trap leads to heart defects and embryonic lethality in mice. Genes Devel 1994; 8:2293-301.

178. Miner JH, Miller JB, Wold BJ. Skeletal muscle phenotypes initiated by ectopic MyoD in transgenic mouse heart. Development 1992; 114:853-60.

179. Rychter Z, Rychterová V, Lemez L. Formation of the heart loop and proliferation structure of its wall as a base for ventricular septation. Herz 1979; 4:86-90.

180. Aranega A, Contreras JA, Alvarez L, et al. Experimental analysis of the formation of the interventricular septum. In:Aranega A, Pexieder T (eds): Correlations Between Experimental Cardiac Embryology and Teratology and Congenital Cardiac Defects. Granada:Universidad de Granada, 1989: 93145.

181. Wessels A, Vermeulen JLM, Verbeek FJ et al. Spatial distribution of "tissue-specific" antigens in the developing human heart and skeletal muscle. III. An immunohistochemical analysis of the distribution of the neural tissue antigen G1N2 in the embryonic heart; implications for the development of the atrioventricular conduction system. Anat Rec 1992; 232:97-111.

182. Ikeda T, Sakai H, Shimokawa I, et al. Expression of Leu-7 antigen in the embryonic heart—with special reference to development of the conduction system. In:Clark EB, Takao A (eds): Developmental Cardiology: Morphogenesis and Function. Mount Kisco, New York:Futura Publishing Company Inc, 1990: 237-256.

183. Aoyama N, Tamaki H, Kikawada R et al. Development of the conduction system in the rat heart as determined by Leu-7 (HNK-1) immunohistochemistry and computer graphics reconstruction. Lab Invest 1995; 72:355-66.

184. Luider TM, Bravenboer N, Meijers C et al. The distribution and characterization of HNK-1 antigens in the developing avian heart. Anat Embryol 1993; 188:307-16.

185. Chan-Thomas PS, Thompson RP, Robert B et al. Expression of the homeobox genes *Msx-1* (Hox-7) and *Msx-2* (Hox-8) during cardiac development in the chick. Dev Dynam 1993; 197:203-16.

186. Virágh S, Challice CE. The development of the conduction system in the mouse embryo heart. I. The first embryonic A-V conduction pathway. Dev Biol 1977; 56:382-96.

187. Virágh S, Challice CE. The development of the conduction system in the mouse embryo heart II. Histogenesis of the atrioventricular node and bundle. Dev Biol 1977; 56:397-411.

188. Vuillemin M, Pexieder T. Normal stages of cardiac organogenesis in the mouse: I. Development of the external shape of the heart. Am J Anat 1989; 184:104-13.

189. Sedmera D, Pexieder T. SEM and image analysis in quantitative evaluation of embryonic myocardial architecture. J Cell Biol 1995; (In press)

190. Huynh DP, Nechiporuk T, Pulst SM. Differential expression and tissue distribution of type I and type II neurofibromins during mouse fetal development. Dev Biol 1994; 161:538-51.

191. Gutmann DH, Geist RT, Rose K et al. Expression of two new protein isoforms of the neurofibromatosis type 1 gene product, neurofibromin, in muscle tissues. Dev Dynam 1995; 202:302-11.

192. Brannan CI, Perkins AS, Vogel KS et al. Targeted disruption of the neurofibromatosis type-1 gene leads to developmental abnormalities in heart and various neural crest-derived tissues. Genes Devel 1994; 8:1019-29.

193. Markwald RR, Mjaatvedt CH, Krug EL. Induction of endocardial cushion tissue formation by adheron-like, molecular complexes derived from the myocardial basement membrane. In:Clark EB, Takao A (eds): Developmental Cardiology: Morphogenesis and Function. Mount Kisco, New York:Futura Publishing Company Inc, 1990: 191-204.

194. Eisenberg LM, Markwald RR. Molecular regulation of valvuloseptal morphogenesis. Circ Res 1995; 1:1-6.

195. Bernanke DH, Markwald RR. Migratory behavior of cardiac cushion tissue cells in a collagen-lattice culture system. Dev Biol 1982; 91:235-45.

196. Krug EL, Runyan RB, Markwald RR. Protein extracts from early embryonic hearts initiate cardiac endothelial cytodifferentiation. Dev Biol 1985; 112:414-26.

197. Krug EL, Mjaatvedt CH, Markwald RR. Extracellular matrix from embryonic myocardium elicits an early morphogenetic event in cardiac endothelial differentiation. Dev Biol 1987; 120:348-55.
198. Mjaatvedt CH, Markwald RR. Induction of an epithelial-mesenchymal transition by an in vivo adheron-like complex. Dev Biol 1989; 136:118-28.
199. Mjaatvedt CH, Lepera RC, Markwald RR. Myocardial specificity for initiating endothelial-mesenchymal cell transition in embryonic chick heart correlates with a particulate distribution of fibronectin. Dev Biol 1987; 119:59-67.
200. Rezaee M, Isokawa K, Halligan N et al. Identification of an extracellular 130-kDa protein involved in early cardiac morphogenesis. J Biol Chem 1993; 268:14404-11.
201. Nakamura A. Cardiac hyaluronidase activity of chick embryos at the time of endocardial cushion formation. J Mol Cell Cardiol 1980; 12:1239-47.
202. Potts JD, Runyan RB. Epithelial-mesenchymal cell transformation in the embryonic heart can be mediated, in part, by transforming growth factor β. Dev Biol 1989; 134:392-401.
203. Potts JD, Dagle JM, Walder JA et al. Epithelial-mesenchymal transformation of embryonic cardiac endothelial cells inhibited by a modified antisense oligodeoxynucleotide to transforming growth factor β3. Proc Natl Acad Sci USA 1991; 88:1516-20.
204. Potts JD, Vincent EB, Runyan RB et al. Sense and antisense TGFβ3 mRNA levels correlate with cardiac valve induction. Dev Dynam 1992; 193:340-5.
205. Nakajima Y, Krug EL, Markwald RR. Myocardial regulation of transforming growth factor-β expression by outflow tract endothelium in the early embryonic chick heart. Dev Biol 1994; 165:615-26.
206. Mahmood R, Flanders KC, Morriss-Kay GM. Interactions between retinoids and TGF βs in mouse morphogenesis. Development 1992; 115:67-74.
207. Lyons KM, Pelton RW, Hogan BLM. Organogenesis and pattern formation in the mouse: RNA distribution patterns suggest a role for Bone Morphogenetic Protein 2 (BMP-2A). Development 1990; 109:833-44.
208. Jones CM, Lyons KM, Hogan BLM. Involvement of Bone Morphogenic Protein-4(BMP-4) and *Vgr-1* in morphogenesis and neurogenesis in the mouse. Development 1991; 111:531-42.
209. Robert B, Sassoon D, Jacq B et al. *Hox-7*, a mouse homeobox gene with a novel pattern of expression during embryogenesis. EMBO J 1989; 8:91-100.
210. Hill RE, Jones PF, Rees AR et al. A new family of mouse homeobox-containing genes: molecular structure, chromosomal location, and development expression of *Hox7.1*. Genes Devel 1989; 3:26-37.
211. Lyons GE, Houzelstein D, Sassoon D et al. Multiple sites of Hox-7 expression during mouse embryogenesis: Comparison with retinoic acid receptor mRNA localization. Mol Repro Dev 1992; 32:303-14.

212. Candia AF, Hu J, Crosby J et al. *Mox-1* and *Mox-2* define a novel homeobox gene subfamily and are differentially expressed during early mesodermal patterning in mouse embryos. Development 1992; 116: 1123-36.

213. Wang Y, Benezra R, Sassoon DA. Id expression during mouse development: A role in morphogenesis. Dev Dynam 1992; 194:222-30.

214. Evans SM, O'Brien TX. Expression of the helix-loop-helix factor Id during mouse embryonic development. Dev Biol 1993; 159:485-99.

215. Song K, Wang Y, Sassoon D. Expression of *Hox 7.1* in myoblasts inhibits terminal differentiation and induces cell transformation. Nature 1992; 360:477-81.

216. Catron KM, Zhang H, Marshall SC et al. Transcriptional repression by Msx-1 does not require homeodomain DNA-binding sites. Mol Cell Biol 1995; 15:861-71.

217. Noden DM. Origins and patterning of avian outlfow tract endocardium. Development 1991; 111:867-76.

218. Schlaeger TM, Qin Y, Fujiwara Y et al. Vascular endothelial cell lineage-specific promoter in transgenic mice. Development 1995; 121:1089-98.

219. Dumont DJ, Gradwohl G, Fong G et al. Dominant-negative and targeted null mutations in the endothelial receptor tyrosine kinase, *tek*, reveal a critical role in vasculogenesis of the embryo. Genes Devel 1994; 8:1897-909.

220. Korhonen J, Polva A, Partanen J et al. The mouse *tie* receptor tyrosine kinase gene: expression during embryonic angiogenesis. Oncogene 1994; 9:395-403.

221. Kuzuoka M, Takahashi T, Guron C et al. Murine homeobox-containing gene, Msx-1: Analysis of genomic organization, promoter structure, and potential autoregulatory *cis*-acting elements. Genomics 1994; 21:85-91.

222. Jones FS, Crossin KL, Cunningham BA et al. Identification and characterization of the promoter for the cytotactin gene. Proc Natl Acad Sci USA 1990; 87:6497-501.

223. Catron KM, Iler N, Abate C. Nucleotides flanking a conserved TAAT core dictate the DNA binding specificity of three murine homeodomain proteins. Mol Cell Biol 1993; 13:2354-65.

224. Edelman GM. Morphoregulation. Dev Dynam 1992; 193:2-10.

225. Satokata I, Maas R. *Msx1* deficient mice exhibit cleft palate and abnormalities of craniofacial and tooth development. Nat Gen 1994; 6:348-56.

226. Vassall-Adams PR. The development of the atrioventricular bundle and its branches in the avian heart. J Anat 1982; 134:169-83.

227. Arrechedera H, Strauss M, Argüello C et al. Ultrastructural study of the myocardial wall of the atrio-ventricular canal during the development of the embryonic chick heart. J Mol Cell Cardiol 1984; 16:885-95.

228. Virágh S, Challice CE. The development of the early atrioventricular conduction system in the embryonic heart. Can J Physiol Pharmacol 1983; 61:775-92.

229. Truex RC, Smythe MQ. Comparative morphology of the cardiac conduction tissue in animals. Ann N Y Acad Sci 1965; 127:19-35.

230. Gittenberger-de-Groot AC, Wenink ACG. The specialized myocardium in the foetal heart. In:Van Mierop LHS, Oppenheimer-Dekker A, Bruins CLDC (eds): Embryology and Teratology of the Heart and the Great Arteries. The Hague:Leiden University Press, 1978: 15-24.

231. Yokouchi Y, Ohsugi K, Sasaki H et al. Chicken homeobox gene *Msx-1*: structure, expression in limb buds and effect of retinoic acid. Development 1991; 113:431-44.

232. Anderson RH. The present-day place of correlations between embryology and anatomy in the understanding of congenitally malformed hearts. In: Aránega A, Pexieder T (eds): Correlations Between Experimental Cardiac Embryology and Teratology and Congenital Cardiac Defects. Granada: University de Granada, 1989: 265-295.

233. Liu Y, Ma L, Wu L et al. Regulation of the Msx2 homeobox gene during mouse embryogenesis: a transgene with 439 bp of 5' flanking sequence is expressed exclusively in the apical ectodermal ridge of the developing limb. Mech Develop 1994; 48:187-97.

234. Spence SG, Argraves WS, Walters L et al. Fibulin is localized at sites of epithelial-mesenchymal transitions in early avian embryo. Dev Biol 1992; 151:473-84.

235. Hurle JM, Kitten GT, Sakai LY et al. Elastic extracellular matrix of the embryonic chick heart: An immunohistological study using laser confocal microscopy. Dev Dynam 1994; 200:321-32.

236. Swiderski RE, Daniels KJ, Jensen KL et al. Type II collagen is transiently expressed during avian cardiac valve morphogenesis. Dev Dynam 1994; 200:294-304.

237. Witte DP, Aronow BJ, Dry JK et al. Temporally and spatially restricted expression of apolipoprotein J in the developing heart defines discrete stages of valve morphogenesis. Dev Dynam 1994; 201:290-6.

238. Zhang H, Chu M, Pan T et al. Extracellular matrix protein fibulin-2 is expressed in the embryonic endocardial cushion tissue and is a prominent component of valves in adult heart. Dev Biol 1995; 167:18-26.

239. Kirby ML. Cellular and molecular contributions of the cardiac neural crest to cardiovascular development. Trends Cardiovasc Med 1993; 3:18-23.

240. Hood LC, Rosenquist TH. Coronary artery development in the chick: origin and deployment of smooth muscle cells, and the effects of neural crest ablation. Anat Rec 1992; 234:291-300.

241. Sumida H, Akimoto N, Nakamura H. Distribution of the neural crest cells in the heart of birds: a three dimensional analysis. Anat Embryol 1989; 180:29-35.

242. Takamura K, Okishima T, Ohdo S et al. Association of cephalic neural crest cells with cardiovascular development, particularly that of the semilunar valves. Anat Embryol 1990; 182:263-72.

243. Fukiishi Y, Morriss-Kay GM. Migration of cranial neural crest cells to the pharyngeal arches and heart in rat embryos. Cell Tissue Res 1992; 268:1-8.

244. Tomita H, Connuck DM, Leatherbury L et al. Relation of early hemodynamic changes to final cardiac phenotype and survival after neural crest ablation in chick embryos. Circulation 1991; 84:1289-95.

245. Aiba S, Creazzo TL. Calcium currents in hearts with persistent truncus arteriosus. Am J Physiol 1992; 262:H1182-90.

246. Creazzo TL, Burch J, Redmond S et al. Myocardial enlargement in defective heart development. Anat Rec 1994; 239:170-6.

247. Chisaka O, Capecchi MR. Regionally restricted developmental defects resulting from targeted disruption of the mouse homeobox gene *hox-1.5*. Nature 1991; 350:473-9.

248. Franz T. Persistent truncus arteriosus in the Splotch mutant mouse. Anat Embryol 1989; 180:457-64.

249. Shaner RF. Comparative development of the bulbus and ventricles of the vertebrate heart with special reference to Spitzer's theory of heart malformations. Anat Rec 1962; 142:519-29.

250. de la Cruz MV, Giménez-Ribotta MA, Saravalli O et al. The contribution of the inferior endocardial cushion of the atrioventricular canal to cardiac septation and to the development of the atrioventricular valves: Study in the chick embryo. Am J Anat 1983; 166:63-72.

251. García-Peláez I, Díaz-Góngora G, Martínez MA. Contribution of the superior atrioventricular cushion to the left ventricular infundibulum. Acta anat 1984; 118:224-30.

252. Icardo JM, Arrechedera J, Colveé E. The atrioventricular valves of the mouse I. A scanning electron microscope study. J Anat 1993; 182:87-94.

253. Lamers WH, Virágh S, Wessels A et al. Formation of the tricuspid valve in the human heart. Circulation 1995; 91:111-21.

254. Rychter Z. Experimental morphology of the aortic arches and the heart loop in chick embryos. Advances in Morphogenesis 1962; 2:333-71.

255. Colvee E, Hurle JM. Malformations of the semilunar valves produced in chick embryos by mechanical interference with cardiogenesis. Anat Embryol 1983; 168:59-71.

256. Hurle JM, Colveé E. Changes in the endothelial morphology of the developing semilunar heart valves. A TEM and SEM study in the chick. Anat Embryol 1983; 167:67-83.

257. Gorski DH, Patel CV, Walsh K. Homeobox transcription factor regulation in the cardiovascular system. Trends Cardiovasc Med 1993; 3:184-90.

258. Gaunt SJ, Sharpe PT, Duboule D. Spatially restricted domains of homeogene transcripts in mouse embryos: relation to a segmented body plan. Development 1988; 104:169-79.

259. Fishman MC, Stainier DY. Cardiovascular development. Prospects for a genetic approach. Circ Res 1994; 74:757-63.

260. Gannon M, Bader D. Initiation of cardiac differentiation occurs in the absence of anterior endoderm. Development 1995; 121:2439-50.

261. Lyons I, Parsons LM, Hartly L, Li R, Andrews JE, Robb L, Harvey RP. Myogenic and morphogenetic defects in the heart tubes of murine embryos lacking the homeo box gene *Nkx2-5*. Genes Devel 1995; 9:1654-66.

262. Yutzey KE, Bader D. Diversification of cardiomyogenic cell lineages during early heart development. Circ Res 1995; 77:216-9.

=========== CHAPTER 3 ===========

CONTRACTILE PROTEIN GENES AND CARDIAC MUSCLE DEVELOPMENT

OVERVIEW

In certain respects the fetal heart resembles a small version of the adult heart. It fulfills similar functions in that it is required to fill, develop pressure, expel blood, and relax, thereby ensuring continuous circulation of blood. At the cellular and molecular level there are, however, significant differences. It can be demonstrated, for example, that muscle from adult heart has a greater ability to generate tension compared to fetal;[1,2] adult and fetal cardiac muscle fibers show differing responses to intracellular calcium,[3] differing resistance to acidosis[4,5] and differing responses to adrenergic stimulation.[6,7] Such developmental differences in contractility have important clinical implications in the understanding and treatment of cardiac disease in neonates and infants.

Experiments with isolated cells have demonstrated that much of the developmental difference in contractility can be attributed to the cardiac myocyte. Isolated fetal and adult cardiac myocytes show differences similar to those seen between intact fetal and adult heart.[2,8] Within the cardiac myocyte, components of the contractile apparatus and calcium handling machinery may, in turn, be largely responsible. Intracellular calcium acts as the molecular signaling mechanism which initiates muscle contraction and relaxation. Maintaining and regulating cyclical changes in calcium concentration, which result in the rhythmic contractions observed in myocytes, requires a highly organized and flexible ion transport system. Similarly, the production of force involves a complex array of contractile proteins in a highly organized structure, the sarcomere.

Molecular Biology of Cardiac Development and Growth, by Paul J.R. Barton, Kenneth R. Boheler, Nigel J. Brand, Penny S. Thomas. © 1995 R.G. Landes Company.

Developmental alterations in calcium handling and contractile proteins probably account for the majority of changes in contractility observed during cardiac development. In this chapter the developmental pattern of contractile protein gene expression during fetal development is described within the general context of cardiac function. Much of the overall developmental program centers on the regulation of gene expression. This is thought to be achieved largely through transcriptional control mechanisms although post-transcriptional mechanisms, including developmentally regulated alternative RNA splicing, also play an important role.

THE CONTRACTILE APPARATUS

The contractile apparatus of striated muscle consists of a highly organized structure arranged within the sarcomere, the basic force producing unit (Fig. 3.1). Force is generated through the interaction of thick and thin filaments via cross bridge formation in response to increased intracellular calcium concentration.[9] The contractile potential of the muscle is therefore determined largely by the structure and organization of the thick and thin filaments including the calcium sensing component, the troponin complex. During development the contractile apparatus undergoes developmental changes in its composition due to changes in the expression of genes encoding its various components. Most components are encoded not by single genes but by members of multigene families where each member specifies a different protein isoform. In certain cases alternative RNA splicing further increases the number of protein isoforms. Alterations in the pattern of expression of these gene families contributes to developmental changes in the contractile characteristics of the myocardium.

Myosin, which makes up the thick filament of striated muscle, is a hexamer of polypeptide chains; two myosin heavy chains (MHC) and four associated myosin light chains (MLC) (see Table 3.1 and Fig. 3.1). Myosin has a globular head region and a coiled tail region where the two MHCs are coiled around one another. The head region contains an ATPase site and the binding site for actin which participates in acto-myosin cross bridge formation. The enzymatic cleavage of ATP, to ADP and inorganic phosphate (Pi), provides the energy for mechanical shortening and force production.[9] The steps involved in the cycle of reactions leading to the generation of mechanical force through this interaction are illustrated in Figure 3.2. On its own, myosin can act to hydrolyse ATP but this activity is greatly enhanced by the interaction of the myosin head with actin. Interestingly, it is the release of ADP from the ATP binding site which is the rate limiting step and not the binding of ATP to myosin or the cleavage of ATP. It is the rate of release which is enhanced through acto-myosin interaction.

Different isoforms of MHC exhibit different ATPase activities and there is a direct correlation between the actin-activated ATPase activity of myosin and the unloaded maximal velocity of shortening of muscle fibers[10-14] (see Table 3.2). It follows that, in general, muscle fibers

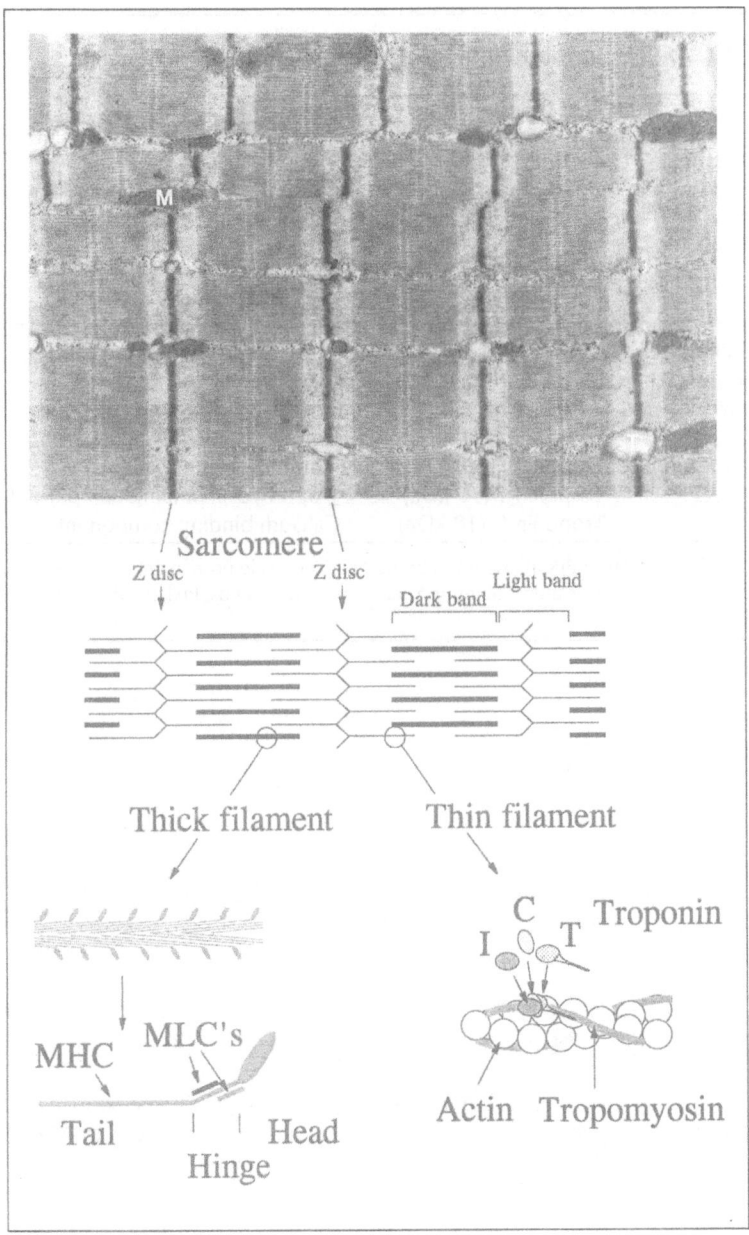

Fig. 3.1. Structural components of the sarcomere. Upper panel: Electron micrograph of a section through cardiac muscle showing classic pattern of striations. Also visible are T-tubules (T) and mitochondria (M).
Lower panel: Schematic representation of the organization and components of the thick and thin filaments. Note that myosin is composed of two myosin heavy chains with the tail regions forming a coiled structure. A single myosin heavy chain is illustrated here with associated myosin light chains. Reproduced from Barton PJR, Bhavsar PK, Brand NJ. Contractile protein genes in development and disease. In: Annual of Cardiac Surgery. Yacoub M, Pepper J eds. Philadelphia: Academic Press. 1995 (in press). Electron micrograph courtesy of Professor N. Severs.

Table 3.1. Major components of the contractile apparatus

Protein	Subunits	Comments
Thick Filament		
Myosin:	2 Myosin heavy chains (220 kDa each)	Major thick filament component which interacts with actin to produce mechanical force
	4 Myosin light chains (20 kDa each)	Involved in regulation by phosphorylation (MLC2)
Thin Filament		
Filamentous Actin	(Actin monomer 42 kDa)	Major thin filament protein
Tropomyosin	2 tropomyosin (32 kDa each)	Lies coiled along the actin filament
Troponin Complex	Troponin I (30 kDa)	Inhibitory component
	Troponin T (30 kDa)	Tropomyosin-binding component
	Troponin C (18 kDa)	Calcium binding component

Reproduced from: Barton PJR, Bhavsar PK, Brand NJ. Contractile protein genes in development and disease. In: Annual of Cardiac Surgery. Yacoub M, Pepper J eds. Philadelphia: Academic Press. 1995 (in press).

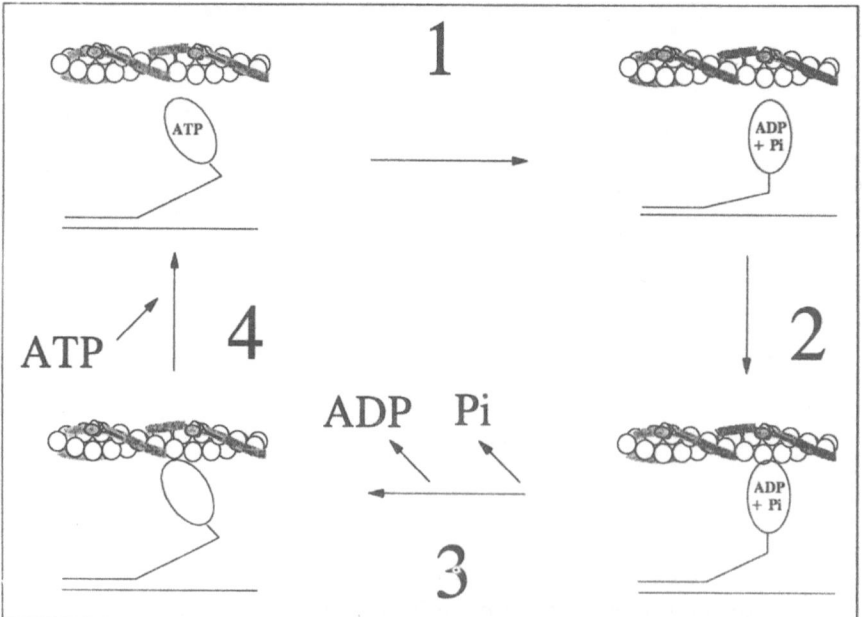

Fig. 3.2. The contractile cycle. Step 1: The myosin head is activated by the hydrolysis of ATP to ADP and Pi. The energy produced being 'stored' in the head (relaxed, energized state). Step 2: during excitation, calcium binds to the troponin complex and a conformational change results which allows myosin and actin to interact (active state). Step 3: The stored energy is released in the power stroke accompanied by the release of Pi and ADP (rigor complex). Step 4: Dissociation of actin and myosin and binding of ATP to myosin (relaxed state).

containing MHC isoforms with a high actin-activated ATPase activity exhibit a fast velocity of contraction and those with low ATPase activity have low velocity of contraction. This is of importance in the developing heart, as the two cardiac isoforms of MHC (α-MHC and β-MHC) have different ATPase activities and undergo developmental alterations in expression (see below). As might be predicted, there is also a direct relation between the rate of muscle contraction and the energy consumption to perform a given workload: the more rapidly a muscle fiber contracts the more energy will be consumed to produce the same amount of work. Slow muscle contraction is therefore energetically more favorable than fast contraction.

Myosin light chains (MLCs) can be grouped into two classes: MLC1 light chains (also known as the alkali light chains) and MLC2 light chains (also known as the regulatory or phosphorylatable light chains).[15] Both are found in the hinge region of myosin with MLC2 light chains wrapped around the head/rod junction region and MLC1 light chains located slightly more towards the head. Recent evidence suggests that the presence of MLC1 type light chains on the myosin molecule is required in order to obtain physiological speeds of muscle shortening.[16,17] The MLC2 (or regulatory) light chains participate in the phosphorylation-mediated regulation of contractility (reviewed by Sweeney et al[18]). MLC2 phosphorylation is mediated by a Ca^{2+}/calmodulin-dependent myosin light chain kinase. In skeletal muscle this phosphorylation correlates with both increased maximal isometric force and rate of force development. The mechanism probably involves altering the position of the myosin head relative to actin, thereby increasing calcium sensitivity and resulting in a leftward shift in the force-calcium curve. In cardiac muscle the exact role of MLC2-phosphorylation has not been determined but it is likely to be similar to that seen in skeletal muscle.

Table 3.2. Myosin heavy chain (MHC) characteristics and expression in rat ventricular muscle

	α-MHC	β-MHC
Contractile characteristics	High ATPase activity Fast shortening velocity Low efficiency of force production	Low ATPase activity Slow shortening velocity High efficiency of force production
Effect of:		
Increased thyroid hormone	↑ expression	↓ expression
Increased workload	↑ expression	↓ expression
Aging	↓ expression	↑ expression

Adapted from Nadal-Ginard B, Mahdavi V. Molecular basis of cardiac contractility. In: Annual of Cardiac Surgery. Yacoub M, Pepper J eds. London: Academic Press 1991; 13-20.

The thin filament of striated muscle is composed of a backbone of coiled filamentous actin with associated tropomyosin and troponin proteins (see Fig. 3.1). These are arranged such that each seven actin monomers are associated with a single tropomyosin molecule and a troponin complex made of three proteins : troponin I (TnI), troponin T (TnT) and troponin C (TnC). The troponin-tropomyosin component of the thin filament acts to regulate contraction through altering its conformation in response to alterations in intracellular calcium concentration. Calcium binds to TnC and, as a result, the troponin-tropomyosin complex moves away from the myosin binding site of the thin filament[19] (see Fig. 3.3). In this situation actin and myosin can interact, the ATPase activity of myosin is enhanced and muscle contraction occurs. When the intracellular calcium concentration decreases, calcium is released from troponin C, the complex returns to its inhibitory state, the acto-myosin interaction is inhibited and muscle relaxation begins.

CARDIAC MYOSIN GENE EXPRESSION

Cardiac myosin is composed of a dimer of myosin heavy chains (MHCs) and associated myosin light chains (MLCs). In mammals there are two cardiac MHC isoforms, α-MHC and β-MHC, which give rise

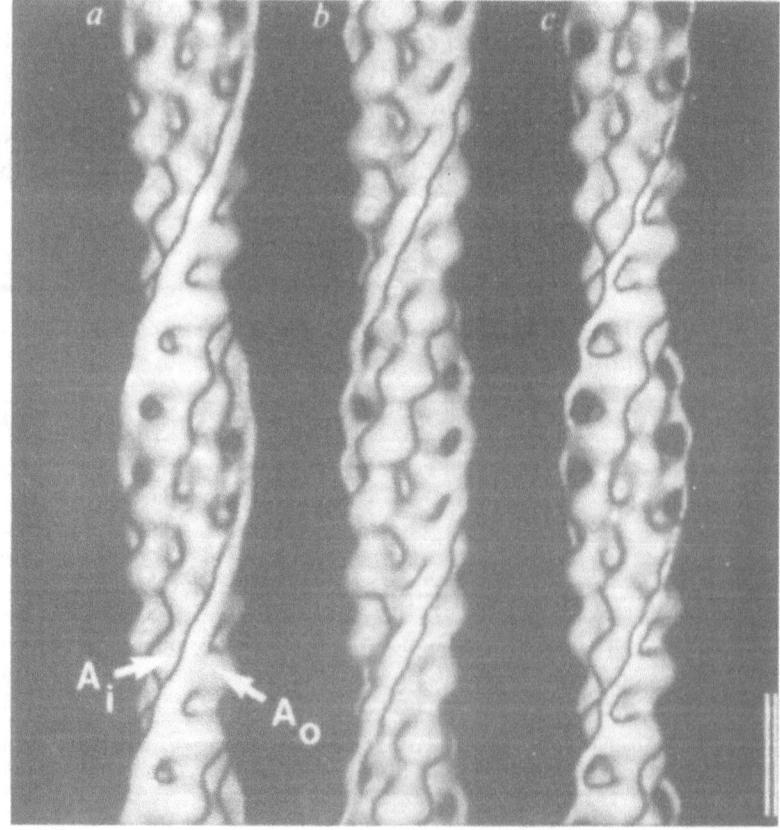

Fig. 3.3. Thin filament regulation. Three dimensional reconstructions of electron micrographs of thin filaments in presence of (a) EGTA (Ca²⁺ depleted) or (b) Ca²⁺. Note the difference in location of the tropomyosin molecule effected by calcium. Ai and Ao denote inner and outer domains of the actin monomer. (c) Reconstruction of filaments in EGTA at different contour level to that in (a) illustrating the contact with Ao. (Reprinted with permission from Lehman et al, Nature 1994; 368:65-67, copyright 1994, Macmillan Magazines Limited)

to three cardiac myosin types : V1 (α/α homodimers), V2 (α/β heterodimers) and V3 (β/β homodimers). Due to structural differences between the α- and β- MHC subunits, V1, V2 and V3 myosin can be distinguished on non-denaturing acrylamide gels.[20] This, together with the use of isoform-specific antibodies allowed the principal patterns of expression of these isoforms to be determined. More recently, molecular cloning of the corresponding genes from a variety of species (see Emerson and Bernstein[21]) has allowed α- and β-MHC gene expression to be examined through analyzing isoform-specific mRNA accumulation. The α- and β-MHC isoforms have significantly different ATPase activities[12,14] and, as a consequence, the corresponding myosins show different ATPase and contractile characteristics. V1 myosin (α/α) has a high actin-activated ATPase activity and produces a fast rate of contraction, V3 myosin (β/β) has a lower actin-activated ATPase activity and produces a slower rate of muscle contraction and V2, which is a heterodimer of α- and β-MHC, has an intermediate ATPase activity and produces an intermediate rate of contraction.

Expression of α- and β-MHC isoforms is subject to both regional and temporal control in the heart. It should be noted, however, that there are significant differences in the pattern of their expression between species and, in general, this correlates with body weight and frequency of heart beat. In small mammals (low body weight, fast heart rate) α-MHC is the predominant isoform in both atrial and ventricular muscle of adult heart[22-25] (see Table 3.3). In larger mammals (higher body weight, lower heart rate), including man, β-MHC is the predominant isoform of the adult ventricle[25-29] and α-MHC is restricted principally to atrial muscle although it is also expressed in parts of the ventricular conduction system.[30] As a consequence, the pattern of MHC expression during development varies between species. In ventricular muscle of small mammals there is a developmental switch from expression of β-MHC, which is the predominant fetal form, to that of α-MHC around the time of birth. In contrast, most larger mammals, including man, express β-MHC as the predominant isoform in ventricular muscle at all developmental stages. In most mammals, α-MHC is the predominant isoform present in atrial muscle throughout fetal development and in the adult. This correlates with the functional differences between atrial and ventricular muscle, in particular with the velocity muscle fiber contraction, and with the developmental changes in ventricular muscle. In man, adult ventricular muscle is a slow contracting muscle relative to that of the atrium. This slow contractile characteristic is shared with slow skeletal muscle fibers. Indeed the same β-MHC gene is expressed in slow skeletal muscle and ventricular muscle in man and other mammals. In small mammals there is a developmental increase in velocity of contraction of ventricular myocytes as β-MHC expression is replaced by that of α-MHC (reviewed by Swynghedauw[10]).

Table 3.3. Developmental changes in contractile protein gene expression in the myocardium of rodents

| | Atrial Muscle | | Ventricular Muscle | | |
	Fetal	Adult	Fetal	Neonatal	Adult
α-actin	α-skeletal	α-cardiac α-cardiac	α-skeletal α-cardiac	α-skeletal	α-cardiac
Myosin heavy chains (MHC)	α-MHC (β-MHC)	α-MHC	β-MHC α-MHC	α-MHC β-MHC	α-MHC (β-MHC)
Myosin light chains (MLC)	MLC1A MLC1V MLC2A	MLC1A MLC2A	MLC1A MLC1V MLC2A MLC2V	MLC1V MLC2V	MLC1V MLC2V
Troponin I (TnI)	TnIs	TnIc	TnIs	TnIc TnIs	TnIc
Troponin T (TnT)	cTnTemb*	cTnTad*	cTnTemb*	cTnTemb* cTnTad*	cTnTad*
Troponin C (TnC)	cTnC	cTnC	cTnC	cTnC	cTnC
Tropomyosin (TM)	α-TM β-TM	α-TM (β-TM)	α-TM β-TM	α-TM β-TM	α-TM (β-TM)

Annotations: Myosin light chains: MLC1A & MLC2A, atrial isoforms; MLC1V & MLC2V, ventricular isoforms. Troponin I, TnIc, cardiac isoform; TnIs, slow skeletal muscle isoform. cTnT, cardiac troponin T. cTnC, cardiac troponin C. Isoforms in parentheses are present in low amounts only.

*Note that multiple cardiac TnT isoforms have been described which are derived by alternative splicing of a single gene. "ad" and "emb" therefore denote multiple embryonic and adult isoforms respectively.

Adapted from Barton PJR, Bhavsar PK, Brand NJ. Contractile protein genes in development and disease. In: Annual of Cardiac Surgery. Yacoub M, Pepper J eds. Philadelphia: Academic Press. 1995 (in press).

Detailed studies on the regional expression of myosin genes during cardiac development have been conducted in several species using immunocytochemistry and in situ hybridization (see Fig. 3.4). In early development in the mouse, α- and β- MHC genes are co-expressed in both the presumptive atrial and presumptive ventricular myocyte populations of the heart tube[22,23] such that, by 8 days post coitum (8 dpc†), α-MHC and β-MHC mRNAs appear equally abundant in both cell types (see Table 3.3). This pattern of expression soon changes to produce asymmetric expression with β-MHC expression predominating in ventricular myocytes and exclusive expression of α-MHC in atrial myocytes. β-MHC expression is lost from atrial myocytes early in development with only low levels detectable at 9.5 dpc and none present by 10.5 dpc. This contrasts with the low but detectable levels of α-MHC mRNA which persist in ventricular myocytes throughout fetal life. After birth the level of β-MHC expression declines in ventricular

†*See page 31 for a discussion of embryonic/fetal staging.*

muscle and α-MHC becomes the predominant mRNA. In the developing rat heart immunocytochemical localization of α-MHC and β-MHC isoforms demonstrates a similar pattern.[31] Co-expression of α and β-MHC isoforms has been demonstrated throughout the tubular heart (10-11 dpc) except for the left free wall of the atrium. Transition to the restricted expression of α-MHC in atrial muscle occurs between 12-13 dpc and that of β-MHC in ventricular muscle slightly later between 12-15 dpc. It should be noted that, in general, development in the rat is retarded relative to that in mouse. Unlike the situation in man, α-MHC expression was not detected in the developing ventricular conduction system (see below).

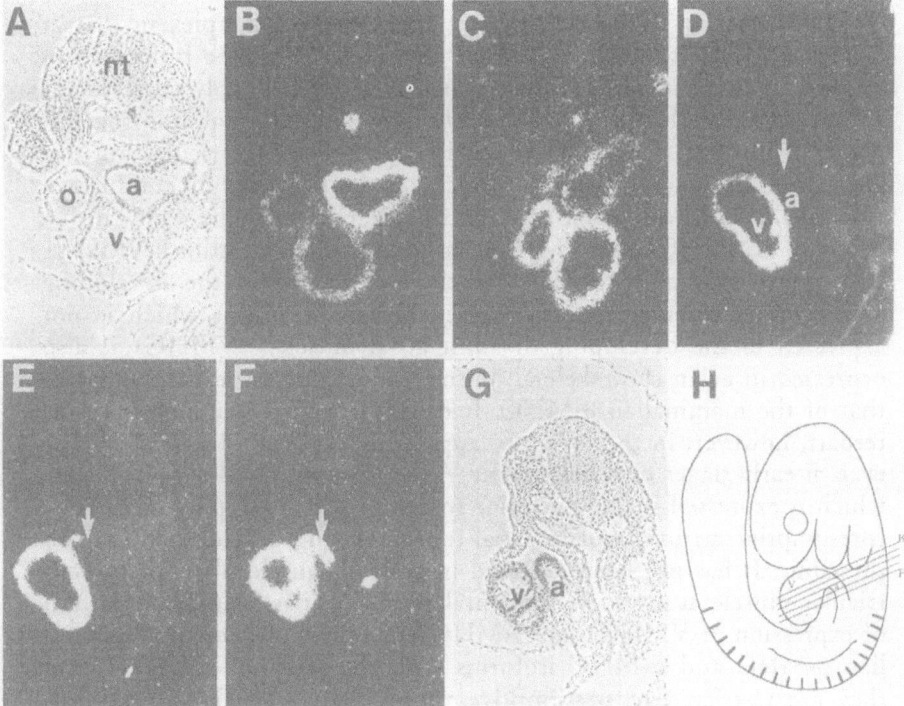

Fig. 3.4. Detection of myosin gene expression in the developing mouse heart. The figure illustrates results of in situ hybridization of radioactively labeled RNA probes to sections of a 9-day mouse embryo. The results are viewed under dark field illumination such that silver grains in the overlying photographic emulsion (which occur in regions where probe was bound) refract light and appear light. (A) Phase contrast image of transverse section shown indicating positions of neural tube (nt) ventricle (v), atrium (a) and outflow tract (o). (B-F) Serial sections hybridized with probes for mRNA encoding myosin heavy chains a-MHC, β-MHC and myosin light chains MLC2V, MLC1V and MLC1A respectively and viewed under dark field optics. (G) phase contrast image of section shown in F. (H) figure showing approximate plane of sections in B-F. Arrow points to atrial muscle. Note the reciprocal patterns of α-MHC and β-MHC and the differing patterns observed with the myosin light chain probes. Only MLC1A is abundant throughout the myocardium at this stage. All other probes are more abundant in one region relative to others. (Reprinted by permission of the publisher from In situ analysis of the cardiac muscle gene program during embryogenesis. Gary E. Lyons, Trends in Cardiovascular Medicine, 4., 70-77. Copyright 1994 by Elsevier Science Inc.)

Immunocytochemical studies of early stages of human cardiac development[26] show that α-MHC and β-MHC isoforms are likewise co-expressed in ventricular muscle at early stages of development. At Carnegie stage 14, (31-35 days gestation) α-MHC is expressed in atrial, ventricular and outflow tract myocardium and β-MHC is expressed in ventricular and outflow tract myocardium. The onset and early stages of cardiac development co-expression of α- and β- MHC isoforms may, therefore, be common to ventricular myocardial cells of all mammals.[33] Subsequently there is specialization such that developing atrial and ventricular myocytes express predominantly α- and β-MHC isoforms respectively. It is only late in fetal life that expression patterns begin to significantly diverge between different mammals, with β-MHC switching to α-MHC in rodents.

The situation in the avian heart appears more complex in that at least five MHC isoforms have been identified.[34-42] Three isoforms can be distinguished in atrial muscle[40] one of which probably corresponds to the recently cloned AMHC1.[41] Different, additional isoforms include an isoform (VMHC1) which is expressed in ventricular but not atrial muscle[41] and an isoform which is specific to adult conduction system fibers of both atrial and ventricular muscle.[35] The regional distribution of AMHC1 and VMHC1 has been partially described: VMHC1 is a ventricular-specific MHC which is expressed in the developing and adult ventricular muscle of the chicken heart but which is not expressed in the developing or adult atrial muscle.[42] VMHC1 is also expressed in avian slow skeletal muscle and in this respect is similar to that of the mammalian β-MHC. It differs from its mammalian counterpart, however, in that it is not apparently expressed in atrial muscle even at early stages of development. AMHC1 is an atrial-specific MHC which is expressed in the posterior region of the developing heart tube (presumptive atrium) and in atrial muscle during development and in the adult. It has not been detected in skeletal muscle fibers or in ventricular muscle at any developmental stage.[41] Superficially, the pattern of expression of VMHC1 and AMHC1 resembles that of the mammalian β-MHC and α-MHC isoforms with the possible difference that they are restricted to presumptive ventricular and atrial muscle lineages earlier in development, at the stage of tube formation.

MYOSIN LIGHT CHAIN GENE EXPRESSION

As with the α- and β-MHC genes, isoforms exist for each of the different classes of myosin light chains. In the adult heart, atrial and ventricular specific isoforms of both MLC1 and MLC2 can be identified (MLC1A, MLC1V, MLC2A and MLC2V respectively). As with myosin heavy chains, MLC1 isoforms are not unique to the myocardium but are also expressed in skeletal muscle. MLC1A, the adult atrial myosin alkali light chain, is expressed in most (and perhaps all) skeletal muscles during embryonic and early fetal development (where it is also known as MLC1emb)[15,43-46] and is one of the earliest markers of skeletal muscle myogenesis in the developing somites of the embryo.[22,23] Interestingly MLC1A, which is not expressed in the majority of adult skeletal muscles, is expressed in certain adult facial muscles including the masseter.[47] These are unlike the majority of skeletal muscles, in that they do not arise from the somites but are derived from head mesenchyme. Other contractile protein isoforms which are primarily restricted to cardiac muscle in the adult can also be found in these facial muscles, including α-MHC.[48] The expression of MLC1V, the (adult) ventricular myosin alkali light chain, resembles that of β-MHC: it is expressed in both adult ventricular and slow skeletal muscle and throughout the heart tube during early development.[22,23]

The expression patterns of MLC2 isoforms are similar to those of MLC1 isoforms. MLC2A is expressed in both atrial and ventricular muscle during development[49] but is restricted to atrial muscle in the adult. MLC2A is not expressed, however, in embryonic or fetal skeletal muscle. MLC2V (like MLC1V and β-MHC) is expressed in both ventricular and slow skeletal muscle in the adult. However, expression of MLC2V is restricted to ventricular (and outflow tract) myocardium early in development; it is the only myosin gene known to show this pattern of restricted expression throughout cardiac development. This feature makes this gene particularly attractive for the analysis of chamber-specific gene regulation.[50]

ACTIN GENE EXPRESSION

In mammals six different actin isoforms have been identified[51-53] which are expressed in a tissue-specific manner. Two, cytoplasmic β- and γ-actin, are co-expressed in all nonmuscle cells studied. Two, α-smooth muscle actin (found in vascular smooth muscle cells and often referred to as α-vascular actin) and γ-smooth muscle actin (found in enteric smooth muscle and often referred to as γ-enteric actin), are found in smooth muscle cells. Two, α-cardiac and α-skeletal actin (known collectively as the α-sarcomeric actins) are expressed in striated muscle and are both expressed in heart. As is the case with myosin gene expression, α-cardiac and α-skeletal actin expression in the heart is subject to variation between species and during development. Their expression is influenced by a variety of factors including thyroid hormone[54] and pressure overload.[55-57]

The different actin isoforms are remarkably similar to each other and α-cardiac and α-skeletal muscle actins differ by only four amino acids out of a total of 375 residues (the skeletal isoform differs from the cardiac by Glu→Asp substitutions at residues 2 and 3 and Met→Leu substitutions at positions 299 and 358, the latter substitutions also being found in smooth muscle and non-muscle actins). The high degree of similarity makes it extremely difficult to distinguish α-cardiac and α-skeletal actin using standard biochemical or immunological methods, although monoclonal antibodies which can distinguish between smooth muscle and sarcomeric muscle isoforms are available.

Determination of the α-cardiac and α-skeletal actin content of striated muscle has relied instead on methods of distinguishing N-terminal peptide fragments (see Vandekerckhove et al[58]). Although highly informative these methods are complex and cannot be adapted for analysis in situ. Detailed analysis of temporal and spatial pattern of α-cardiac and α-skeletal actin genes in the heart has required molecular cloning techniques to provide isoform-specific probes. These have relied principally on the use of 3' and 5'-nontranslated regions of the genes, which show significant sequence differences both between species and between genes within a species. A large number of actin cDNAs and genes have been cloned[59] including those of man, rodents, and chicken[60-74] and their developmental pattern of expression in the heart determined. There is a reasonable correlation between the relative abundance of actin mRNAs and their corresponding isoforms where this has been assessed.[58] Analysis of expression in chick, mouse, rat and man show that α-cardiac and α-skeletal actin genes are both expressed in cardiac muscle but that their relative proportion varies between species, with development and with senescence. It has also been noted that during early stages development (of the embryo) γ-smooth muscle actin is transiently expressed in the developing heart.

In rodents α-cardiac and α-skeletal actins are co-expressed in the developing heart although α-cardiac actin is always predominant.[58,60,72,75,76] At 8 dpc in the mouse α-cardiac and α-skeletal actin mRNAs are detectable in the heart tube although α-skeletal actin is less abundant.[75] Co-expression continues throughout fetal development of the mouse heart but in the adult α-skeletal actin is a minor component. Studies in the rat[60,62,76,77] have shown a similar pattern of fetal expression. However, α-skeletal actin mRNA represents 28% of sarcomeric actin mRNA at 17-19 dpc and 40% at three weeks postnatal development, subsequently decreasing to 20% at one month and less than 5% in adult hearts.[76] No significant differences were observed between male and female rats in these studies and there was no detectable difference in adult rats with senescence. Studies on smooth muscle and non-muscle actin have revealed that, in addition to the sarcomeric actins, α-smooth muscle actin mRNA is expressed during cardiac development and is detectable at low levels in adult myocardium.[62,77] Studies using antibodies have shown that this expression is detectable throughout the atrial and ventricular myocardium.[78] In contrast, γ-smooth muscle actin has not been detected in cardiac muscle at any stage.[62,77]

The pattern of expression in chick[79-81] is similar to that seen in rodents. There is co-expression of α-skeletal and α-cardiac actin isoforms during fetal development but in the adult α-cardiac actin is the predominant isoform detected.[79,82] Also similar to rodents, α-smooth muscle actin is expressed during the early stages of cardiac development. In chick it has been demonstrated that α-smooth mRNA is detectable before the onset of expression of α-cardiac or α-skeletal actin,[79,81] and in this respect the molecular phenotype of early myocardium resembles that of smooth muscle. Other smooth muscle genes are not expressed in embryonic myocardium. For example, the smooth muscle myosin heavy chain gene is not expressed in developing rat heart.[83]

In larger animals, including man, the pattern of α-actin gene expression appears to be different from that seen in rodents and birds.[58,64,84,85] Studies on mRNA accumulation[85] have shown that during fetal life there is co-expression of α-cardiac and α-skeletal isoforms with α-cardiac actin representing 80% of the sarcomeric actin. There is a continual increase in α-skeletal actin expression after birth such that it reaches about 50% of sarcomeric actin within the first decade after birth and $\geq$ 60% in adult life. In man, therefore, α-skeletal actin mRNA is the major actin transcript detectable in adult life.

The functional significance of the two α-sarcomeric actins is not fully understood. Their similarity (only four amino acid changes in a total of 375) would suggest little structural or functional difference between the two isoforms although two of the amino acid differences lie within the myosin-binding region of the actin molecule (see refs. 10 and 59). Yet, these differences have been highly conserved through evolution[86,87] and the two genes show distinct patterns of expression. Evidence in support of functional differences between α-cardiac and α-skeletal actin has recently been obtained, however, from the analysis of cardiac function in mice with abnormal patterns of actin expression. In BALB/c mice there is a structural rearrangement of the α-cardiac actin gene such that the promoter and first three exons are duplicated within a 9.5 kb direct repeat.[88] Through a mechanism which probably involves disruption of the transcriptional machinery, the duplication results in reduced expression of α-cardiac actin and a concomitant increase in α-skeletal actin expression. As a result, the hearts of BALB/c mice contain about 50% γ-skeletal actin compared to less than 5% in mice without the gene duplication.[88,89] Subsequent analysis has demonstrated that variable levels of α-skeletal actin mRNA expression (26% to 54%) are observed between individual BALB/c hearts and that this correlates well with their individual contractile performance.[90] In particular, isolated perfused BALB/c hearts with higher α-skeletal actin expression show an increased maximal rate of contraction and decreased time to peak pressure. No correlation was found with differences in rate of relaxation between different BALB/c hearts. As α-skeletal actin expression levels are, in general, higher in fetal heart, these data would suggest that α-skeletal actin expression contributes to the increased contractility observed in fetal heart compared to adult.

TROPONIN GENE EXPRESSION

Isoforms exist for the thin filament proteins troponin I (TnI), troponin T (TnT) and troponin C (TnC) and these are subject to altered expression during cardiac development.[91]

Separate TnC genes encode the fast skeletal and the slow skeletal/ cardiac isoforms.[92] The slow skeletal/cardiac isoform is expressed in both atrial and ventricular myocardium and is one of the few contractile proteins which does not undergo isoform switching in the heart during development.

In the case of troponin I, two isoforms are expressed during cardiac development. During fetal life the predominant isoform is that of slow skeletal muscle TnI (TnIs) and it is replaced by the adult cardiac isoform (TnIc) in the early postnatal period in both rodents and in man.[93-100] Detailed analysis of the regional accumulation of TnI proteins has further demonstrated that both the onset of TnIc protein accumulation and the loss of TnIs occur earlier in atrial muscle than in ventricular muscle in rats[101] although not in birds.[94] The cardiac and skeletal isoforms of troponin I differ in that the cardiac isoform has an extended N-terminal amino acid sequence which contains serine residues which become phosphorylated in response to β-adrenergic stimulation. Phosphorylation of cardiac troponin I at these sites alters the calcium binding characteristics of the troponin complex and, as a consequence, the contractile characteristics of the myocardium is altered.[102-106] Phosphorylation of TnIc has been shown to decrease the myofilament calcium sensitivity in intact ventricular myocardium[103] and skinned muscle fibers[105] which would act to increase the rate of muscle relaxation. This forms part of the overall phosphorylation-mediated response to adrenergic stimulation. The switch from expression of TnIs to that of TnIc would therefore be expected to have an effect on the ability of the heart to respond to adrenergic stimulation. Alteration of troponin I expression may have other effects, including contributing to the increased resistance to acidosis of fetal heart.[4,5,97,107] Studies in normal, euthyroid and hypothyroid adult and neonatal rats all demonstrate that neonatal myocardium exhibits a less pronounced change in calcium sensitivity as a result of acidosis and that this difference correlates with the presence of the TnIs isoform.[97,108]

A single troponin T (TnT) gene is expressed during cardiac development but multiple isoforms are produced by alternative splicing of the RNA transcript. This represents a second major level for the regulation of gene expression during cardiac development. Initial studies identified two isoforms in rodents and chick which have been fully characterized at the molecular level.[109-111] These differ by ten amino acid residues in the N-terminal region. During fetal life the splicing pattern results in the inclusion of these amino acid residues, but in the adult they are omitted. Recent studies suggest that multiple cardiac TnT isoforms can be identified with a variable number being detected in different species.[112,113] Molecular cloning has demonstrated that in rat,[110,114] rabbit[115] and man[116-120] the cardiac TnT gene is, in fact, subject to complex patterns of alternative splicing. Alternative splicing has been demonstrated in both 5' and 3' regions, resulting in protein isoforms with small differences in both N-terminal and C-terminal regions. Several studies suggest that differences in the N-terminal region may influence the calcium binding characteristics of the troponin complex. In particular, transition from the fetal to adult pattern of cardiac TnT expression in rabbit has been correlated with a developmental decrease in calcium sensitivity.[121,122] Similarly, two bovine isoforms, which differ by the presence or absence of 5 amino acid residues in the N-terminal region, confer different calcium sensitivities to reconstituted troponin in vitro.[123] Studies conducted in man have shown that one of the fetal isoforms can be detected in variable amounts in adult failing hearts and that the proportion of this isoform correlates with the observed decrease in myofibrillar ATPase activity.[124] Similar alternative splicing has been described for the 3'-regions of cardiac TnT of rat, rabbit and man. The functional significance of this splicing has not yet been determined.

TROPOMYOSIN GENE EXPRESSION

Multiple tropomyosin (TM) isoforms have been described and are the products of a limited number of genes which exhibit complex patterns of alternative splicing.[125] There are four distinct TM genes (α-TM, β-TM, TM30 and TM4) in vertebrates,[126-131] of which α-TM and β-TM are expressed in striated muscle, including cardiac muscle. The α-TM gene (see Fig. 3.5) encodes at least nine different isoforms in rat which are generated by alternative splicing and which include isoform(s) expressed in cardiac and skeletal muscle.[131-133] The β-TM gene encodes alternatively spliced isoforms which are expressed in striated as well as smooth muscle and nonmuscle specific forms in rat[130,134] and in chicken.[135,136] Early experiments suggested that only α-TM is present in rodent heart.[137] Recent studies in mouse have demonstrated, however, that both α-TM and β-TM genes are expressed throughout fetal development and in the adult, at least at the level of mRNA accumulation.[138-140] α-TM mRNA is always predominant, with expression of β-TM mRNA decreasing with development such that the α-TM/β-TM ratio increases from about 4.5 at 11 dpc to 18.5 at 2 days neonatal and 60.5 in adult heart.[140] The accumulation of β-TM protein during rodent development has not been described. Detailed studies using S1 nuclease protection analysis capable of distinguishing different isoforms of α-TM and β-TM mRNAs have also been performed in mouse.[140] These show that the predominant mRNA detectable in mouse heart is that corresponding to the striated muscle α-TM type (see Fig. 3.5). Partially protected fragments of the α-TM probe were detected in these experiments, indicating the presence of smooth muscle α-TM mRNA as well as a second, minor striated muscle mRNA (α-TM str'). Similarly the predominant β-TM mRNA detected corresponds to that of striated muscle, but partial protection corresponding to non-muscle/smooth muscle mRNA were also detected. The spatial distribution of

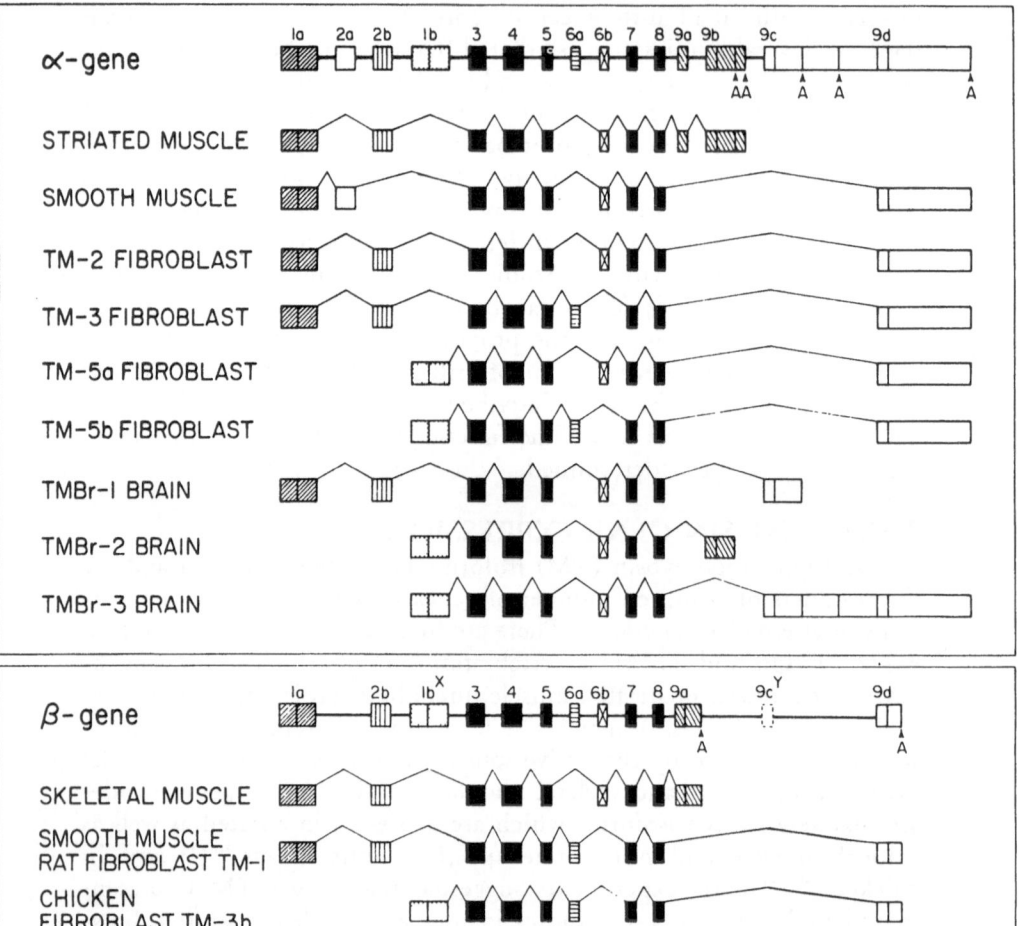

Fig. 3.5. *Alternative splicing of α-TM and β-TM genes. The intron-exon organization of vertebrate TM genes is shown together with alternative splicing combinations observed in muscle and non-muscle cells. Exons are shown as boxes and intervening sequence as horizontal lines. The α-TM gene is that of rat;[131-133] the β-TM gene is a composite of rat and chick.[130,135,136] The striated muscle isoform of α-TM is expressed in both skeletal and cardiac muscle. Note that the isoform marked as 'skeletal muscle' for the β-TM skeletal muscle form is also expressed in cardiac muscle.[140] (From Lees-Millar and Helfman, BioEssays 1991; 13:429-437).*

these isoforms within the heart remains unknown. In the bovine and human heart both α-TM and β-TM isoforms have been described at the protein level[141] although again α-TM is the predominant type detected. Detailed analysis of the alternative splicing of the human genes has not been reported.

GENE EXPRESSION IN THE CONDUCTION SYSTEM

The cardiac conduction system, including the sinoatrial (SA) node, atrioventricular (AV) node, bundle branches and Purkinje fibers, consists of cardiac myocytes which show patterns of gene expression which distinguish them from the rest of the myocardium. Different parts of the conduction system have been shown, for example, to express connexins,[142-146] intermediate filament proteins,[147,148] neurofilament proteins,[149] dystrophin isoforms,[150] inositol 1,4,5-triphosphate receptors,[151] angiotensin II binding sites,[152] Na^+/K^+-ATPase isoforms,[153] polysialated NCAM[154] and other, unidentified, epitopes[32] which distinguish them from the rest of the myocardium. Some of these differences probably relate to the functional requirements of the conduction system for transmission of impulses. Less easily explained is the distinct pattern of contractile protein gene expression. Patterns of expression of myosin heavy chain genes,[30,155-157] myosin light chains[46] and troponin I[101] have been described which distinguish certain conduction system myocytes from other cardiac myocytes. In some cases this represents expression of isoforms not seen in the rest of the myocardium at any stage of development, as with certain myosin heavy chains.[155,157] In other cases, however, isoforms which are expressed transiently throughout the myocardium during development are persistently expressed in the conduction system. For example, myosin light chain MLC1A is expressed in both atrial and ventricular myocardium during development but is restricted in the adult to atrial muscle and the ventricular conduction system.[15,46] The slow skeletal muscle isoform of troponin I is expressed throughout the myocardium during development but becomes restricted to conduction system myocytes after birth.[101] The functional significance of differing contractile protein gene expression in conduction system myocytes is unclear. Expression of these particular isoforms may be necessary for the physiology of the conduction system myocytes but this has yet to be tested experimentally.

REGULATION OF CONTRACTILE PROTEIN GENE EXPRESSION

During early development, when cardiac myocytes are formed, genes encoding the components of the contractile apparatus become activated. Later, during fetal life, their expression is modulated, resulting in the pattern of expression associated with mature atrial and ventricular muscle (see Table 3.3). Many of these genes are expressed in both cardiac and skeletal muscle during development and some are also expressed in both in the adult. For example β-MHC, the major isoform of ventricular muscle in man, is the same as found in slow skeletal muscle. In the case of actin both α-cardiac and α-skeletal actin isoforms are expressed in both cardiac and skeletal muscle during development. These patterns of expression have led to the speculation that the regulation of contractile protein gene expression during development might be shared by cardiac and skeletal muscle. This somewhat simplistic view does not, however, take into account significant differences in both the developmental biology of these two striated muscle types and in their cellular structure and function. Cardiac muscle is derived from anterior lateral plate mesoderm, whereas skeletal muscle is derived from the dorsal paraxial mesoderm which becomes segmented along the neural tube to form the somites.[158] The timing of cardiac and skeletal muscle formation is, as a consequence, different. Cells become committed to the cardiac lineage soon after gastrulation. Cardiac myocyte differentiation (that is, the formation of contractile myocyte cells) occurs soon after. In the avian heart, for example, expression of the ventricular myosin heavy chain VMHC1 can be detected in the cardiogenic mesoderm as early Hamburger Hamilton[159] stage 7[160] (see also chapter 2). This precedes both the formation of the heart tube and the formation of somites. Given the differences in physical location and timing of formation of their respective precursors it is likely that the molecular and cellular signals responsible for programming cardiac and skeletal muscle determination and differentiation are different. A further difference lies in the timing of withdrawal from the cell-cycle. During skeletal muscle development, terminal differentiation of precursor myoblasts involves cell fusion to form multi-nucleated myotubes, the activation of muscle-specific gene expression and irreversible withdrawal from the cell-cycle. Recent evidence that the myogenic factor MyoD interacts with elements of the cell cycle machinery including the retinoblastoma gene product Rb, p107, cyclin D1 and cdk kinase[161-165] provides a link between these events at the molecular level. During cardiac muscle development the activation of muscle-specific gene expression occurs at the time of formation of the cardiac tube (see chapter 2), yet cardiac myocytes continue to undergo cell division during fetal life and, in mammals, withdraw from the cell cycle after birth. Subsequent increase in cardiac mass is believed to occur primarily by trophic growth of myocytes without significant myocyte cell division. The molecular events associated with the process of muscle-specific gene expression and withdrawal from the cell cycle are, therefore, separate events during cardiac muscle development (see Table 3.4).

Little is currently known of the molecular mechanisms responsible for the activation of cardiac muscle gene expression at the onset of cardiac myocyte differentiation. This situation contrasts with that of skeletal muscle differentiation where myogenic regulatory factors (MyoD, myogenin, Myf-5 and MRF4) have been identified (see chapter 4). One reason for this discrepancy is the lack of cardiac myocyte cell lines. During development cardiac myocytes cease dividing at around the time of birth and appear unable to re-enter the cell cycle in vivo or in vitro. In contrast, skeletal muscle myoblast cells isolated from fetal or adult skeletal muscle are capable both of continuous replication and of differentiation into functional skeletal muscle fibers in cell culture. There is, however, accumulating data on the molecular mechanisms responsible for maintaining or modulating gene expression during cardiac development. In many cases muscle gene regulation is achieved at the level of transcription (discussed in detail in chapter 4). This is

Table 3.4. Events in the formation of cardiac muscle

Stage	Molecular Events
Post-gastrulation	Determination: Commitment of mesodermal cells to the myogenic lineage. Molecular mechanisms unknown.
Embryonic/Early Fetal	Differentiation: Activation of muscle gene expression through transcriptional mechanisms resulting in a population of (dividing) myocytes.
Late Fetal	Modulation of muscle gene expression through transcriptional mechanisms (e.g., α-actin and myosin genes) and post transcriptional mechanisms (e.g., alternative RNA splicing of cardiac troponin T gene).
Neonatal	Terminal differentiation: Establishment of "adult" pattern of gene expression (e.g., switch in troponin I gene expression) and cessation of cell division.
Adult	Pattern of gene expression characteristic of mature myocardium.Subject to alterations in response to hemodynamic load and aging.

supported by numerous studies on mRNA accumulation (both through Northern blot analysis of tissue extracts and by in situ hybridization to tissue sections) and corresponding studies on protein accumulation, analyzed by electrophoretic and/or immunocytochemical techniques. In these studies parallel mRNA and protein accumulation suggests that the overall pattern of expression is governed by the accumulation of the corresponding mRNA. That this may be regulated at the level of transcription is supported by increasing numbers of studies which show that cloned cardiac muscle genes are regulated like the endogenous gene when introduced (as naked DNA) into cardiac cells in culture or into whole animals through transgenic techniques. Definitive proof that cardiac gene expression is transcriptionally regulated requires, however, direct measurement of gene transcription. In spite of its obvious importance few such studies have been conducted. Boheler et al[166] have determined the transcriptional activity of the α-MHC compared to that of β-MHC genes and of the α-cardiac actin gene relative to that of α-skeletal actin in intact nuclei isolated from myocytes of newborn rats. The results showed that the level of mRNA accumulation was directly paralleled by the relative transcriptional activity of the corresponding genes. Similar results have been obtained in skeletal muscle nuclei[167] confirming that, in striated muscle, a major level of regulation is that of gene transcription.

A further level of complexity is that of heterogeneity in the patterns of gene expression within regions of myocardium. Attention is often focused, as in this chapter, on the developmental pattern of expression relative to adult atrial and ventricular muscle phenotypes. It will be evident from discussion of early development presented in chapter 2 that simple segmentation of the developing heart is not sufficient to describe the events leading from a tubular structure to the four chambered adult organ. Patterns of expression in the outflow tract myocardium as well as in junctional regions (in particular the atrioventricular junction) are particularly important (see chapter 2). It is also important to consider regional differences within a given chamber after the heart is septated. For example, there is a transitory transmural gradient of β-MHC expression during postnatal development in the rat.[168,169] There also exists some evidence for left-right differences in myosin gene expression.[168,170-173] Transmural gradients have also been described for muscle creatine kinase expression with highest levels in the epicardial layer,[174] for myosin light chain MLC1A[22] and myoglobin.[175] These patterns raise intriguing questions concerning the regulation of gene expression which may relate to gradients in wall stress or pressure loading.[23,169]

Considerable effort is currently centered on identifying transcription factors involved in activating gene expression during cardiogenesis (see chapter 2), in determining cardiac-specific gene expression (see chapter 4) and in altering transcriptional activity in response to altered workload (see chapter 5). One of the better understood mechanisms is that of transcriptional regulation by thyroid hormone, particularly in relation to myosin gene expression during development. Thyroid hormone has a major influence on cardiac myocytes and has a direct effect on the transcriptional activity of α-MHC and β-MHC genes. Studies in hypothyroid rats have shown that lack of thyroid hormone represses expression of α-MHC and enhances β-MHC expression and that thyroid hormone replacement restores normal expression pattern.[20,24,29,176-183] This regulation also accounts for the developmental switch from β-MHC to α-MHC expression observed after birth in rodents and which coincides with the postnatal surge in circulating levels of thyroid hormone.[179] The identification of the thyroid hormone receptor as a ligand-dependent DNA-binding protein and the determination of the DNA sequence to which it binds (the thyroid hormone response element or TRE) is, in itself, a paradigm of transcriptional regulation. Thyroid hormone receptors, products of the *c-erbA* proto-oncogenes (see ref. 184) are nuclear proteins which interact with thyroid hormone (T_3). Multiple isoforms of receptor are produced as a result of alternative splicing of at least two genes. Receptor binding to the TRE requires the presence of T_3. In the presence of thyroid hormone the receptor binds to the TRE and interacts with the transcriptional machinery. Of particular interest is the ability of different MHC genes to respond differently to thyroid hormone depending on the gene and tissue concerned.[183] In rat, the α-MHC gene is positively regulated by thyroid hormone in ventricular muscle (but shows only minimal response in atrial muscle) whereas β-MHC gene expression is negatively regulated in ventricular muscle. As a consequence, thyroidectomy results in loss of α-MHC expression and concomitant expression of β-MHC. However, thyroidectomy does not affect β-MHC expression in atrial muscle or in slow skeletal muscle. The differences between α-MHC and β-MHC may be explained by the organization of the TRE elements, which is different between the two genes. Repression of expression is probably due to interference of bound receptor with essential binding factors as has been described for other genes.[185,186] Differing responses between tissues may be accounted for by differences in receptor isoform expression between tissues and by variable expression of other thyroid hormone binding proteins.[187] For a more detailed discussion of MHC transcriptional regulation, see chapter 4.

CONTRACTILE PROTEIN GENE ORGANIZATION

Multiple isoforms exist for all the contractile proteins. Many of the genes encoding these isoforms have been isolated and their structure and organization within the genome determined. Three types of gene organization can be identified (see Table 3.5): (i) Isolated individual genes encoding single isoforms. This is the case for the majority of contractile protein genes. It is exemplified by the actin genes where the sarcomeric α-cardiac and α-skeletal actins, like the smooth muscle and non-muscle isoforms, are encoded by separate genes each of which produces a single protein isoform. These genes are not closely linked and, in those species where this has been determined, are located on separate chromosomes.[188-190] Similarly, neither the myosin light chain genes expressed in atrial and ventricular muscle[188,191-193] nor the troponin I genes[194-196] are linked in the genome. (ii) Closely linked genes each encoding single isoforms. This is the case for the myosin heavy chain genes where individual genes encode cardiac α-MHC and β-MHC and the different skeletal muscle isoforms. These are arranged into two gene clusters located on two separate chromosomes. One cluster contains the cardiac (α-MHC and β-MHC) genes, the other contains skeletal muscle genes including those encoding the developmentally regulated embryonic neonatal and adult fast skeletal muscle MHC isoforms.[188,197-202] (iii) Individual genes each encoding multiple isoforms. This is exemplified by the troponin T genes. Multiple cardiac TnT isoforms are encoded by a single cardiac troponin T gene through the use of alternative RNA splicing. Two other separate genes[203,204] encode the multiple isoforms found in fast and slow skeletal muscle respectively.[110,111,114,117-120,199,203-206]

The only contractile protein gene family known to show close linkage are the myosin heavy chain genes. The cardiac α-MHC and β-MHC isoforms are encoded by two separate but closely linked genes. MHC polypeptides are large (ca. 200,000 Daltons) and the mRNA and genes which encode them are correspondingly large. The α- and β-MHC

Table 3.5. *Chromosomal location of sarcomeric protein genes*

Protein	Isoform	Gene Locus	Chromosomal Location	Refs
Troponin T	Cardiac (cTnT)	TNNT2	1q	117
(TnT)	Slow skeletal (sTnT)	TNNT1	19q13.4	203,204
	Fast Skeletal (fTnT)	TNNT3	Unknown	
Troponin I	Cardiac (TnIc)	TNNI3	19p13.2-q13.2	194
(TnI)	Slow skeletal (TnIs)	TNNI1	1q32	195,196
	Fast Skeletal (TnIf)	TNNI2	10p13.2-q13.2	117
Troponin C	Cardiac/slow (cTnC)	TNNC1	Unknown	
(TnC)	Fast skeletal (TnCf)		Unknown	
Myosin Heavy Chain	MHC skeletal genes (≥ 6 genes)	MYH1-MYH6	17p31.1	200,210,221,222 198,223-225
	α-MHC	MYH 6	14q12	198,201,207,226
	β-MHC	MYH 7	14q12	198,227
Myosin light chain 1	Atrial (MLC1A)	MYL4	17	192,228
(MLC1)	Ventricular (MLC1V)	MYL3	3p	192,229
	Fast Skeletal (MLC1F/MLC3F)	MYL1	2q32-qter	230,231
Myosin light chain 2	Ventricular (MLC2V)		12q23-q24.2	193
(MLC2)	Atrial (MLC2A)		Unknown	
	Fast skeletal (MLC2F)		Unknown	
α-actin	α-cardiac	ACTC	15q11-qter	189,232
	α-skeletal	ACTSK	1q42	189,233
	α-smooth muscle	ACTSA	10q22-q24	234
Tropomyosin	α-Tm (Cardiac)	TPM1	15q22	235
(Tm)	β-Tm(Cardiac)		Unknown	
	α-Tm(Non-Muscle)	TPM3	1q22q23	236

Reproduced from Barton PJR, Bhavsar PK, Brand NJ. Contractile protein genes in development and disease. In: Annual of Cardiac Surgery. Yacoub M, Pepper J eds. Philadelphia:Academic Press. 1995 (in press).

mRNAs are approximately 7000 nucleotides in length and the combined α- and β-MHC genes cover over 50 kb of genomic DNA. In man, the genes are located on chromosome 14q12 and are separated by 4.5 kb of genomic DNA.[198,201,207] Each gene is comprised of 43 exons and spans approximately 22 kb of DNA. It is unclear why the myosin genes are organized into two clusters in this way. Sequence and structure comparisons would argue that most of the contractile protein gene families arose through gene duplication mechanisms. In this scheme an ancestral gene, probably resembling most closely the modern-day non-muscle gene, would give rise to new genes through duplication. These would subsequently diverge both in sequence and pattern of expression resulting, ultimately, in the generation of tissue-specific adaptations and patterns of expression. Such mechanisms have, for example, been proposed for actin[87] and myosin alkali light chain genes.[15] In most cases gene duplication has been accompanied by dispersion in the genome such that, for example, the different actin and myosin light chain genes are not genetically linked and are often located on different chromosomes. The myosin heavy chain genes have, in contrast, remained closely linked in two genomic clusters. It is possible that this arrangement plays a role in their regulation. Other gene families which show a sequential expression during development and which are closely linked in the genome include the α- and β- globin gene clusters[208] where the organization of the genes reflects their pattern of expression during hematopoesis. This is believed to be achieved through *cis*-acting elements located at the 5'-end of the gene cluster.[208] Another striking example is that of the *Hox* homeobox gene clusters (reviewed by Krumlauf[209]) where the genomic organization of the genes reflects their pattern of expression along the anterior-posterior axis of the embryo. Genes located further towards the 5' end of a given cluster are expressed in more posterior regions of the embryo. In the case of myosin heavy chain genes there no such obvious correlation. Although closely linked, the skeletal muscle genes are not arranged in order of their sequential expression during skeletal muscle embryogenesis[200,210] and the α-MHC and β-MHC genes show different developmental patterns depending on the species concerned. It remains possible that their organization in two clusters has some other functional significance.

An important aspect of identifying the chromosomal location of genes is to determine whether there is genetic linkage to inherited disease loci (see chapter 6). This has proved to be the case for the contractile protein genes and inherited forms of familial hypertrophic cardiomyopathy (FHC). The definition of a genetic locus for FHC on chromosome 14q1[211] in a Canadian pedigree raised the question of MHC involvement, as the cardiac MHC gene cluster had previously been mapped to this region of chromosome 14. Subsequent analysis identified a mutation in the β-MHC gene responsible for the disease in this family. A large number of different β-MHC mutations have now been identified (for reviews see Schwartz et al[212] and Watkins[213]) in different families with FHC, but not all FHC mutations map to chromosome 14q.[214-216] Separate additional loci have been identified on chromosome 1q3,[217] 11p13-q13,[218] and 15q2.[219] Two of these have now been defined at the molecular level[220] and involve mutation in the cardiac troponin T gene (chromosome 1q locus) and α-tropomyosin gene (chromosome 15q2 locus). The molecular basis of the chromosome 11 FHC locus is currently unknown. Together these data show that many, and perhaps all, of the familial cases of hypertrophic cardiomyopathy are due to mutations in genes encoding contractile proteins.

REFERENCES

 1. Friedman WF. The intrinsic physiologic properties of the developing heart. Progress in Cardiovascular Diseases 1972; 15:87-111.
 2. Anderson PAW, Glick KL, Manring A. Developmental changes in cardiac contractility in fetal and postnatal sheep: In vitro and in vivo. Am J Physiol 1984; 247:H3718-98.
 3. Fabiato A. Calcium release in skinned cardiac cells: variations with species, tissues and development. Federation Proceedings 1982; 41:2238-44.
 4. Solaro RJ, Lee JA, Kentish JC et al. Effects of acidosis on ventricular muscle from adult and neonatal rats. Circ Res 1988; 63:779-87.
 5. Solaro RJ, Kumar P, Blanchard EM et al. Differential effects of pH on calcium activation of myofilaments of adult and perinatal dog hearts. Evidence for developmental differences in thin filament regulation. Circ Res 1986; 58:721-9.
 6. Park IS, Michael LH, Driscoll DJ. Comparative response of the developing canine myocardium to inotropic agents. Am J Physiol 1982; 242:H13-8.
 7. Artman M, Kithas PA, Wike JS et al. Inotropic responses change during postnatal maturation in rabbit. Am J Physiol 1988; 255:H335-42.
 8. Nassar R, Reedy MC, Anderson PA. Developmental changes in the ultrastructure and sarcomere shortening of the isolated rabbit ventricular myocyte. Circ Res 1987; 61:465-83.
 9. Spudich JA. How molecular motors work. Nature 1994; 372:515-8.
10. Swynghedauw B. Developmental and functional adaptation of contractile proteins in cardiac and skeletal muscles. Physiol Rev 1986; 66:710-71.
11. Barany M. ATPase activity of myosin correlated with speed of muscle shortening. J Gen Physiol 1967; 50:Suppl:197-218.
12. Schwartz K, Lecarpentier Y, Martin JL et al. Myosin isoenzyme distribution correlates with speed of myocardial contraction. J Mol Cell Cardiol 1981; 13:1071-5.
13. Lompré A, Schwartz K, D'Albis A et al. Myosin isoenzyme redistribution in chronic heart overload. Nature 1979; 282:105-7.
14. Pope B, Hoh JF, Weeds A. The ATPase activities of rat cardiac myosin isoenzymes. FEBS Lett 1980; 118:205-8.
15. Barton PJR, Buckingham M. The myosin alkali light chain proteins and their genes. Biochem J 1985; 231:249-61.
16. Lowey S, Waller GS, Trybus KM. Skeletal muscle light chains are essential for physiological speeds of shortening. Nature 1993; 365:454-6.
17. Trybus KM. Role of myosin light chains. J Musc Res Cell Mot 1994; 15:587-94.
18. Sweeney HL, Bowman BF, Stull JT. Myosin light chain phosphorylation in vertebrate striated muscle: regulation and function. Am J Physiol 1993; 264:C1085-95.
19. Lehman W, Craig R, Vibert P. Ca(2+)-induced tropomyosin movement in Limulus thin filaments revealed by three-dimensional reconstruction. Nature 1994; 368:65-7.

20. Hoh JFY, McGrath PA, Hale PT. Electrophoretic analysis of multiple forms of rat cardiac myosin: Effects of hypophysectomy and thyroxine replacement. J Mol Cell Cardiol 1978; 10:1053-76.

21. Emerson CPJ, Bernstein SI. Molecular genetics of myosin. Annu Rev Biochem 1987; 56:695-726.

22. Lyons GE, Schiaffino S, Sassoon D et al. Developmental regulation of myosin expression in mouse cardiac muscle. J Cell Biol 1990; 111:2427-37.

23. Lyons GE. In situ analysis of the cardiac muscle gene program during embryogenesis. Trends Cardiovasc Med 1994; 4:70-7.

24. Schwartz K, Lompré AM, Bouveret P et al. Comparisons of rat 'cardiac' myosins at fetal stages in young animals and in hypothyroid adults. J Biol Chem 1981; 257:14412-8.

25. Lompré AM, Mercadier JJ, Wisnewsky C et al. Species and age-dependent changes in the relative amounts of cardiac myosin isoenzymes in mammals. Dev Biol 1981; 84:286-90.

26. Wessels A, Vermeulen JLM, Virgh S et al. Spatial distribution of "tissue-specific" antigens in the developing human heart and skeletal muscle. II. An immunocytochemical analysis of myosin heavy chain isoform expression patterns in the embryonic heart. Anat Rec 1991; 229:355-68.

27. Mercadier JJ, Bouveret P, Gorza L et al. Myosin isoenzymes in normal hypertrophied human ventricular myocardium. Circ Res 1983; 53:52-62.

28. Gorza L, Mercadier JJ, Schwartz K et al. Myosin types in the human heart. An immunofluorescence study of normal and hypertrophied atrial and ventricular myocardium. Circ Res 1984; 54:694-702.

29. Chizzonite RA, Everett AW, Prior G et al. Comparison of myosin heavy chains in atria and ventricles from hyperthyroid, hypothyroid, and euthyroid rabbits. J Biol Chem 1984; 259:15564-71.

30. Kuro-o M, Tsuchimochi H, Ueda S et al. Distribution of cardiac myosin isozymes in human conduction system. Immunohistochemical study using monoclonal antibodies. J Clin Invest 1986; 77:340-7.

31. De Groot IJM, Lamers WH, Moorman AFM. Isomyosin expression patterns during rat heart morphogenesis: An immunohistochemical study. Anat Rec 1989; 224:365-73.

32. Wessels A, Vermeulen JLM, Verbeek FJ et al. Spatial distribution of "tissue-specific" antigens in the developing human heart and skeletal muscle. III. An immunohistochemical analysis of the distribution of the neural tissue antigen G1N2 in the embryonic heart; implications for the development of the atrioventricular conduction system. Anat Rec 1992; 232:97-111.

33. Moorman AFM, Lamers WH. Molecular anatomy of the developing heart. Trends Cardiovasc Med 1994; 4:257-64.

34. Sartore S, Pierobon-Bormioli S, Schiaffino S. Immunohistochemical evidence for myosin polymorphism in the chicken heart. Nature 1978; 274:82-3.

35. Gonzalez-Sanchez A, Bader D. Immunochemical analysis of myosin heavy chains in the developing chicken heart. Dev Biol 1984; 103:151-8.

36. de Jong F, Geerts WJC, Lamers WH et al. Isomyosin expression pattern during formation of the tubular chicken heart: A three-dimensional immunohistochemical analysis. Anat Rec 1990; 226:213-27.

37. de Jong F, Geerts WJC, Lamers WH et al. Isomyosin expression patterns in tubular stages of chicken heart development: a 3-D immunohistochemical analysis. Anat Embryol 1987; 177:81-90.

38. Sanders E, Moorman AFM, Los JA. The local expression of adult chicken heart myosins during development. I. The first three days embryonic chicken heart. Anat Embryol 1984; 169:185-91.

39. Sweeney LJ, Zak R, Manasek FJ. Transitions in cardiac isomyosin expression during differentiation of the embryonic chick heart. Circ Res 1987; 61:287-295.

40. Evans D, Miller JB, Stockdale FE. Developmental patterns of expression and coexpression of myosin heavy chains in atria and ventricles of the avian heart. Dev Biol 1988; 127:376-83.

41. Yutzey KE, Rhee JT, Bader D. Expression of the atrial-specific myosin heavy chain AMHC1 and the establishment of anteroposterior polarity in the developing chicken heart. Development 1994; 120:871-83.

42. Bisaha JG, Bader D. Identification and characterization of a ventricular-specific avian myosin heavy chain, VMHC1: expression in differentiating cardiac and skeletal muscle. Dev Biol 1991; 148:355-64.

43. Barton PJR, Robert B, Fiszman M et al. The same myosin alkali light chain gene is expressed in adult cardiac atria and in fetal skeletal muscle. J Mus Res Cell Motil 1985; 6:461-75.

44. Barton PJR, Robert B, Cohen A et al. Structure and sequence of the myosin alkali light chain gene expressed in adult cardiac atria and fetal striated muscle. J Biol Chem 1988; 263:12669-76.

45. Whalen RG, Sell SM. Myosin from fetal hearts contains the skeletal muscle embryonic light chain. Nature 1980; 286:731-3.

46. Whalen RG, Sell SM, Eriksson A et al. Myosin subunit types in skeletal and cardiac tissues and their developmental distribution. Dev Biol 1982; 91:478-84.

47. Soussi-Yanicostas N, Barbet P, Laurent-Winter C et al. Transition of myosin isozymes during development of human masseter muscle. Persistence of developmental myosin isoforms. Development 1990; 108:239-49.

48. Bredman JJ, Wessels A, Weijs WA et al. Demonstration of 'cardiac-specific' myosin heavy chain in masticatory muscles of human and rabbit. Histochemical Journal 1991; 23:160-70.

49. Hailstones D, Barton PJR, Chan-Thomas P et al. Differential regulation of the atrial isoforms of the myosin light chains during striated muscle development. J Biol Chem 1992; 32:23295-300.

50. Kubalak SW, Miller-Hance WC, O'Brien TX et al. Chamber specification of atrial myosin light chain-2 expression precedes septation during murine cardiogenesis. J Biol Chem 1994; 269:16961-70.

51. Vandekerckhove J, Weber K. The complete amino acid sequence of actins from bovine aorta, bovine heart, bovine fast skeletal muscle, and rab-

bit slow skeletal muscle. A protein chemical analysis of muscle actin differentiation. Differentiation 1979; 14:123-33.

52. Vandekerckhove J, Weber K. Mammalian cytoplasmic actins are the products of at least two genes and differ in primary structure in at least 25 identified positions from skeletal muscle actins. Proc Natl Acad Sci USA 1978; 75:1106-10.

53. Vandekerckhove J, Weber K. At least six different actins are expressed in a higher mammal: an analysis based on the amino acid sequence of the amino-terminal tryptic peptide. J Mol Biol 1978; 126:783-802.

54. Winegrad S, Wisnewsky C, Schwartz K. Effect of thyroid hormone on the accumulation of mRNA for skeletal and cardiac α-actin in hearts from normal and hypophysectomized rats. Proc Natl Acad Sci USA 1990; 87:2456-60.

55. Schiaffino S, Samuel JL, Sassoon D et al. Nonsynchronous accumulation of α-skeletal actin and β-myosin heavy chain mRNAs during early stages of pressure-overload-induced cardiac hypertrophy demonstrated by in situ hybridization. Circ Res 1989; 64:937-48.

56. Schwartz K, de la Bastie D, Bouveret P et al. α-Skeletal actin mRNAs accumulate in hypertrophied adult rat hearts. Circ Res 1986; 59:551-5.

57. Izumo S, Nadal-Ginard B, Mahdavi V. Protooncogene induction and reprogramming of cardiac gene expression produced by pressure overload. Proc Natl Acad Sci USA 1988; 85:339-43.

58. Vandekerckhove J, Bugaisky G, Buckingham M. Simultaneous expression of skeletal muscle and heart actin proteins in various striated muscle tissues and cells. J Biol Chem 1986; 261:1838-43.

59. Hennessey ES, Drummond DR, Sparrow JC. Molecular genetics of actin function. Biochem J 1993; 291:657-71.

60. Mayer Y, Czosnek H, zeelon PE et al. Expression of the genes coding for the skeletal muscle and cardiac actins in the heart. Nuc Acids Res 1984; 12:1087-100.

61. Zakut R, Shani M, Givol D et al. Nucleotide sequence of the rat skeletal muscle actin gene. Nature 1982; 298:857-9.

62. McHugh KH, Lessard JL. The developmental expression of the rat α-vascular and γ-enteric smooth muscle isoactins: Isolation and characterization of a rat γ-enteric actin cDNA. Mol Cell Biol 1988; 8:5224-31.

63. Miwa T, Kamada S. The nucleotide sequence of a human smooth muscle (enteric type) γ-actin cDNA. Nuc Acids Res 1990; 18:4263

64. Gunning P, Ponte P, Blau H et al. α-Skeletal and α-cardiac actin genes are coexpressed in adult human skeletal muscle and heart. Mol Cell Biol 1983; 3:1985-95.

65. Gunning P, Ponte P, Okayama H et al. Isolation and characterization of full-length cDNA clones for human α-, β-, and γ-actin mRNAs: skeletal but not cytoplasmic actins have an amino-terminal cysteine that is subsequently removed. Mol Cell Biol 1983; 3:787-95.

66. Minty AJ, Caravatti M, Robert B et al. Mouse actin messenger RNAs. J Biol Chem 1981; 256:1008-14.

67. Ordahl CP, Cooper TA. Strong homology in promoter and 3'-untranslated regions of chick and rat α-actin genes. Nature 1983; 303:348-9.
68. Reddy S, Ozgur K, Lu M et al. Structure of the human smooth muscle α-actin gene. J Biol Chem 1990; 265:1683-7.
69. Ueyama H, Hamada H, Battula N et al. Structure of a human smooth muscle actin gene (aortic type) with a unique intron site. Mol Cell Biol 1984; 4:1073-8.
70. Hu MC, Sharp SB, Davidson N. The complete sequence of the mouse skeletal α-actin gene reveals several conserved and inverted repeat sequences outside of the protein-coding region. Mol Cell Biol 1986; 6:15-25.
71. Hanauer A, Levin M, Heilig R et al. Isolation and characterization of cDNA clones for human skeletal muscle α actin. Nuc Acids Res 1983; 11:3503-3506.
72. Minty AJ, Alonso S, Caravatti M et al. A fetal skeletal muscle actin mRNA in the mouse and its identity with cardiac actin mRNA. Cell 1982; 30:185-92.
73. Gunning P, Ponte P, Blau H et al. alpha-skeletal and alpha-cardiac actin genes are coexpressed in adult human skeletal muscle and heart. Mol Cell Biol 1983; 3:1985-95.
74. Paterson BM, Eldridge JD. α-Cardiac actin is the major sarcomeric isoform expressed in embryonic avian skeletal muscle. Science 1984; 224:1436-8.
75. Sassoon DA, Garner I, Buckingham M. Transcripts of α-cardiac and α-skeletal actins are markers for myogenesis in the mouse embryo. Development 1988; 104:155-64.
76. Carrier L, Boheler KR, Chassagne C et al. Expression of the sarcomeric actin isogenes in the rat heart with development and senescence. Circ Res 1992; 70:999-1005.
77. McHugh KM, Crawford K, Lessard JL. A comprehensive analysis of the developmental and tissue-specific expression of the isoactin multigene family in the rat. Dev Biol 1991; 148:442-58.
78. Sawtell NM, Lessard JL. Cellular distribution of smooth muscle actins during mammalian embryogenesis: Expression of the α-vascular but not the γ-enteric isoform in differentiating striated myocytes. J Cell Biol 1989; 109:2929-47.
79. Ruzicka DL, Schwartz RJ. Sequential activation of α-actin genes during avian cardiogenesis: vascular smooth muscle α-actin gene transcripts mark the onset of cardiac differentiation. J Cell Biol 1988; 107:2575-86.
80. Hayward LJ, Schwartz RJ. Sequential expression of chicken actin genes during myogenesis. J Cell Biol 1986; 102:1485-93.
81. Sugi Y, Lough J. Onset of expression and regional deposition of alpha-smooth and sarcomeric actin during avian heart development. Dev Dynam 1992; 193:116-24.
82. Ordahl CP. The skeletal and cardiac α-actin genes are coexpressed in early embryonic striated muscle. Dev Biol 1986; 117:488-92.
83. Miano JM, Cserjesi P, Ligon KL et al. Smooth myscle myosin heavy chain exclusively marks the smooth muscle lineage during mouse embryogenesis. Circ Res 1994; 75:803-12.

84. Bennetts BH, Burnett L, dos Romedios CG. Differential co-expression of α-actin genes within the human heart. J Mol Cell Cardiol 1986; 18:993-6.

85. Boheler KR, Carrier L, de la Bastie D et al. Skeletal actin mRNA increases in the human heart during ontogenic development and is the major isoform of control and failing adult hearts. J Clin Invest 1991; 88:323-30.

86. Kovilur S, Jacobson JW, Beach RL et al. Evolution of the chordate muscle actin gene. J Mol Evol 1993; 36:361-8.

87. Alonso S, Minty A, Bourlet Y et al. Comparison of three actin-coding sequences in the mouse; evolutionary relationships between the actin genes of warm-blooded vertebrates. J Mol Evol 1986; 23:11-22.

88. Garner I, Minty AJ, Alonso A et al. A 5' duplication of the α-cardiac actin gene in BALB/c mice is associated with abnormal levels of α-cardiac and α-skeletal actin mRNAs in adult cardiac tissue. EMBO J 1986; 5:2559-67.

89. Garner I, Sassoon D, Vandekerckhove J et al. A developmental study of the abnormal expression of α-cardiac and α-skeletal actins in the striated muscle of a mutant mouse. Dev Biol 1989; 134:236-45.

90. Hewett TE, Grupp IL, Grupp G et al. α-Skeletal actin is associated with increased contractility in the mouse heart. Circ Res 1994; 74:740-6.

91. Schiaffino S, Gorza L, Ausoni S. Troponin isoform switching in the developing heart and its functional consequences. Trends Cardiovasc Med 1993; 3:12-7.

92. Parmacek MS, Leiden JM. Structure, function and regulation of troponin C. Circulation 1991; 84:991-1003.

93. Saggin L, Gorza L, Ausoni S et al. Troponin I switching in the developing heart. J Biol Chem 1989; 264:16299-302.

94. Sabry MA, Dhoot GK. Identification and pattern of expression of a developmental isoform of troponin I in chicken and rat cardiac muscle. J Muscle Res Cell Motil 1989; 10:85-91.

95. Ausoni S, De Nardi C, Moretti P et al. Developmental expression of rat cardiac troponin I mRNA. Development 1991; 112:1041-51.

96. Murphy AM, Jones L, Sims HF et al. Molecular cloning of rat cardiac troponin I and analysis of troponin I isoform expression in developing rat heart. Biochemistry 1991; 30:707-12.

97. Martin AF, Ball K, Gao LZ et al. Identification and functional significance of troponin I isoforms in neonatal rat heart myofibrils. Circ Res 1991; 69:1244-52.

98. Hunkeler NM, Kullman J, Murphy AM. Troponin I isoform expression in human heart. Circ Res 1991; 69:1409-14.

99. Bhavsar PK, Dhoot GK, Cumming DVE et al. Developmental expression of troponin I isoforms in the fetal human heart. FEBS Lett 1991; 292:5-8.

100. Sasse S, Brand NJ, Kyprianou P et al. Troponin I gene expression during human cardiac development and in end-stage heart failure. Circ Res 1993; 72:932-8.

101. Gorza L, Ausoni S, Merciai N et al. Regional differences in troponin I isoform switching during rat heart development. Dev Biol 1993; 156:253-64.

102. Liao R, Wang CK, Cheung HC. Time-resolved tryptophan emission study of cardiac troponin I. Biophys J 1992; 63:986-95.

103. Solaro RJ, Moir AJ, Perry SV. Phosphorylation of troponin I and the inotropic effect of adrenaline in the perfused rabbit heart. Nature 1976; 262:615-7.

104. Robertson SP, Johnson JD, Holroyde MJ et al. The effect of troponin I phosphorylation on the Ca^{2+}-binding properties of the Ca^{2+}-regulatory site of bovine cardiac troponin. J Biol Chem 1982; 257:260-3.

105. Mope L, McClellan GB, Winegrad S. Calcium sensitivity of the contractile system and phosphorylation of troponin in hyperpermeable cardiac cells. J Gen Physiol 1980; 75:271-82.

106. Winegrad S. Regulation of cardiac contractile proteins. Correlations between physiology and biochemistry. Circ Res 1984; 55:565-74.

107. el Saleh SC, Solaro RJ. Troponin I enhances acidic pH-induced depression of Ca2+ binding to the regulatory sites in skeletal troponin C. J Biol Chem 1988; 263:3274-8.

108. Dieckman LJ, Solaro RJ. Effect of thyroid status on thin-filament Ca2+ regulation and expression of troponin I in perinatal and adult rat hearts. Circ Res 1990; 67:344-51.

109. Saggin L, Ausoni S, Gorza L et al. Troponin T switching in the developing rat heart. J Biol Chem 1988; 263:18488-92.

110. Jin JP, Lin JJC. Isolation and characterisation of cDNA clones encoding embryonic and adult isoforms of rat cardiac troponin T. J Biol Chem 1989; 264:14471-7.

111. Cooper TA, Ordahl CP. A single cardiac troponin T gene generates embryonic and adult isoforms via developmentally regulated alternate splicing. J Biol Chem 1985; 260:11140-8.

112. Anderson PAW, Oakeley AE. Immunological identification of five troponin T isoforms reveals an elaborate maturational troponin T profile in rabbit myocardium. Circ Res 1989; 65:1087-93.

113. Malouf NN, McMahon D, Oakeley AE et al. A cardiac troponin T epitope conserved across phyla. J Biol Chem 1992; 267:9269-74.

114. Jin J, Huang Q, Yeh HI et al. Complete nucleotide sequence and structural organization of rat cardiac troponin T gene. J Mol Biol 1992; 227:1269-76.

115. Greig A, Hirschberg Y, Anderson PAW et al. Molecular basis of cardiac troponin T isoform heterogeneity in rabbit heart. Circ Res 1994; 74:41-7.

116. Mesnard L, Samson F, Espinasse I et al. Molecular cloning and developmental expression of human cardiac troponin T. FEBS Lett 1983; 328:139-44.

117. Townsend PJ, Farza H, MacGeoch C et al. Human cardiac troponin T: identification of fetal isoforms and assignment of the TNNT2 locus to chromosome 1q. Genomics 1994; 21:311-6.

118. Farza H, Townsend PJ, Yacoub MH, et al. Human cardiac troponin T gene expression in developing and failing heart. In:Hauson S, Kjeldsen K (eds): Proceedings of the International Society for Heart Research XV European Section Meeting. 1995: 565-569.

119. Anderson PAW, Greig A, Mark TM et al. Molecular basis of human cardiac troponin T isoforms expressed in the developing, adult and failing heart. Circ Res 1995; 76:681-6.

120. Mesnard L, Logeart D, Taviaux S et al. Human cardiac troponin T: Cloning and expression of new isoforms in the normal and failing heart. Circ Res 1995; 76:687-92.

121. McAuliffe JJ, Gao L, Solaro RJ. Changes in myofibrillar activation and troponin C Ca²⁺ binding associated with troponin T isoform switching developing rabbit heart. Circ Res 1990; 66:1204-16.

122. Nassar R, Malouf NN, Kelly MB et al. Force-pCa relation and troponin T isoforms of rabbit myocardium. Circ Res 1991; 69:1470-5.

123. Tobacman LS, Lee R. Isolation and functional comparison of bovine cardiac troponin T isoforms. J Biol Chem 1987; 262:4059-64.

124. Anderson PAW, Malouf NN, Oakeley AE et al. Troponin T isoform expression in humans: a comparison among normal and failing heart, fetal heart, and adult and fetal skeletal muscle. Circ Res 1991; 69:1226-33.

125. Lees-Miller J, Helfman DM. The molecular basis for tropomyosin isoform diversity. BioEssays 1991; 13:429-37.

126. Clayton L, Reinach FC, Chumbley GM et al. Organization of the hTMnm gene implications for the evolution of muscle and non-muscle tropomyosins. J Mol Biol 1988; 201:507-15.

127. Mak A, Smillie LB, Stewart GR. A comparison of the amino acid sequences of rabbit skeletal muscle α and β tropomyosin. J Biol Chem 1980; 255:3647-53.

128. Ruiz-Opazo N, Weinberger J, Nadal-Ginard B. comparison of α tropomyosin sequences from smooth and striated muscle. Nature 1985; 315:67-70.

129. Cummins P, Perry SV. Chemical and immunochemical characteristics of tropomyosins from striated and smooth muscle. Biochem J 1974; 141:43-9.

130. Helfman DM, Cheley S, Kuismanen E et al. Nonmuscle and muscle tropomyosin isoforms are expressed from a single gene by alternative splicing and polyadenylation. Mol Cell Biol 1986; 6:3582-5.

131. Ruiz-Opazo N, Nadal-Ginard B. α-Tropomyosin gene organization. J Biol Chem 1987; 262:4755-65.

132. Lees-Miller JP, Goodwin LO, Helfman DM. Three novel brain tropomyosin isoforms are expressed from the rat α-tropomyosin gene through the use of alternative promoters and alternative RNA processing. Mol Cell Biol 1990; 10:1729-42.

133. Wieczorek D, Smith C, Nadal-Ginard B. The α rat tropomyosin gene generates a minimum of six different mRNAs coding for striated, smooth and nonmuscle isoforms by alternative splicing. Mol Cell Biol 1988; 8:679-94.

134. Yamawaki-Kataoka Y, Helfman DM. Rat embryonic fibroblast tropomyosin 1. J Biol Chem 1985; 260:14440-5.

135. Forry-Schaudies S, Maihle NJ, Hughes SH. Generation of skeletal, smooth and low molecular weight non-muscle tropomyosin isoforms from the chicken tropomyosin 1 gene. J Mol Biol 1990; 211:321-30.

136. Libri D, Lemonnier M, Meinnel T et al. A single gene encodes for the β-subunits of smooth and skeletal muscle tropomyosin in the chicken. J Biol Chem 1989; 264:2935-44.

137. Cummins P, Perry S. The subunits and biological activity of polymorphic forms of tropomyosin. Biochem J 1973; 133:765-77.

138. Wang SM, Rubenstein PA. Choice of 3' cleavage/polyadenylation site in β-tropomyosin RNA processing is differentiation-dependent in mouse BC3H1 muscle cells. J Biol Chem 1992; 267:2728-36.

139. Wang YC, Rubenstein PA. Splicing of two alternative exon pairs in β-tropomyosin pre-mRNA is independently controlled during myogenesis. J Biol Chem 1992; 267:12004-10.

140. Muthuchamy M, Pajak L, Howles P et al. Developmental analysis of tropomyosin gene expression in embryonic stem cells and mouse embryos. Mol Cell Biol 1993; 13:3311-23.

141. Humphreys JE, Cummins P. Regulatory proteins of the myocardium. Atrial and ventricular tropomyosin and troponin-I in the developing and adult bovine and human heart. J Mol Cell Cardiol 1984; 16:643-57.

142. Gourdie RG, Green CR, Severs NJ et al. Immunolabelling patterns of gap junction connexins in the developing and mature rat heart. Anat Embryol 1992; 185:363-78.

143. van Kempen MJA, Fromaget C, Gros D et al. Spatial distribution of connexin-43, the major cardiac gap junction protein, in the developing and adult rat heart. Circ Res 1991; 68:1638-51.

144. Gourdie RG, Green CR, Severs NJ et al. Evidence for a distinct gap-junctional phenotype in ventricular conduction tissues of the developing and mature avian heart. Circ Res 1993; 72:278-89.

145. Kanter HL, Laing JG, Beyer EC et al. Multiple connexins colocalize in canine ventricular myocyte gap junctions. Circ Res 1993; 73:344-50.

146. Kanter HL, Laing JG, Beau SL et al. Distinct patterns of connexin expression in canine Purkinje fibers and ventricular muscle. Circ Res 1993; 72:1124-31.

147. Kjîrell U, Thornell LE, Lehto V et al. A comparative analysis of intermediate filament proteins in bovine heart Purkinje fibres and gastric smooth muscle. Eur J Cell Biol 1987; 44:68-78.

148. Vincent M, Levasseur S, Currie RW et al. Persistence of an embryonic intermediate filament-associated protein in the smooth muscle cells of elastic arteries and in Purkinje fibres. J Mol Cell Cardiol 1991; 23:873-82.

149. Vitadello M, Matteoli M, Gorza L. Neurofilament proteins are co-expressed with desmin in heart conduction system myocytes. J Cell Sci 1990; 97:11-21.

150. Bies RD, Friedman D, Roberts R et al. Expression and localization of dystrophin in human cardiac purkinje fibers. Circulation 1992; 86:147-53.

151. Gorza L, Schiaffino S, Volpe P. Inositol 1,4,5-trisphosphate receptor in heart: evidence for its concentration in Purkinje myocytes of the conduction system. J Cell Biol 1993; 121:345-53.

152. Saito K, Gutkind JS, Saavedra JM. Angiotensin II binding sites in the conduction system of rat hearts. Am J Physiol 1987; 253:H1618-22.

153. Zahler R, Brines M, Kashgarian M et al. The cardiac conduction system in the rat expresses the α2 and α3 isoforms of the Na⁺,K⁺-ATPase. Proc Natl Acad Sci USA 1992; 89:99-103.

154. Watanabe M, Timm M, Fallah-Najmabadi H. Cardiac expression of polysialylated NCAM in the chicken embryo: correlation with the ventricular conduction system. Dev Dynam 1992; 194:128-41.

155. Gorza L, Saggin L, Sartore S et al. An embryonic-like myosin heavy chain is transiently expressed in nodal conduction tissue of the rat heart. J Mol Cell Cardiol 1988; 20:931-41.

156. Gorza L, Sartore S, Thornell L et al. Myosin types and fiber types in cardiac muscle. III. Nodal conduction tissue. J Cell Biol 1986; 102:1758-66.

157. Gonzalez-Sanchez A, Bader D. Characterization of a myosin heavy chain in the conductive system of the adult and developing chicken heart. J Cell Biol 1985; 100:270-5.

158. Buckingham M. Making muscle in mammals. Trends Genet 1992; 8:144-8.

159. Hamburger V, Hamilton HL. A series of normal stages in the development of the chick embryo. J Morphol 1951; 38:49-92.

160. Gonzalez-Sanchez A, Bader D. In vitro analysis of cardiac progenitor cell differentiation. Dev Biol 1990; 139:197-209.

161. Gu W, Schneider JW, Condorelli G et al. Interaction of myogenic factors and the retinoblastoma protein mediates muscle cell commitment and differentiation. Cell 1993; 72:309-24.

162. Schneider JW, Gu W, Zhu L et al. Reversal of terminal differentiation mediated by p107 in Rb⁻/⁻ muscle cells. Science 1994; 264:1467-71.

163. Rao SS, Chu C, Kohtz DS. Ectopic expression of cyclin D1 prevents activation of gene transcription by myogenic basic helix-loop-helix regulators. Mol Cell Biol 1994; 14:5259-67.

164. Shapek SX, Rhee J, Spicer DB et al. Inhibition of myogenic differentiation in proliferating myoblasts by cyclin D1-dependent kinase. Science 1995; 267:1022-4.

165. Halevy O, Norvitch BG, Spicer DB et al. Correlation of terminal cell cycle arrest of skeletal muscle with induction of p21 by MyoD. Science 1995; 267:1018-21.

166. Boheler KR, Chassagne C, Martin X et al. Cardiac expressions of α- and β-myosin heavy chains and sarcomeric α-actins are regulated through transcriptional mechanisms. J Biol Chem 1992; 267:12979-85.

167. Cox RD, Garner I, Buckingham ME. Transcriptional regulation of actin and myosin genes during differentiation of a mouse muscle cell line. Differentiation 1990; 43:183-91.

168. Gorza L, Pauletto P, Pessina AC et al. Isomyosin distribution in normal and pressure overloaded rat ventricular myocardium: An immunocytochemical study. Circ Res 1981; 49:1003-9.

169. Dechesne CA, Leger JOC, Leger JJ. Distribution of α- and β-myosin heavy chains in the ventricular fibres of the postnatal developing rat. Dev Biol 1987; 123:169-78.

170. Mercadier JJ, Lompré A, Wisnewsky C et al. Myosin isoenzyme changes in several models of rat cardiac hypertrophy. Circ Res 1981; 49:525-32.

171. Satore S, Gorza L, Pierobon-Bormioli S et al. Myosin types and fibre types in cardiac muscle. I. ventricular myocardium. J Cell Biol 1981; 88:226-33.

172. Litten RZ, Martin BJ, Buchthal RH et al. Heterogeneity of myosin isozyme content of rabbit heart. Circ Res 1985; 57:406-14.

173. Kelly R, Alonso S, Tajbakhsh S, Cossu G, Buckingham M. Myosin light chain 3F regulatory sequences confer regionalised cardiac and skeletal muscle expression in transgenic mice. J Cell Biol 1995; (in press).

174. Lyons GE, Mühlebach S, Moser A et al. Developmental regulation of creatine kinase gene expression by myogenic factors in embryonic mouse and chick skeletal muscle. Development 1991; 113:1017-29.

175. Parsons W, Richardson J, Graves K et al. Gradients of transgene expression directed by the human myoglobin promoter in the developing mouse heart. Proc Natl Acad Sci U S A 1993; 90:1726-30.

176. Everett AW, Clark WA, Chizzonite RA et al. Change in synthesis rates of α- and β-myosin synthesis by thyroid hormone. J Biol Chem 1983; 258:2421-5.

177. Holubarsch C, Goulette RP, Litten RZ et al. The economy of isometric force development, myosin isoenzyme pattern and myofibrillar ATPase activity in normal and hypothyroid rat myocardium. Circ Res 1985; 56:78-86.

178. Litten RZ, Low BJ, Alpert NR. Altered myosin isozyme patterns from pressure overloaded and thyrotoxic hypertrophied rabbit hearts. Circ Res 1982; 50:856-64.

179. Lompré A, Nadal-Ginard B, Mahdavi V. Expression of the cardiac ventricular α- and β-myosin heavy chain genes is developmentally and hormonally regulated. J Biol Chem 1984; 259:6437-46.

180. Gustafson TA, Markham BE, Morkin E. Effects of thyroid hormone on α-actin and myosin heavy chain gene expression in cardiac and skeletal muscles of the rat: Measurement of mRNA content using synthetic oligonucleotide probe. Circ Res 1986; 59:194-201.

181. Chizzonite RA, Zak R. Regulation of myosin isoenzyme composition in fetal and neonatal rat ventricle by endogenous thyroid hormones. J Biol Chem 1984; 259:12628-32.

182. Rohrer DK, Hartong R, Dillmann WH. Influence of thyroid hormone and retinoic acid on slow SR Ca-ATPase and myosin heavy chain α expression in cardiac myocytes. J Biol Chem 1991; 266:8638-48.

183. Izumo SM, V., Mahdavi V, Nadal-Ginard B. All members of the MHC multigene family respond to thyroid hormone in a highly tissue-specific manner. Science 1986; 231:231-597.

184. Evans RM. Steroid and thyroid hormone receptors as transcriptional regulators of development and physiology. Science 1988; 240:889-95.

185. Chatterjee V, Lee J, Rentoumis A et al. Negative regulation of the thyroid stimulating hormone α gene by thyroid hormone: Receptor interaction adjacent to the TATA box. Proc Natl Acad Sci USA 1989; 86:9114-8.

186. Crone DE, Kim H, Spindler SR. α and β thyroid hormone receptors bind immediately adjacent to the rat growth hormone gene TATA in a negatively hormone-responsive region. J Biol Chem 1990; 265:10851-6.
187. Osty J, Rappaport L, Samuel JL et al. Characterization of a cytosolic triiodothyronine binding protein in atrium and ventricle of rat heart with different sensitivity toward thyroid hormone levels. Endocrinology 1988; 122:1027-33.
188. Robert B, Barton PJR, Minty A et al. Investigation of genetic linkage between myosin and actin genes using an interspecific mouse back-cross. Nature 1985; 314:181-3.
189. Gunning P, Ponte P, Kedes L et al. Chromosomal location of the co-expressed human skeletal and cardiac actin genes. Proc Natl Acad Sci USA 1984; 81:1813-7.
190. Erba HP, Eddy R, Shows T et al. Structure, chromosome location, and expression of the human gamma-actin gene: differential evolution, location, and expression of the cytoskeletal beta- and gamma-actin genes. Mol Cell Biol 1988; 8:1775-89.
191. Cohen-Haguenauer O, Barton PJR, Van Cong N et al. Assignment of the human fast skeletal muscle myosin alkali light chain gene (MLC1F/MLC3F) to 2q32.1-2qter. Hum Gen 1988; 78:65-70.
192. Cohen-Haguenauer O, Barton PJR, Van Cong N et al. Chromosomal assignment of the myosin alkali light chain genes encoding the ventricular/slow skeletal isoform and the atrial/fetal muscle isoform. Hum Gen 1989; 81:278-82.
193. Macera MJ, Szabo P, Sadgaonkar R et al. Localization of the gene coding for ventricular myosin regulatory light chain (MYL2) to human chromosome 12q23-q24.3. Genomics 1992; 13:829-31.
194. MacGeoch C, Barton PJR, Vallins WJ et al. The human cardiac troponin I locus: assignment to chromosome 19p13.2-19q13.2. Hum Genet 1991; 88:101-4.
195. Wade R, Eddy R, Shows TB et al. cDNA sequence, tissue-specific expression, and chromosomal mapping of the human slow-twitch skeletal muscle isoform of troponin I. Genomics 1990; 7:346-57.
196. Eyre HJ, Akkari PA, Meredith C et al. Assignment of the human slow skeletal muscle troponin gene (TNNI1) to 1q32 by fluorescence in situ hybridisation. Cytogenet Cell Genet 1993; 62:181-2.
197. Weydert A, Daubas P, Lazaridis I et al. Genes for skeletal muscle myosin heavy chains are clustered and are not located on the same mouse chromosome as a cardiac myosin heavy chain gene. Proc Natl Acad Sci USA 1985; 82:7183-7.
198. Saez LJ, Gianola KM, McNally EM et al. Human cardiac myosin heavy chain genes and their linkage in the genome. Nuc Acids Res 1987; 15:5443-59.
199. Gahlmann R, Troutt AB, Wade RP et al. Alternative splicing generates variants in important functional domains of human slow skeletal troponin T. J Biol Chem 1987; 262:16122-6.

200. Yoon SJ, Seiler SH, Kucherlapati R et al. Organization of the human skeletal myosin heavy chain gene cluster. Proc Natl Acad Sci USA 1992; 89:12078-82.

201. Matsuoka R, Yoshida MC, Kanda N et al. Human cardiac myosin heavy chain gene mapped within chromosome region 14q11.2→q13. Am J Med Genet 1989; 32:279-84.

202. Edwards YH, Parkar M, Povey S et al. Human myosin heavy chain genes assigned to chromosome 17 using a human cDNA clone as probe. Ann Hum Genet 1985; 49:101-9.

203. Samson F, de Jong PJ, Trask BJ et al. Assignment of the human slow skeletal troponin T gene to 19q13.4 using somatic cell hybrids and fluorescence in situ hybridization analysis. Genomics 1992; 13:1374-5.

204. Novelli G, Gennarelli M, Rocchi M et al. Assignment of the slow troponin T (TNNT1) gene to chromosome 19 using polymerase chain reaction. Hum Genet 1992; 88:697-8.

205. Breitbart RE, Nguyen HT, Medford RM et al. Intricate combinatorial patterns of exon splicing generate multiple regulated troponin T isoforms from a single gene. Cell 1985; 41:67-82.

206. Breitbart RE, Nadal-Ginard B. Complete nucleotide sequence of the fast skeletal troponin T gene: alternative spliced exons inhibit unusual interspecies divergence. J Mol Biol 1986; 188:313-24.

207. Matsuoka R, Chambers A, Kimura M et al. Molecular cloning and chromosomal localization of a gene coding for human cardiac myosin heavy-chain. Am J Med Genet 1988; 29:369-76.

208. Crossley M, Orkin SH. Regulation of the β-globin locus. Curr Opin Genet Develop 1993; 3:232-7.

209. Krumlauf R. Hox genes in vertebrate development. Cell 1994; 78:191-201.

210. Soussi-Yanicostas N, Whalen RG, Petit C. Five skeletal myosin heavy chain genes are organized as a multigene complex in the human genome. Human Molecular Genetics 1993; 2:563-9.

211. Jarcho JA, McKenna W, Pare JA et al. Mapping a gene for familial hypertrophic cardiomyopathy to chromosome 14q1. N Engl J Med 1989; 321:1372-8.

212. Schwartz K, Carrier L, Guicheney P et al. Molecular basis of familial cardiomyopathies. Circulation 1995; 91:532-40.

213. Watkins H. Multiple disease genes cause hypertrophic cardiomyopathy. Br Heart J 1994; 72:S4-9.

214. Solomon SD, Jarcho JA, McKenna W et al. Familial hypertrophic cardiomyopathy is a genetically heterogeneous disease. J Clin Invest 1990; 86:993-9.

215. Schwartz K, Beckmann J, Dufour C et al. Exclusion of cardiac myosin heavy chain and actin gene involvement in hypertrophic cardiomyopathy of several French families. Circ Res 1992; 71:3-8.

216. Epstein ND, Fananapazir L, Lin HJ et al. Evidence of genetic heterogeneity in five kindreds with familial hypertrophic cardiomyopathy. Circulation 1992; 85:635-47.

217. Watkins H, MacRae C, Thierfelder L et al. A disease locus for familial hypertrophic cardiomyopathy maps to chromosome 1q3. Nat Gen 1993; 3:333-57.

218. Carrier L, Hengstenberg C, Beckmann JS et al. Mapping of a novel gene for familial hypertrophic cardiomyopathy to chromosome 11. Nat Gen 1993; 4:311-3.

219. Thierfelder L, MacRae C, Watkins H et al. A familial hypertrophic cardiomyopathy locus maps to chromosome 15q2. Proc Natl Acad Sci U S A 1993; 90:6270-4.

220. Thierfelder L, Watkins H, MacRae C et al. Alpha-tropomyosin and cardiac troponin T mutations cause familial hypertrophic cardiomyopathy: a disease of the sarcomere. Cell 1994; 77:701-12.

221. Schwartz CE, McNally E, Leinwand L et al. A polymorphic human myosin heavy chain locus is linked to an anonymous single copy locus (D17S1) at 17p13. Cytogenet Cell Genet 1986; 43:117-20.

222. Karsch-Mizrachi I, Feghali R, Shows TB et al. Generation of a full-length human perinatal mysoin heavy-chain-encoding cDNA. Gene 1990; 89:298-4.

223. Stedman HH, Eller M, Jullian EH et al. The human embryonic myosin heavy chain. Complete primary structure reveals evolutionary relationships with other developmental isoforms. J Biol Chem 1990; 265:3568-76.

224. Eller M, Stedman HH, Sylvester JE et al. Human embryonic myosin heavy chain cDNA. Interspecies sequence conservation of the myosin rod, chromosomal locus and isoform specific transcription of the gene. FEBS Lett 1989; 256:21-8.

225. Rappold GA, Vosberg HP. Chromosomal localization of human myosin heavy chain gene by in situ hybridization. Hum Genet 1983; 65:195-7.

226. Epp TA, Dixon IM, Wang HY et al. Structural organization of the human cardiac alpha-myosin heavy chain gene (MYH6). Genomics 1993; 18:505-9.

227. Qin H, Kemp J, Yip MY et al. Localization of human cardiac beta-myosin heavy chain gene (MYH7) to chromosome 14q12 by in situ hybridization. Cytogenet Cell Genet 1990; 54:74-6.

228. Seharaseyon J, Bober E, Hsieh CL et al. Human embryonic/atrial myosin alkali light chain gene characterization, sequence, and chromosomal location. Genomics 1990; 7:289-93.

229. Fodor WL, Darras B, Seharaseyon J et al. Human ventricular/slow twitch myosin alkali light chain gene characterization, sequence, and chromosomal location. J Biol Chem 1989; 264:2143-9.

230. Cohen Haguenauer O, Barton PJR, Nguyen VC et al. Assignment of the human fast skeletal muscle myosin alkali light chains gene (MLC1F/MLC3F) to 2q32.1-2qter. Hum Genet 1988; 78:65-70.

231. Seidel U, Bober E, Winter B et al. Alkali myosin light chains in man are encoded by a multigene family that includes the adult skeletal muscle, the embryonic or atrial, and nonsarcomeric isoforms. Gene 1988; 66:135-46.

232. Kramer PL, Luty JA, Litt M. Regional localisation of the gene for cardiac actin (ACTC) on chromosome 15q. Genomics 1992; 13:904-5.
233. Akkari PA, Eyre HJ, Wilton SD et al. Assignment of the human skeletal muscle alpha actin gene (ACTA1) to 1q42 by fluorescence in situ hybridisation. Cytogenet Cell Genet 1994; 65:265-7.
234. Ueyama H, Bruns G, Kanda N. Assignment of the vascular smooth muscle actin gene ACTSA to human chromosome 10. Jinrui Idengaku Zasshi 1990; 35:145-50.
235. Eyre H, Akkari PA, Wiltron SD et al. Assignment of the human skeletal muscle alpha-tropomyosin gene (TPM1) to band 15q22 by fluorescence in situ hybridization. Cytogenet Cell Genet 1995; 69:15-7.
236. Wilton SD, Eyre H, Akkari PA et al. Assigment of the human alpha-tropomyosin gene TPM3 to 1q22→q23 by fluorescence in situ hybridisation. Cytogenet Cell Genet 1994; 68:122-4.

CHAPTER 4

REGULATORS OF MUSCLE GENE EXPRESSION

INTRODUCTION

Specific interactions between transcription factors (TF) and RNA polymerase II are the basis of the regulation of gene transcription and play a central role in many developmental processes. The timing of these interactions and the cells or tissues in the embryo in which they take place are controlled by a well-ordered cascade of TF gene expression. For example, the early stages of embryogenesis are characterized by the formation of the three germ layers (ectoderm, endoderm and mesoderm) and the laying down of a primary body plan with the establishment of the anterior (head)-posterior (tail) and dorsal-ventral axes. Mutations within genes encoding TFs (such as homeobox or zinc finger genes) which disrupt their expression or the function of the proteins they encode can have profound consequences on development and normal morphogenetic processes. These events have been analyzed in detail in the fruit fly *Drosophila melanogaster*[1] and it is clear that many of the developmental mechanisms underlying fly development have their counterparts at the molecular level in vertebrate mammalian development. For example, the genetic organization of the *HOM/hox* gene clusters in fly and vertebrates suggest that they have evolved from a common ancestral group of genes.[2,3] Skeletal muscle development is studied as a well-defined model of tissue specialization and terminal differentiation. Recent years have brought a considerable advance in understanding how genes are expressed in a tissue-specific manner, and have allowed the characterization of how key molecular markers of muscle differentiation, such as contractile protein genes, are activated. This has come about chiefly through the study of a well-defined myoblast to myotube transition in skeletal muscle-derived cell

Molecular Biology of Cardiac Development and Growth, by Paul J.R. Barton, Kenneth R. Boheler, Nigel J. Brand, Penny S. Thomas. © 1995 R.G. Landes Company.

lines which differentiate in culture and the molecular cloning of cDNAs encoding TFs central to skeletal muscle differentiation. These include the myogenic basic/helix-loop-helix (bHLH) factors MyoD,[4] myogenin,[5,6] *myf-5*[7] and MRF4.[8] These proteins, members of the bHLH family of transcription factors (see chapter 1), are restricted in expression to skeletal muscle and have been proposed as "master regulators", indispensable for tissue-specific gene expression. In contrast to the well-established situation in skeletal muscle, our understanding of how contractile protein gene expression is regulated in cardiac muscle remains vague. Although the expression of certain genes transcribed in both cardiac and skeletal muscle have been shown to depend upon myogenic bHLH factors in skeletal muscle, these factors have not been detected in cardiac muscle, though reportedly related factors have been detected in pre-cardiac mesoderm of early chick and mouse embryos.[9]

Advances in the field of skeletal muscle molecular biology are of vital importance in molecular cardiology. For example, members of the MEF-2 family of TFs are expressed in both cardiac and skeletal muscle.[10,11] The mechanisms by which MEF-2 proteins bind DNA and activate gene expression in skeletal muscle can be predicted to be true for cardiac muscle as well. Table 4.1 lists a number of factors expressed in skeletal and cardiac muscle which have been identified by molecular cloning or through their interactions with particular muscle genes. Three things are apparent from the table. First, it includes members of all the major transcription factor gene families described in chapter 1. Secondly, several different factors usually interact to activate the expression of a particular gene. For example, high-level, thyroid hormone-inducible expression of the rodent α-MHC gene requires the binding of members of the GATA and MEF-2 families in addition to the thyroid hormone receptor.[12-14] Third, only a few TFs, such as the myogenic bHLH proteins, are restricted in expression to a particular tissue or cell lineage. Many are distributed more widely or are expressed ubiquitously. This argues against the general use of single, tissue-restricted factors as master regulators for the cell-specific activation (or repression) of genes and the development of a differentiated cell phenotype. In this chapter we look at how genes are regulated in a tissue-specific manner by the binding of multiple transcription factors. We focus on two distinct pathways for activating tissue-specific gene expression in the two striated muscle types. Pathways regulated by myogenic bHLH factors, which are used extensively in skeletal muscle, are reviewed in the first part of the chapter. Though there is no data currently to support such a program of gene expression in cardiac muscle, the example of myogenic bHLH-controlled pathways and the importance of interactions between different TFs in skeletal muscle provides an insight into how gene expression may be regulated in cardiac muscle. In particular, we look at MyoD and its properties as a DNA-binding protein and transcriptional activator. In the second part we illustrate how largely bHLH-independent pathways are used in cardiac muscle, from which myogenic bHLH proteins are absent.

THE MYOGENIC bHLH FAMILY: MASTER REGULATORS OF SKELETAL MUSCLE-SPECIFIC GENE EXPRESSION

bHLH FACTORS BIND TO E-BOXES

The MyoD gene was identified originally by its ability, when introduced and expressed in the mouse embryonic fibroblast cell line $10T_{1/2}$, to transform them into myoblasts expressing normal markers

Table 4.1. Transcription factors expressed in striated muscle and examples of gene promoters they bind to

Factor	Key Domain	Expression	Binding site	Promoter
MyoD, myogenin, myf-5, MRF4	bHLH	skeletal muscle	CANNTG	myogenin, MCK
			α-cardiac actin	
E2A[a]	bHLH	ubiquitous	CANNTG	-
Id[b]	HLH	widespread	-	-
Nkx-2.5/csx	homeodomain	myocardium + others	unknown	unknown
Mhox	homeodomain	striated muscle and uterus	$T(^A/_T)ATAAT(^A/_T)$	MCK
MEF-2 (A,B,C,D)	MADS	skeletal/cardiac muscle + others	$CTA(^A/_T)_4TA(^A/_G)$	myogenin myoglobin (human) MCK
SRF/CArG	MADS	ubiquitous	$CC(^A/_T)_6GG$	α-cardiac actin c-fos
GArC	unknown	widespread	$G(A)_{4-5}TCT$	α-MHC (human)
GATA-4	GATA (zfp[c])	myocardium/ endoderm	$^A/_TGATA^A/_G$	α-MHC (rat) cardiac TnC enhancer (mouse)
MCBF/TEF-1	TEA domain	skeletal/cardiac muscle	$CATTCC^A/_T$	cTnT (chicken)
MEF3	unknown	ubiquitous?	$(^T/_G)(^G/_C)_2TCAGG$	skeletal TnC enhancer (mouse) aldolase A (rat)
HF-1a	unknown	ubiquitous	GGTCATG	MLC2V (rat)
HF-1b	C_2H_2 zinc finger	skeletal/cardiac muscle + brain	GGGTTATTTTA	MLC2V (rat)
CBP40	unknown	skeletal/cardiac + others	CCCCACCCC	myoglobin (human)
MNF	HNF-3/forkhead	"	CCCCACCCC	"
Thyroid hormone receptors	C_4 zinc twist	widespread	$(AGGTCA)_2$	α-MHC
jun/fos (AP-1)	bZIP	widespread	$TGA^G/_CT^C/_A$	ANP

[a] E2A proteins are ubiquitous dimerization partners (also known as Class A proteins) for tissue-specific bHLH factors such as MyoD.

[b] Id lacks the basic domain and dimerizes with Class A bHLH factors, preventing their binding to tissue-restricted bHLH factors to form competent DNA-binding heterodimers.

[c] GATA proteins have a novel zinc finger structure, distinct from the C_2H_2 zinc finger or C_4 zinc twist.

of skeletal myogenesis.[4,15] This property is shared by three related genes called myogenin, *myf-5* and MRF4 (also known as *myf-6* or Herculin).[16,17] Like all bHLH proteins, they bind DNA as dimers. The HLH domain, composed of two α-helices separated by a loop, mediates protein dimerization, leaving the adjoining basic helices free to contact DNA. The dimer binds a short partially-palindromic DNA sequence known as the E-box (Table 4.1). This site, found in the enhancers of genes which are targets for regulation by bHLH proteins, has the consensus sequence CANNTG, where N is any nucleotide. The activation of genes coding for many skeletal muscle markers, including contractile proteins and enzymes, is achieved largely, but not exclusively, by the binding of myogenic bHLH factors to E-boxes present in target genes. The bHLH gene family is an extensive one and contains genes whose protein products play central roles in regulating gene expression in cell lineages other than skeletal muscle (Fig. 4.1). These include the *Drosophila achaete-scute* genes, which function as determinants of neuronal development in fly, and their rat homologues MASH1 and MASH2.[18] However, the human E12 and E47 bHLH proteins,[19] encoded by the E2A gene and produced by alternative splicing of the E2A primary transcript, are expressed ubiquitously, as are their homologues from rat Pan-1 and Pan-2.[20] (These are often called the Class A bHLH proteins: see Fig. 4.1). Functional heterodimers are formed in vivo by the association of tissue-restricted bHLH factors with ubiquitously expressed ones: heterodimers of MyoD and E47 bind DNA over ten times more efficiently than MyoD homodimers.[21-23] The formation of heterodimers is a property that has been conserved during evolution. In *Drosophila* the *achaete-scute* neurogenic regulators probably dimerize with the product of the ubiquitously expressed daughterless gene,[9,22] whereas MASH1 from mouse forms heterodimers with human E12.[24] However, MyoD from *Drosophila* is unable to dimerize with E2A proteins when expressed in mammalian cells and will not *trans*-activate gene expression,[9] suggesting that the formation of a heterodimer between a tissue-restricted bHLH factor and a ubiquitous partner is precise and a prerequisite for accurate and tight DNA binding and *trans*-activation (see below). The formation of active heterodimers also increases the number of potential regulatory dimers which can be assembled in the cell from a limited repertoire of bHLH genes and their protein products.

MyoD Binding the E-box; Implications for *Trans*-Activation

The bHLH domain consists of two α-helices separated by a loop which varies in length among bHLH family members (Fig. 4.1). These helices contain a number of conserved hydrophobic or charged residues which participate in protein dimerization. Contiguous with the N-terminal end of the first helix is the basic helix which is rich in positively charged residues such as arginine and lysine (see also Fig. 1.5).

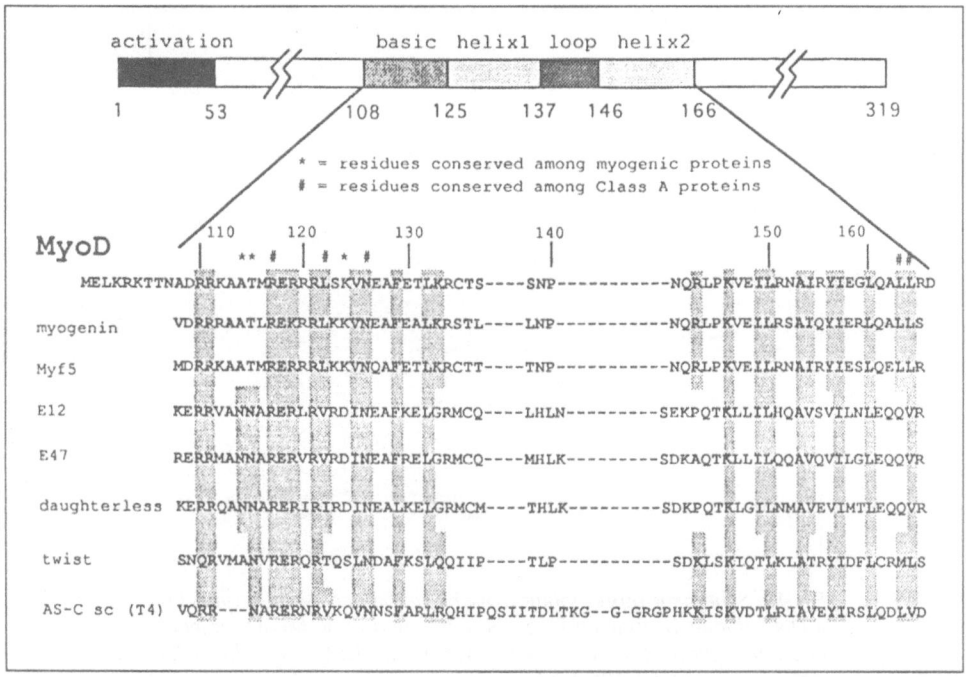

Fig. 4.1. (Top) schematic structure of MyoD protein and homology with other bHLH proteins. MyoD has an N-terminal transcriptional activation domain (black) and a central basic/helix-loop-helix (bHLH) domain which mediates DNA-binding and dimerization. (Bottom) amino acid identity between MyoD bHLH domain and those of other bHLH proteins is indicated by shading. Class A proteins are the ubiquitously expressed bHLH proteins such as E12, E47 and daughterless. (From Ma et al, Cell 1994;77:451-9)

The structure of a homodimer of the MyoD bHLH domain bound to an E-box has been established by X-ray crystallography.[25] The bHLH dimer forms a parallel, four-helix bundle with the two basic helices contacting the DNA across the double helix, rather as a pair of scissors would be positioned to cut a rope (Fig. 4.2). A similar arrangement is found in bZIP proteins such as members of the *c-fos* and *c-jun* families of proto-oncogenes (see chapter 1) which also bind DNA as homo- or heterodimers. The structure is very similar to those derived by crystallography for homodimers of the bHLH domains of E47[26] and Max,[27] the dimerization partner of the proto-oncogene *c-myc*. Remarkably, the backbones of the α-helices of MyoD and Max can be superimposed upon one another almost exactly, despite only about 25% similarity in amino acid sequence between the two bHLH domains,[25] indicating the high degree of conservation of tertiary structure of bHLH domains.

How can different bHLH dimers differentiate between genes containing similar E-box sequences, yet maintain the specificity of binding essential to regulating such diverse developmental processes as

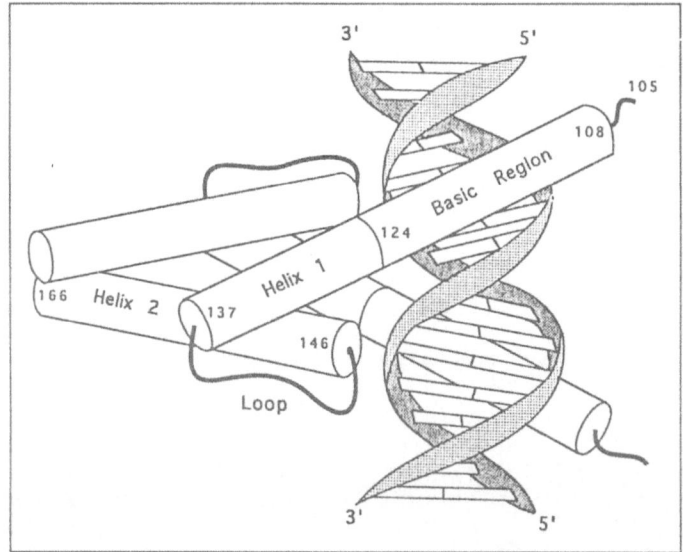

Fig. 4.2. Schematic representation of a homodimer of MyoD contacting a DNA duplex bearing an E-box. Numbering on one MyoD molecule refers to amino acids residues of MyoD: for clarity, only the bHLH domains (residues 108-166) are depicted. (From Ma et al, Cell 1994;77:451-9)

myogenesis and neurogenesis? All bHLH proteins share homology within the bHLH domain, but differ considerably outside of this region, even within a sub-family such as that comprising the myogenic bHLH proteins. Other parts of the protein act as transcriptional activation domains (TAD), which make protein-protein interactions with other factors and/or RNA polymerase II when bound to the gene. For example, MyoD has a TAD in its N-terminal region[22] (Fig. 4.1). Solving the crystal structure for the MyoD bHLH domain bound to the E-box has shed more light on how bHLH proteins work in vivo. The myogenic bHLH proteins differ from other bHLH factors in that they contain a pair of amino acid residues in the basic helix, an alanine and a threonine residue, that have been conserved through evolution and are essential for *trans*-activation from a muscle enhancer (see Fig. 4.1). Several lines of evidence indicate that these two residues specify myogenic activation. Mutation of these residues in myogenin (Ala86-Thr87) abolishes *trans*-activation and their replacement by two asparagine residues, the equivalent amino acids in the E12 basic helix, produces a protein which, though able to form dimers with E12 and to bind DNA avidly, cannot *trans*-activate from the MCK muscle-specific enhancer.[28] The equivalent residues in MyoD are Ala114-Thr115 and it was assumed that these residues would make specific contacts with the E-box. Surprisingly, the crystal structure of the MyoD bHLH/E-box complex reveals that these two residues do not make direct interactions with the DNA.[25] Rather, the compact and neutral character of their side-chains and their positions in the basic helix appears to influence the relative position of a conserved arginine residue with respect to the DNA. The side chain of this arginine residue (Arg111) protrudes into the major groove of the E-box and contacts the G in one half-site of the E-box. Replacement of alanine or threonine with amino acids having charged or bulkier side chains, such as asparagine, presumably causes local distortion of the basic helix and disrupts the interaction of DNA-contacting groups such as Arg111. It has been suggested that the interaction of Arg111 of MyoD with the E-box triggers transcriptional activation by altering the conformation of the amino-terminal TAD. In the absence of DNA, the TAD may contact the bHLH domain directly, in some way masking TAD function.[9]

The central degenerate nucleotides of the E-box are likely to contribute to specificity of binding. Binding-site selection experiments indicate that MyoD homodimers or MyoD-E47 heterodimers have a preference to bind the sequence CACGTG, and the 3D-structure of the E47 homodimer bound to CACCTG reveals asymmetric contacts to the central bases.[26] The overall conformation of a heterodimer may also be important. For example, transcriptional activation may be determined in terms of what TADs are present on the two proteins and where they sit in relation to each other and to other proteins of the transcriptional machinery bound to the gene, particularly basal transcription factors associated with RNA polymerase II.

Nucleotides adjacent to E-boxes also play a discriminatory role. E12 is an activator of the immunoglobulin heavy chain (IgH) enhancer in B-lymphocytes,[19] which contains three E-boxes. MyoD is unable to activate expression from this enhancer because of the nature of the DNA sequence within and immediately adjacent to one of the E-boxes which has the sequence CACCTG (the preferred binding sequence for E47).[29] The prevention of activation of the IgH enhancer by MyoD in transient expression studies in cultures of mouse 3T3 fibroblasts appears to be directed at the MyoD bHLH domain, possibly through a repressor protein binding the E-box.[29]

PCR-based binding-site selection experiments (CASTing:[30] Cyclic Amplification and Selection of Targets), conducted on partially-purified myogenic bHLH proteins, have been used to identify high-affinity DNA binding sites. CASTing involves mixing protein(s) with thousands of double-stranded oligonucleotides containing a central region of randomized DNA sequence and allowing a TF to "select" its preferred binding sequence from the many sequence combinations present in the mix. Funk and Wright have written a concise review of CASTing and similar techniques developed in other laboratories.[30] The specific protein-DNA complexes are then purified (usually using an antibody to the factor under study), the bound DNA sequences eluted, then amplified by PCR and mixed with protein. After several cycles of enrichment, the amplified sequences are cloned and their DNA sequences determined, giving a group of similar sequences representing high-affinity sites for the factor. Such experiments have shown that E-box sequences selected by myogenin are often flanked by binding sites for MEF-2 proteins.[30] Muscle-specific genes such as MCK or α-MHC also contain E-boxes in close proximity to a MEF-2 site. This may reflect the need for co-operativity in binding between different factors in order to activate transcription. In some cases, antagonism between factors binding to adjacent sequences has been noted. Stimulation of expression of *c-fos* by serum growth factors is mediated through a *cis*-acting sequence in the *c-fos* promoter called the serum response element or SRE which binds a protein called serum response factor (SRF).[31] Muscle cell differentiation in culture is characterized by up-regulation of MyoD expression and a decrease in the expression of *c-fos* and other proto-oncogenes. In culture, MyoD binds an E-box immediately adjacent to the *c-fos* SRE and blocks the normal pathways for responding to mitogenic or other growth signals which are focused through the binding of SRF to the SRE.[32]

MYOGENIC bHLH FACTORS PLAY SPECIFIC ROLES IN MYOGENESIS

From studies on mRNA expression in vivo analyzed by in situ hybridization and protein expression analyzed by immunohistochemistry it appears that the four myogenic bHLH proteins exhibit different patterns of expression during the development of skeletal muscle. The main skeletal muscles of the trunk and limbs have their origins in the somites, which form early in development (somitogenesis) by segmentation of the paraxial mesoderm which lies on either side of the neural tube (Fig. 4.3). Somitogenesis occurs first in the anterior (rostral) paraxial mesoderm, so that somite development progresses in an anterior to posterior direction, with anterior somites being more advanced in their development than those posterior to them. The somites contain three types of cells which differ in their prospective developmental fates. The dermamyotome, which occupies the upper (dorsal) half of the early somite, later develops into two distinct layers, the myotome and dermatome, giving rise to the precursors of skeletal muscle and dermis respectively. The lower (ventral) half of the somite, the schlerotome, gives rise to cartilaginous tissues. Fate-mapping and cell transplantation experiments[9,33] in chick suggest that the developmental fate of pre-somitic (unsegmented) paraxial mesoderm is fixed early on, probably soon after gastrulation and the establishment of the three germ layers.

The development of skeletal muscle is a complex phenomenon involving events of lineage commitment and differentiation. The first stage in skeletal muscle development is determination, as progenitor cells develop into myoblasts. Myoblasts are capable of cell division, but are already irreversibly committed to the skeletal muscle lineage.

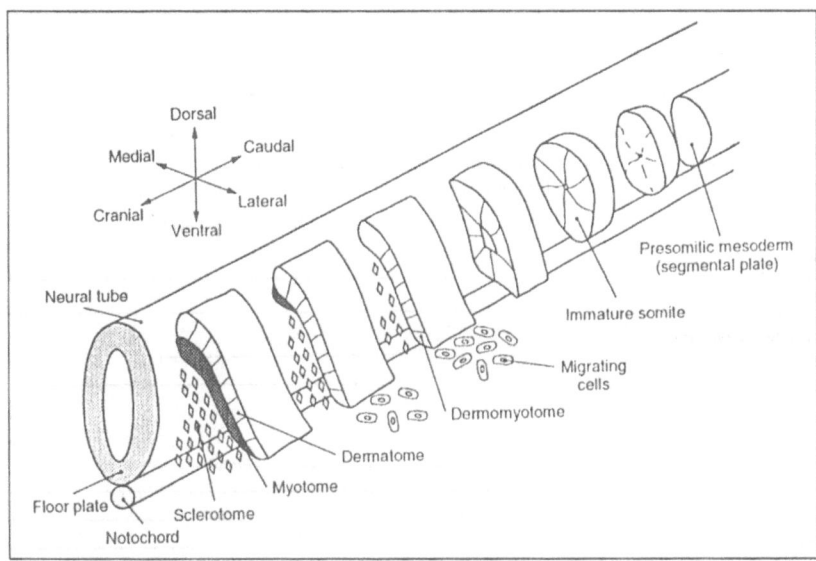

Fig. 4.3. Somitogenesis. Somites are derived from paraxial mesoderm adjacent to the neural tube and notochord and maturation proceeds in the cranial (rostral) to caudal direction such that anterior somites are more mature than posterior ones. (From Buckingham, Cell 1994;78:15-21)

At this point they do not yet express muscle genes (markers). The second stage, differentiation into myocytes which will fuse to form multinucleate myotubes, is characterized by withdrawal of the myoblast from the cell cycle and the expression of muscle-specific genes. The characterization of these events has been aided greatly by the study of stable skeletal muscle-derived cell lines, such as the mouse line C2C12,[34] which can be induced to differentiate and undergo fusion in culture.

The expression of the four myogenic bHLH factors during skeletal muscle development has been extensively reviewed.[9,33,35] The first factor to be detected in the somite is *myf-5*, detectable by in situ hybridization from 8 dpc (days post coitum) in the anterior somites of the mouse embryo (Table 4.2). Myogenin is detected, again in the anterior somites, at about 8.5 dpc and MyoD, which is associated with differentiation and expression of muscle markers, appears two days later. MRF4 (*myf-6*) is expressed transiently during somitogenesis at around 9 dpc and is then down-regulated until it is re-expressed in developing muscle at 15.5-16 dpc. Thereafter, it becomes the major myogenic bHLH factor expressed after birth. Skeletal muscle formation in the forelimbs, in contrast, is marked by *myf-5* expression at 10.5-12 dpc, followed by co-activation of MyoD and myogenin expression at 11 dpc The reason why the four myogenic bHLH proteins are expressed at different times during myogenesis is becoming clearer. It appears

Table 4.2.

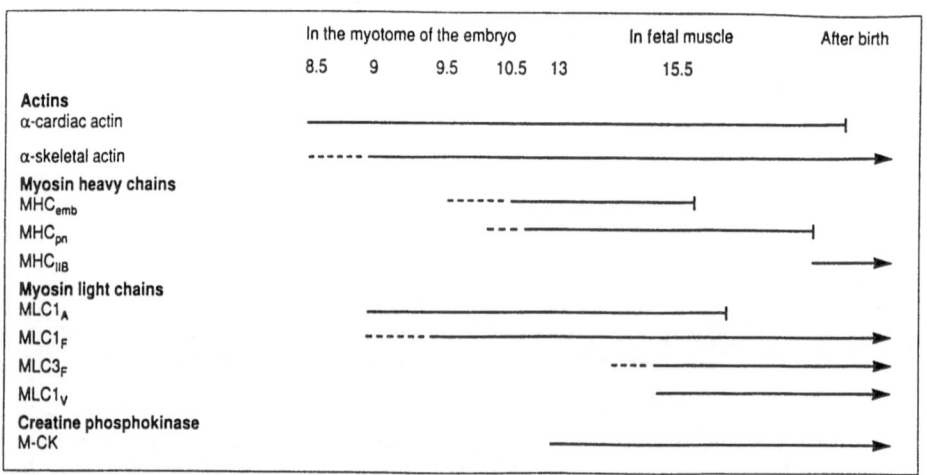

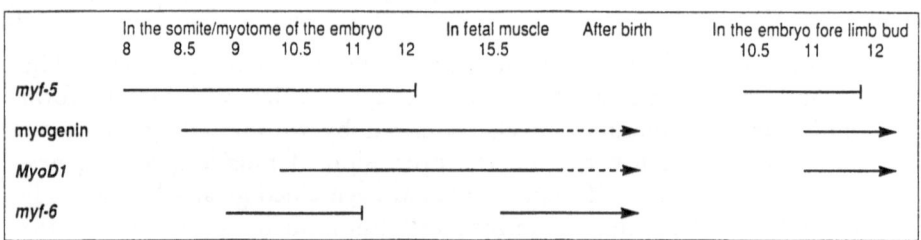

a. Summary of times of expression of various sarcomeric proteins during skeletal muscle development in the myotome of the mouse embryo, in differentiated fetal muscle or in post-natal muscle.

b. Times of expression of myogenic bHLH factors during mouse development in embryonic somite/myotome, fore limb bud, fetal muscle and post-natally.
(From Buckingham, Trends Genet 1992; 8:144-9)

that certain myogenic bHLH factors fulfill particular functions during the progression to a fully differentiated phenotype. Recent evidence shows that myogenin acts downstream of *myf-5* in vivo and is a key player in establishment of the muscle phenotype and activating muscle genes. This is suggested by the observation that "knock-out" mice (see chapter 1), in which the myogenin gene has been inactivated by homologous recombination (and hence myogenin protein is not produced), make normal amounts of MyoD and *myf-5*, yet have greatly reduced amounts of muscle and contain myoblasts in the regions where muscle fibers should be.[35-38] These results are consistent with myogenin being expressed at the stage of determination and being essential for progression to differentiation (also seen in culture), and support the idea that determined myogenic precursors migrate out of the somites to take up position in the body and limbs ready for differentiation. Knock-out mice bearing a homozygous mutation in the *myf-5* or MyoD genes go on to develop muscle, though the appearance of differentiating skeletal myocytes in mice lacking *myf-5* is retarded by a day or two. These results suggest that other myogenic bHLH proteins may compensate for the loss of MyoD or *myf-5* (in the latter, muscle development is presumably retarded until an appropriate factor is expressed). Double knock-out mice, in which both *myf-5* and MyoD loci are disrupted,[39] make no muscle and do not form precursor cells. Taken together, these knock-out data strongly suggest (1) that *myf-5* and myogenin function at the stage of skeletal muscle cell determination, (2) that myogenin is essential for advancing from determination to differentiation and (3) that some myogenic bHLHs can compensate for the function of others, as witnessed by the delay in myogenesis of mice lacking *myf-5*.

REGULATION OF MYOGENIC bHLH PROTEINS: STUDIES IN CULTURED CELLS

How do bHLH proteins respond to extracellular cues or stimuli? Experimental evidence from several sources in vivo and in culture suggests they regulate their own and each others expression.[35,40,41] They are subject to various positive feedback loops, as well as negative regulation by proto-oncogenes, growth factors and the inhibitory HLH factor Id (Fig. 4.4).[41] In myoblast culture systems, removal of growth factors from culture media leads to growth arrest and, at suitable densities,

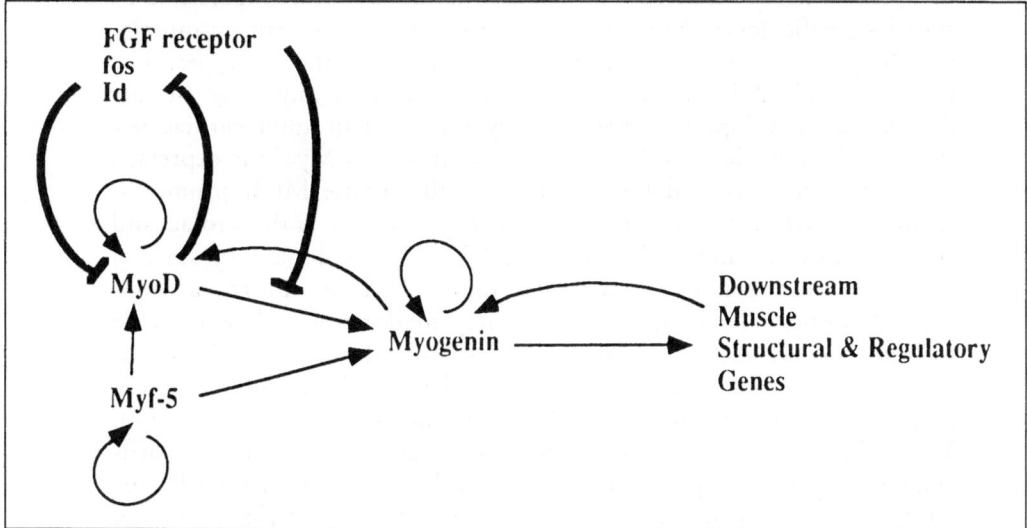

Fig. 4.4. Regulatory pathways and feedback loops governing myogenic bHLH protein expression. Positive regulation is depicted by arrows, repression by thick lines. (From Weintraub, Cell 1993;75:1241-4)

differentiation and cell fusion. Such results indicate that at certain threshold concentrations growth factors may hold the myoblast in a determined state, but prevent differentiation and activation of the muscle program. Evidence from studies on the function of myogenin in cultured skeletal muscle cell lines suggest that the conserved threonine residue (Thr87) in the basic helix of myogenic bHLH proteins is a target for phosphorylation by intracellular protein kinases activated in response to growth hormones.[42] Phosphorylation of this residue abolishes DNA-binding and *trans*-activation. Thus, both direct activation of the myogenic program and inhibition can be channeled through the basic domain. This results in a differentiation program in skeletal muscle that depends upon expression of myogenic bHLH factors and the release of inhibition to bind to DNA through post-translational events associated with the phosphorylation state of the protein.

MYOGENIC bHLH PROTEINS ARE ABSENT FROM THE HEART

Although cardiac and skeletal muscle are both striated muscles which share much in common there are pronounced differences between them. The embryonic origins of the two muscle types are different. Cardiac muscle is derived from lateral plate mesoderm and commitment to the cardiac lineage occurs early in development and probably simultaneously with expression of muscle-specific markers including contractile protein genes (see chapter 2). The result is a population of cardiac myocytes that continues to divide as they mature through fetal life and withdraw from the cell cycle around the time of birth. In contrast, skeletal muscle is derived from paraxial mesoderm and its development is divided into two distinct phases: determination or commitment to the myogenic lineage, followed by differentiation and the expression of muscle-specific genes. As a result, the TFs employed to activate muscle-specific gene expression differ considerably between the cell types. The myogenic bHLH factors have not been detected at any stage in cardiac muscle development, nor are they expressed in adult cardiac pathology. Transgenic mice have been made in which MyoD is expressed ectopically in heart under the control of the mouse MCK promoter. Under normal circumstances, MCK is expressed in both cardiac and skeletal muscle, whereas MyoD would be expressed only in skeletal muscle. Homozygous transgenic mice died in utero around 15.5-18.5 dpc.[43] Their hearts were visibly malformed and, while able to beat, had low ejection volume. Several skeletal muscle markers, not normally expressed in heart, were detected in the hearts of transgenic mice, as was myogenin, which may have mediated their expression. Thus, the ectopic expression of MyoD in cardiac myocytes is sufficient to cause the expression of some markers of skeletal muscle differentiation, including myogenin, and is reminiscent of the activation of the myogenic program by exogenous myogenic bHLH factors in permissive fibroblast cell lines in culture. However, the cardiac myocyte cellular environment is clearly not sufficient to support full expression of the skeletal phenotype, suggesting that myogenic bHLH proteins interact with other TFs in skeletal muscle to bring about the differentiated state.

UBIQUITOUS bHLH FACTORS ARE EXPRESSED IN THE HEART

The absence of equivalents of the myogenic bHLH proteins in cardiac muscle is at odds with the detection of the ubiquitous E2A and inhibitory Id proteins.[44] This observation immediately suggests that tissue-specific dimerization partners may exist in cardiac myocytes. (The search for cardiac-specific TFs in general has been hindered by the lack of a suitable cardiac cell line and the fact that cardiac myocytes become committed to the cardiac myocyte lineage and differentiate much earlier during embryogenesis.) Notably, some genes whose expression is regulated by bHLH proteins in skeletal muscle most likely are expressed in cardiac muscle by E-box-independent pathways. Regulation of the human α-cardiac actin gene may be controlled in this manner.

HUMAN α-CARDIAC ACTIN:
ROLE OF bHLH PROTEINS IN ITS REGULATION

The α-skeletal actin and α-cardiac actin genes are expressed in cardiac muscle (see chapter 3). In the rodent, both α-cardiac and α-skeletal actin are expressed, although α-cardiac actin is always the predominant isoform. There is also a gradual reduction in α-skeletal actin expression in the heart during development. Human development is markedly different with respect to actin isoform expression. Both isoforms are expressed during fetal life with α-cardiac actin representing up to 80% of sarcomeric actin at early stages, but α-skeletal actin expression increases after birth and can exceed 60% of the total sarcomeric actin in the adult human heart.

The α-cardiac actin gene is expressed in developing skeletal muscle as well as myocardium and α-cardiac actin expression has long been used as one of the earliest markers of skeletal myocyte differentiation. Some evidence, chiefly from band-shift assays (see chapter 1), suggests that the human α-cardiac actin gene (HCA) binds a bHLH factor present in isolated neonatal cardiac myocytes.[45,46] A HCA promoter-reporter construct containing -485/+68 of HCA exhibits maximal activity when expressed transiently in cardiac myocytes or C2C12 myotubes in culture: a shorter construct containing -177/+68 exhibits 30% activity by comparison.[46] The latter HCA fragment contains an E-box, which binds MyoD in skeletal myocytes, a binding site for the C_2H_2 zinc finger protein Sp1 and the most proximal of four SREs (often referred to as CArG boxes in the context of muscle genes) which bind SRF.[45] Overexpression of the inhibitory HLH factor Id (Table 4.1 and see below) significantly repressed HCA expression in C2C12 cells and cardiac myocytes.[46] In addition, methylation interference assays (see chapter 1) revealed protein-DNA contacts made over the E-box by a factor present in cardiac myocyte nuclei. Taken together, these results and the observation from band-shift assays using specific antibodies that ubiquitous E2A bHLH factors bind the E-box suggested that HCA expression in cardiac muscle is regulated, in part, by ubiquitous bHLH proteins forming heterodimers with a bHLH factor that is tissue specific.[46]

These findings are at odds with the failure so far to detect a cardiac-specific bHLH factor in embryonic, fetal or adult myocardium. More consistent with these observations are results obtained through the differentiation of primitive embryonal carcinoma (EC) cells. Undifferentiated P19 EC cells can be induced to differentiate into either cardiac or skeletal myocytes under appropriate conditions. Lines of P19 cells have been created in which a construct containing the wild-type HCA promoter and 5' sequences as far as -440 driving expression of the *E. coli* β-galactosidase gene *lacZ* was stably integrated into the genome.[47] Parallel lines were made which contained the same region of the HCA gene, but with the E-box (sequence CAACTG) replaced by the sequence AGATCT. These lines were differentiated subsequently into either skeletal or cardiac myocytes using selective agents. As the lines underwent differentiation they expressed muscle markers, including HCA and myosin heavy chains. The experiments indicated that the E-box in the HCA promoter is essential for HCA expression in P19-derived skeletal muscle, but was not essential for expression in P19-derived cardiac muscle. When PCR was used to amplify bHLH cDNAs from the differentiated cardiac myocytes, several ubiquitous bHLH sequences were detected. Thus, it is possible that in vivo the HCA E-box binds bHLH factors, but only in skeletal muscle would these be heterodimeric proteins formed between ubiquitous and myogenic bHLH proteins able to activate HCA expression. However, in cardiac muscle, although the HCA E-box may bind bHLH dimers, they may not contribute to cardiac-specific expression. An explanation for these observations could lie with the sequence of the E-box and adjacent nucleotides because, as described earlier, the sequence of an E-box in the IgH enhancer discriminates against binding and activation by MyoD, possibly through the binding of an as yet uncharacterized repressor protein.[29]

Several laboratories have sought to identify cardiac-specific bHLH factors without success, by either screening early fetal or embryonic cDNA libraries or using RT-PCR.[40,48] This suggests that either cardiac-specific bHLH factors do not exist per se, and that other types of TF may act as master regulators of cardiac muscle development, or that cardiac-specific bHLH factors might be expressed at an earlier point in development than the stages examined. In support of the latter, it has recently been reported that such a factor has been identified from chicken cardiac muscle.[9]

Id: An Inhibitory HLH Factor that Cannot Bind DNA

Id genes (for Inhibitor of DNA binding) contain an HLH dimerization domain, but lack the DNA-binding basic helix. They inhibit DNA-binding by forming heterodimers with other bHLH proteins (usually ubiquitous ones) in order to prevent DNA-binding.[49] Id genes, of which at least four have been identified to date, are expressed in various undifferentiated tissues and proliferating cell lines, and their transcription is down-regulated in several cell lines upon differentiation. During development Id is expressed extensively in mesoderm and later down-regulated in many tissues, though co-localization of Id expression to regions of the heart where the homeobox gene *Msx-1* and the bHLH factor twist are expressed, such as the endocardial cushions of the AV septal region and outflow tract, has been observed.[50,51]

Could Id proteins antagonise gene expression in the heart by complexing with as yet unidentified cardiac-specific bHLH factors? This is unlikely. Studies on cardiac gene expression in cultured ventricular myocytes reveal that Id expression does not correlate with down-regulation of a number of cardiac genes.[51] One isoform, Id-1, is up-regulated in the rat heart around the time of birth. Measurements of Id-1 expression in isolated rat cardiac myocytes indicates that Id-1 expression rises to a maximum at neonatal day 17, thereafter declining to lower, stable levels.[44] The increase suggests a role for Id proteins in the heart other than down-regulating cardiac gene expression by complexing with cardiac bHLH factors. This may be linked to the withdrawal of the cardiac myocyte from the cell cycle (the transition from hyperplastic to hypertrophic growth) which occurs during the post-natal period. It has been reported that Id-1, 2 and 3 are induced in growth-arrested NIH3T3 fibroblasts in culture in response to serum or growth factor stimulation.[52] Furthermore, transfection of these cells with antisense oligonucleotides (see chapter 6) directed against one or all Id transcripts retards the re-entry of 3T3 cells into the cell cycle. Id mRNA levels increase within 30 minutes of stimulation in culture[51] and also rise in response to α-adrenergic hypertrophic stimulus.[44] Such observations implicate Id genes as important targets for signal transduction in cardiac myocytes which respond to growth factors and mitogenic signals. The signal is probably transmitted to a specific enhancer within Id genes. For example, the broad-spectrum chemotherapy drug Doxorubicin (Dox) stimulates expression of Id-2 via an enhancer in the 5' flanking sequence of the Id-2 promoter.[53] Skeletal muscle cell lines in culture show elevated levels of Id mRNA in response to Dox treatment which interferes with the ability of exogenous MyoD to activate the myogenic program in pluripotent $10T^{1/2}$ cells. Not unexpectedly, this effect is mediated through the ubiquitous bHLH factors, as over-expression of E2A in Dox-treated cells can overcome the block on MyoD-mediated *trans*-activation, presumably through titering out Id proteins.[54] Interestingly, the development of cardiomyopathies are a notable side effect of Dox treatment. As both E2A and Id are expressed in the heart,[44] perturbation by Dox of the levels of their expression may have some consequence for certain pathways of cardiac-specific gene regulation involving Id and/or E2A isoforms.

MEF-2 FACTORS: KEY ACTIVATORS OF CARDIAC AND SKELETAL MUSCLE GENES

MEF-2 IS A MADS DOMAIN PROTEIN

The mammalian MEF-2 gene family of transcription factors (for Myocyte Enhancer Factor 2) comprises four genes called *MEF-2A-D* (Fig. 4.5).[11,54,55] A single *MEF-2* gene has been identified in *Drosophila* (*D-MEF-2*[56]) and two genes called *SL-1* and *SL-2* in *Xenopus*[10] appear to be homologues of mammalian *MEF-2D* and *MEF-2A*, respectively. MEF-2 was initially identified as a nuclear protein from differentiated rodent skeletal myotubes which bound an A/T-rich sequence in the MCK enhancer.[57] Then, cDNA clones for factors related in sequence to the human serum response factor, SRF, which were called RSRF for related to SRF were shown to be indistinguishable from MEF-2 in terms of sequence-specific binding.[58] All MEF-2 proteins exhibit significant amino acid sequence homology over a 55 amino acid DNA-binding and dimerization domain called the MADS domain or MADS box (Fig. 4.5B). MADS is an acronym derived from four transcription factors in which this region of homology was first noted: the yeast mating type factor MCM1, plant homeotic factors Agamous and Deficiens and human SRF). In addition, all MEF-2 factors possess a 29 amino acid region of homology on the C-terminal side of the MADS box which is not present in other MADS factors (Fig. 4.5: the MEF-2 domain). MEF-2 proteins bind as dimers to an A/T rich sequence in the enhancers of target genes which agree with the general consensus sequence $CTA(^A/_T)_4TA(^A/_G)$[58,59] (Table 4.1 and Fig. 4.6). The MEF-2 site resembles other sites that bind transcription factors as dimers in that it is partially palindromic. Each half of the palindrome contains a conserved TA dinucleotide (underlined, Fig. 4.6A) and probably represents a binding half-site for one MEF-2 monomer.

ALTERNATIVE mRNA SPLICING OF *MEF-2* GENES: MANY TRANSCRIPTS BUT FEW PROTEINS

MEF-2 transcripts are expressed in several tissues, but are particularly enriched in cardiac and skeletal muscle, and are developmentally regulated. For example, during the development of skeletal muscle, expression of *MEF-2A, B* and *C* is detected in differentiated skeletal myocytes whereas MEF-2D is expressed only in skeletal myoblasts. Expression of *MEF-2* genes is subject to both alternative mRNA splicing and post-translational modification of their proteins. Some alternate transcripts are widely distributed whilst others are restricted to

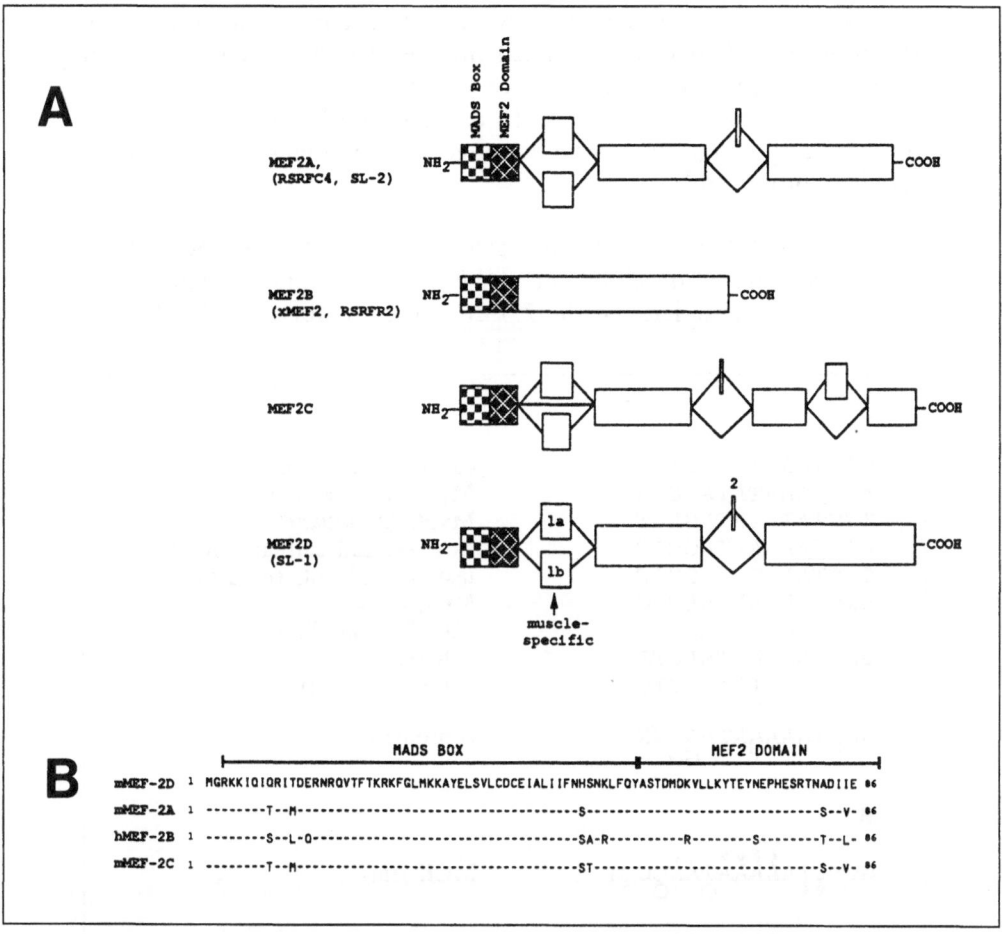

Fig. 4.5. The MEF-2 gene family. (A) schematic representation of the genomic organization of the four mammalian MEF-2 genes showing alternative splicing of their transcripts. Coding exons are depicted by boxes; the N-terminal exons encoding the MADS box and MEF-2 domain are hatched. Alternate names for some isoforms are shown in parentheses: human RSRF isoforms were identified by homology to serum response factor,[58] SL-1, SL-2[10] were isolated from Xenopus. (B) alignment of amino acid sequences within the MADS box and MEF-2 domain for mouse MEF-2A, C and D and human MEF-2B. (From Martin et al, Mol Cell Biol 1994;14:1647-56)

tissues in which MEF-2 protein has been detected. In MEF-2A, an alternate transcript which incorporates a 24 bp exon encoding an eight amino acid sequence comprised largely of glutamic acid residues is expressed in brain, cardiac and skeletal muscles.[11,55] This correlates with tissues in which endogenous MEF-2A binding activity is detected.[11] *MEF-2D* expression is also complex, with a similar small exon (Fig. 4.5A, labeled as "2" on *MEF-2D* splicing scheme) being spliced into some transcripts. The exon immediately downstream of the MEF-2 domain in *MEF-2D* is also alternatively spliced: transcripts containing exon 1a

(Fig. 4.5A) are widely expressed whereas transcripts containing exon 1b are expressed mainly in striated muscles.[55] It has been noted that since MEF-2 proteins bind DNA as dimers and the four genes could generate theoretically as many as 15 different MEF-2 proteins through alternative splicing, an enormous range of functional MEF-2 dimers could be produced.[54]

MEF-2 Regulates Gene Expression in Cardiac Myocytes

During cardiac development, *MEF-2C* transcripts are detected in the precardiac mesoderm of the 7.5 dpc mouse embryo, with *MEF-2A*

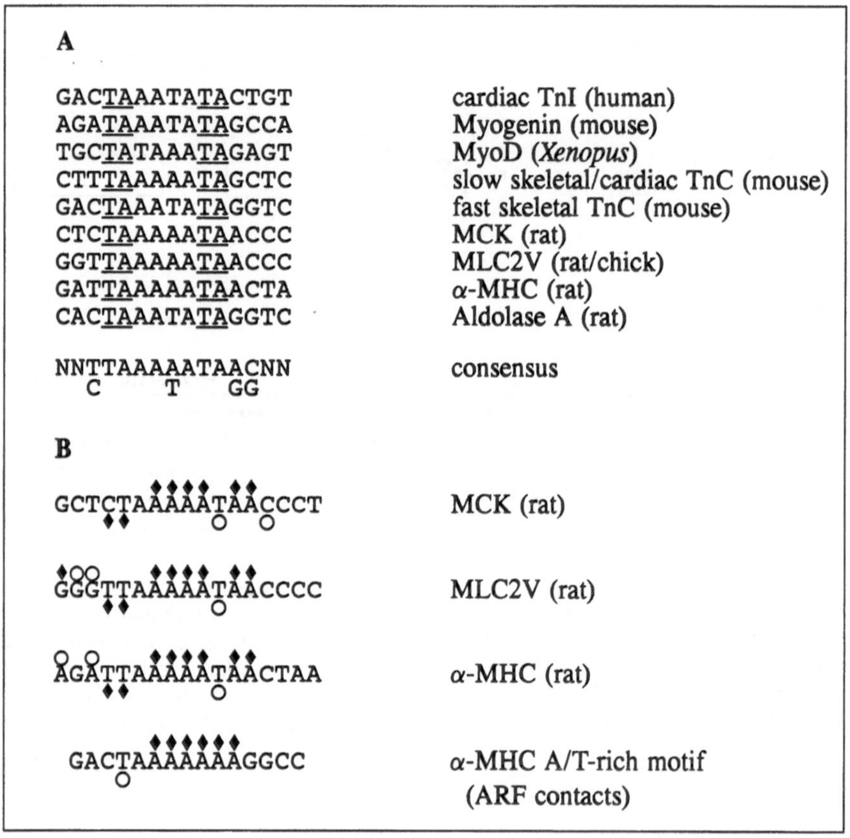

A

GAC**TAA**ATA**TA**CTGT	cardiac TnI (human)
AGA**TAA**ATA**TA**GCCA	Myogenin (mouse)
TGC**TAT**AAA**TA**GAGT	MyoD (*Xenopus*)
CTT**TAA**AAA**TA**GCTC	slow skeletal/cardiac TnC (mouse)
GAC**TAA**ATA**TA**GGTC	fast skeletal TnC (mouse)
CTC**TAA**AAA**TA**ACCC	MCK (rat)
GGT**TAA**AAA**TA**ACCC	MLC2V (rat/chick)
GAT**TAA**AAA**TA**ACTA	α-MHC (rat)
CAC**TAA**ATA**TA**GGTC	Aldolase A (rat)

```
NNTTAAAAATAACNN          consensus
  C       T   GG
```

B

GCTC**TA**A**A**A**A**A**TA**ACCCT   MCK (rat)

GGG**TTA**A**A**A**A**TA**ACCCC   MLC2V (rat)

AGA**TTA**A**A**A**A**TA**ACTAA   α-MHC (rat)

GAC**TA**A**A**A**A**A**A**AGGCC   α-MHC A/T-rich motif
                                    (ARF contacts)

Fig. 4.6. (A) alignment of MEF-2-like sequences from the following genes: human cardiac TnI (P. Bhavsar, N. Brand, P. Barton, unpublished); myogenin,[74,75] slow skeletal/cardiac TnC (skeletal enhancer),[86] and fast skeletal TnC[129] (all mouse); rat MCK,[62,63] MLC2V,[80] cardiac α-MHC[14] and aldolase A[85] and Xenopus MyoD.[73] A consensus sequence based on the alignment is shown at the bottom: N, any nucleotide. (B) protein-DNA contact points, determined by DEPC interference assays, made by MEF-2 proteins binding MEF-2 sequences (MCK,[14] MLC2V[14,61] and α-MHC,[14]) and by a MEF-2-related factor, ARF, binding to the A/T-rich sequence in the rat α-MHC gene.[101] Strong (♦) and weak (o) contacts are indicated. Only the upper (coding) strand of DNA is shown for clarity: contacts shown below the sequence are made to the complementary DNA strand.

and *D* transcripts becoming detectable in developing myocardium about 24 hours later.[60] MEF-2 proteins are therefore likely to be major determinants of cardiac-specific gene expression during these stages. Many genes expressed in cardiac muscle contain MEF-2 sites conforming to a specific binding sequence $(^C/_T)TAAA(^A/_T)ATA(^A/_G)$. These include *MCK*, *α-MHC*, myosin light chain 2V (*MLC2V*) and cardiac troponin C (*cTnC*) (Fig 4.6). The precise contacts made between MEF-2 proteins and their binding sites have been studied using DNAseI footprinting and methylation interference assays and two things are noteworthy. First, interference experiments show that contacts are made predominantly to the A residues within the consensus binding sequence. Also, MEF-2 contacts one of the TA dinucleotides on one DNA strand, and the other TA on the opposite strand (Fig. 4.6B), consistent with each monomer contacting each half-site. Secondly, both footprinting and interference data suggest that MEF-2 proteins make contacts with nucleotides flanking the consensus binding sequence. This may be relevant in determining how MEF-2 dimers discriminate between different genes. For example, the MEF-2 sites of α-MHC and MLC2V of rat are virtually identical, but differ in their flanking sequences. Contacts made to the MEF-2 sites in these genes are essentially equivalent, but unique contacts are made to the flanking nucleotides[14,61] (Fig. 4.6B).

MEF-2 proteins also bind to A/T-rich sequences which do not conform exactly to consensus binding sites. The *MCK* gene contains a 200 bp upstream muscle-specific enhancer in which are located an SRE/CArG box, two E-boxes, a MEF-2 binding site and an A/T-rich sequence[62,63] (Fig. 4.7). The A/T-rich sequence is very similar to the MEF-2 sequence and differs by only two base pairs (see Fig. 4.6A), yet MEF-2 protein binds poorly to it. The A/T-rich site is a binding site for MHox (for Muscle Homeobox), a mesodermally-restricted homeoprotein from mouse, which recognizes the consensus sequence $T(^A/_T)ATAAT(^A/_T)A.$[64] However, it is MEF-2, rather than MHox, which is important in *trans*-activating from this site.[65] This was determined experimentally by looking at the effect of mutations within the A/T-rich site on the ability to bind MEF-2 and MHox (as determined by band shift assays) and to drive expression of a CAT reporter gene when transfected into C2C12 cells in culture. The wild-type A/T-rich sequence binds MHox tightly, but MEF-2 and Oct-1 (a POU-homeodomain factor) weakly, and can be mutated into strong binding sites for MEF-2 or Oct-1 (Fig. 4.7: A/T→MEF2 and A/T→Oct-1, respectively). Other mutated sequences bound MHox tightly, but did not bind MEF-2 at all (Fig. 4.7: A/T→MHox-A and B). Only the sequence mutated to a MEF-2 binding site functioned as an enhancer when expressed in differentiated C2C12 myotubes, although it was only half as active as the wild-type sequence (Fig. 4.7, 54% as opposed to 100% activity obtained with MCK-CAT). These experiments

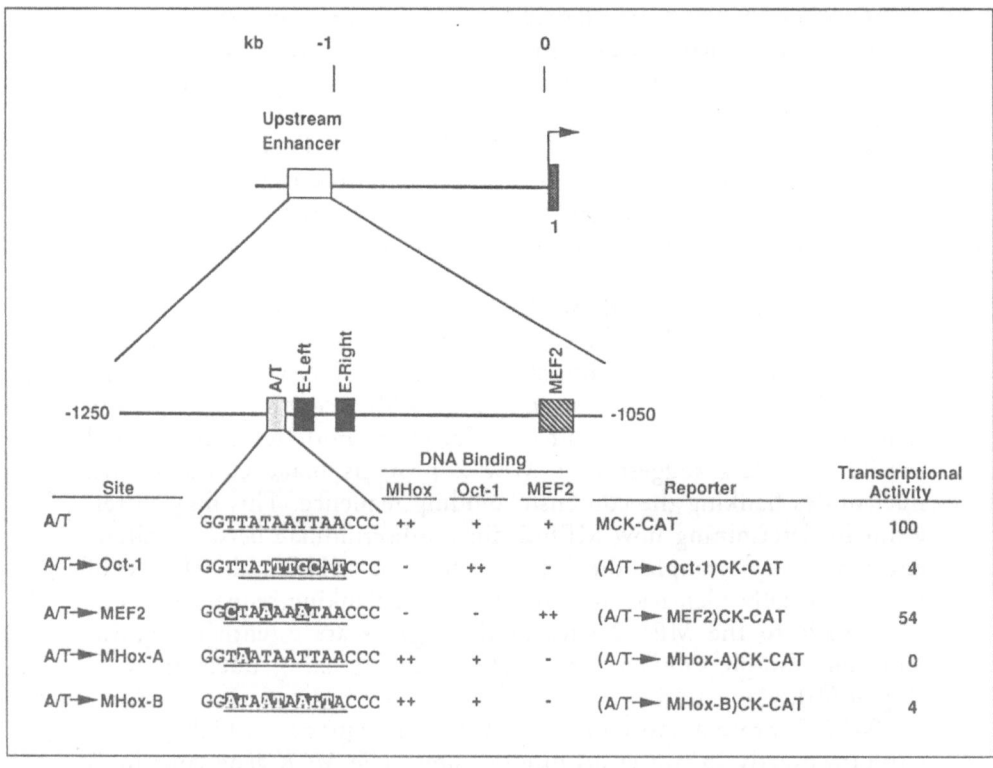

Fig. 4.7. (Top) the upstream enhancer of rat muscle creatine kinase (MCK). The enhancer contains a MEF-2 site preceded by two E-boxes (E-left, E-right) and an A/T rich sequence which binds the homeobox protein MHox. (Bottom) assessment of DNA-binding and trans-activation potential of A/T-rich site. DNA sequences of the wild-type A/T rich sequence (A/T) and four mutated variants are shown, together with DNA-binding of MHox, POU-1 and MEF-2 to those sites (++, strong binding, -, no detectable binding) and transcriptional activity, expressed as percentage of wild-type activity when driving expression of a CAT reporter gene in cultured C2 myotubes. (From Cserjesi et al, J Biol Chem 1994; 269:16740-5)

emphasize the subtle relationship between factor binding and *trans*-activation which underlies the regulation of expression of many genes.

The functional significance of two types of transcription factor recognizing one DNA sequence may have important consequences for controlling gene regulation. Recent studies suggest homeobox-containing factors and other types of transcription factors recognize binding sites for MADS-domain proteins and may stabilize MADS binding. For example, Phox (for Paired-like homeobox, related to the *Drosophila* homeoprotein Paired and the human homologue of MHox), enhances the rate of binding (and dissociation) of SRF to the SRE in the *c-fos* promoter.[66] This binding enhancement is unique to MHox/PHox and other factors belonging to the paired sub-class of homeoproteins and experiments with truncated homeoproteins reveal that the effect is directed through the homeodomain itself.[66] MHox/Phox may be required to recruit a MADS protein to a particular gene in the course of tissue specialization or in response to a particular signal. In this respect it should be noted that SRF and associated proteins are targets for phosphorylation in response to growth factor stimulation in vitro.[31]

Co-operation of two factors at one site may be a widespread mechanism for regulating gene expression. Another homeoprotein, Nkx-2.5,[67,68] which is expressed in cardiac myocytes (see chapter 2), is also reported to bind the SRE/CArG motif.[9] The skeletal α-actin promoter, which is expressed transiently during development in embryonic myocardium and re-induced during hypertrophy in the mature heart, contains three SREs, one of which also binds a novel zinc finger protein called Yin-Yang-1 (YY1) which can function as a transcriptional activator (*c-myc*) or repressor (*c-fos*, α-skeletal actin) depending upon which gene it binds to.[69,70] YY1 competes with SRF for binding to the SRE in both *c-fos* and skeletal α-skeletal actin promoters[70] and the expression of the YY1 and SRF genes shows an inverse relationship during embryonic skeletal muscle differentiation, suggesting YY1 may be a sensitive switch marking the transition from proliferation to differentiation.[69] Under appropriate conditions, YY1 also functions as an initiator of transcription.[71,72]

The *Xenopus* maternally expressed MyoD gene (*XMyoDa*), which is expressed during oogenesis, is regulated through the binding of endogenous MEF-2 to a TATA sequence which resembles a consensus MEF-2 binding site[73] (Fig. 4.6A). Mutation of individual nucleotides within and either side of the TATA motif revealed that the TATA-box binding protein (TBP) and MEF-2 possess distinct binding specificities for the site. Transfection experiments in differentiated muscle or non-muscle cell lines suggest that TBP-binding is essential in the non-muscle lines, but that MEF-2 may substitute for TBP in driving expression from the *XMyoDa* promoter in muscle, possibly in order to stabilize transcription.[73]

INTERACTION OF MEF-2 PROTEINS WITH MYOGENIC bHLH FACTORS

There is compelling evidence to suggest that myogenic bHLH and MEF-2 proteins may regulate and amplify each other's expression, perhaps defining threshold levels of expression of transcription factors and muscle-specific proteins to bring about withdrawal from the cell cycle and the development of a differentiated phenotype. The mouse myogenin gene contains a MEF-2 site within its promoter and, from transgenic experiments, it is apparent that expression of the myogenic bHLHs alone is insufficient to program skeletal muscle differentiation in vivo.[74,75] Furthermore, human MEF-2D is expressed before myogenin, making it a likely candidate for the activation of myogenin expression.[54,55] MEF-2A shares with MyoD and other myogenic bHLH factors the ability to initiate myogenesis when expressed in non-muscle cell lines.[76] These findings are not surprising for, as described earlier, there is considerable feedback and auto-regulation among myogenic bHLH proteins[41] and the expression of MEF-2 factors is up-regulated during differentiation of muscle cell lines in culture.[59] The relationship between bHLH and MEF-2 proteins in myogenesis may be easier to define in *Drosophila* where there exists a single MEF-2 gene (*D-MEF-2*) and one gene for *nautilus*, the fly homologue of MyoD.[56] D-MEF-2 appears to be the key player in myogenesis in the fly. Although *nautilus*-deficient transgenic flies lack some muscle, D-MEF-2-deficient embryos do not form any muscle at all and die. Interestingly, a central role for MEF-2 proteins in smooth and cardiac muscle development has also been shown by generating flys lacking D-MEF-2. During normal development, D-MEF-2 is expressed in all muscle lineages, including the precursors of skeletal, cardiac and smooth muscles. However, transgenic flies in which the *D-MEF-2* gene has been inactivated do not make skeletal muscle (though myoblasts form and migrate to the correct regions), show mid-gut abnormalities characterized by an apparent lack of differentiated visceral muscle cells and exhibit negligible expression of muscle-specific markers in the heart tube.[56] Thus, MEF-2 proteins may be key transcriptional regulators in early events associated with determination in the three muscle lineages, but subsequent activation of the myogenic bHLH proteins in skeletal muscle would be required in order to develop the fully differentiated phenotype.

SUMMARY

The regulation of muscle-specific gene expression is necessarily complex in order to ensure that genes are expressed correctly, at the appropriate time during development, in the correct tissues. It is clear that tissue-specific and developmentally-regulated gene expression is directed through the binding of transcription factors to discreet DNA sequences within the gene. In many muscle-specific genes these are located in the promoter region or 5' flanking sequences but can be found elsewhere in the gene, such as within introns. Once the appropriate factors have bound to their cognate binding sites (which collectively constitute an enhancer element) they interact with one another via protein-protein interactions to bring about high-level, tissue-specific gene expression. Interactions between multiple factors can restrict expression of a gene to a particular tissue or lineage or allow the gene to respond to physiological changes (for example, phosphorylation by activated protein kinases in response to growth factor stimulation). Transcription factor genes can be expressed in a highly restricted manner (e.g., MyoD and the related skeletal muscle-specific bHLH factors), or expressed in cells of a particular lineage (e.g., the MEF-2 factors, enriched in striated muscles) or ubiquitously (e.g., Class A bHLH factors such as Pan or E47). Some tissue-restricted factors such as MyoD and other myogenic bHLH factors activate the myogenic program when expressed ectopically in non-muscle cell lines in culture. This suggests they function as master regulators and can initiate a cascade of gene expression leading to development of the fully differentiated muscle phenotype. However, in vivo analyses using homologous recombination in mice reveal that only some myogenic bHLH proteins, such as myogenin, are truly indispensable for skeletal muscle differentiation and that ectopic expression of MyoD in the heart (from which the endogenous myogenic bHLH factors are absent) does not lead to expression of the full skeletal phenotype. Several factors which play a key role in skeletal muscle gene regulation, such as MEF-2 proteins, are also found in cardiac muscle. This suggests that common pathways for regulating gene expression may exist between the two types of muscle. In the remainder of this chapter we review the regulation of expression of some key cardiac muscle genes, including those for α- and β-myosin heavy chains, ventricular myosin light chain 2 (MLC2V), myoglobin and troponins.

VENTRICULAR MYOSIN LIGHT CHAIN 2: A PARADIGM OF CARDIAC-SPECIFIC GENE EXPRESSION?

The cardiac ventricular myosin light chain 2 gene (*MLC2V*) is expressed in slow skeletal and cardiac muscle (see chapter 3). In the heart it is expressed only in ventricular muscle, first detectable during embryogensis in the developing ventricular region of the heart tube.[77] Molecular dissection of the rat *MLC2V* gene has revealed that approximately 250 bp encompassing the *MLC2V* promoter and immediate upstream sequences are sufficient to recapitulate cardiac expression when introduced into the germline of transgenic mice and therefore constitutes the cardiac enhancer of *MLC2V*.[78] Expression in slow skeletal myotubes requires additional sequences located further upstream in addition to sequences within the promoter region.[78] The core of the cardiac enhancer is a 28 -bp element called HF-1 which drives high levels of expression when linked to a heterologous promoter and transfected into rat neonatal cardiac myocytes in culture.[79] The HF-1 element can be further divided into two adjacent sequences called HF-1a and HF-1b (Fig. 4.8A). The contribution of these individual sequences to ventricular muscle-specific expression has been assessed by creating mutations within them and ascertaining the effect of these on promoter activity when assayed in culture[61,80] or in vivo in transgenic mice.[81] The HF-1a sequence binds a ubiquitously expressed nuclear protein while HF-1b, which is A/T-rich, reportedly binds two different factors. One of these is a member of the MEF-2 family,[80] the other a novel C_2H_2 zinc finger protein (zfp) which has been termed HF-1b zfp.[61] The HF-1b sequence is also conserved in the *MLC2V* gene from chicken and binds a MEF-2-related factor called BBF-1 which is expressed in pre-cardiac mesoderm.[82] Both the HF-1a and HF-1b pro-

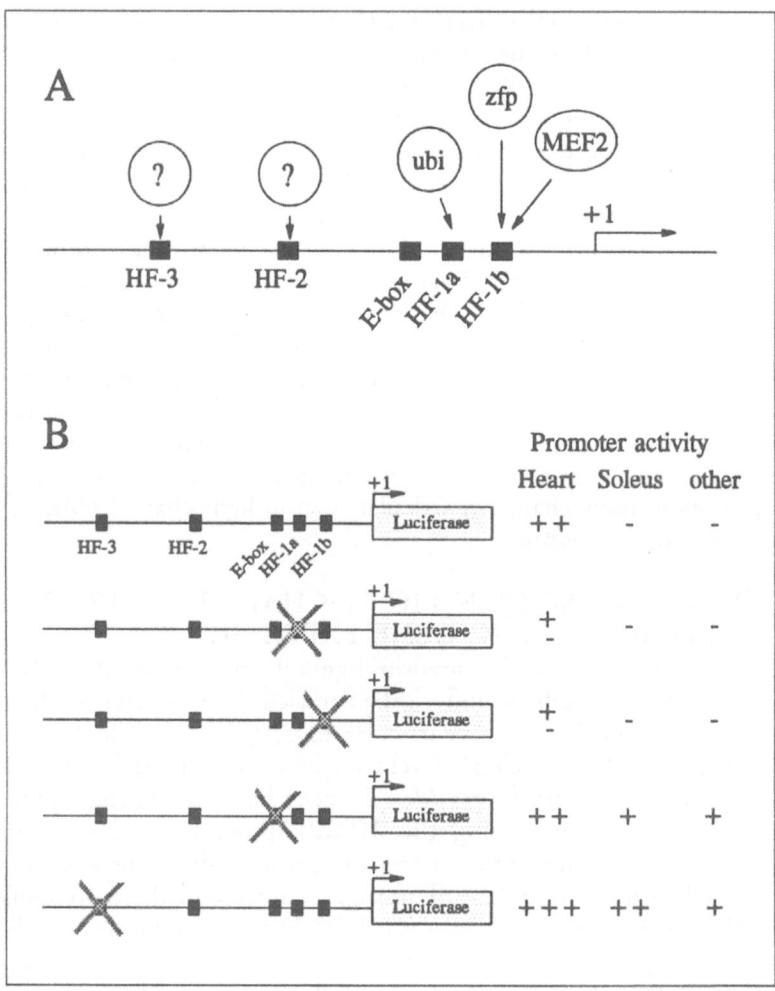

Fig. 4.8. Transcription factors binding the rat MLC2V cardiac enhancer. (A) HF-1a, HF-1b, HF-2, HF-3 and E-box are cis-acting DNA sequences shown to be binding sites for transcription factors essential for cardiac-specific expression of MLC2V (see text). HF-1b binds MEF2 and a novel zinc finger protein (zfp). HF-1a is bound by a ubiquitously expressed factor (ubi). The factors binding HF-2 and HF-3 remain uncharacterized. +1, start-site of transcription. (B) summary of effects of mutations in cis-acting elements upon MLC2V promoter driving expression of a luciferase reporter gene in the tissues of transgenic mice.[81] Promoter activity ranging from strongest (+++) to undetectable (-) in heart, soleus muscle or other tissues is shown. (From Barton et al, Annual of Cardiac Surgery; 1995 (in press).

moter elements are essential for high-level expression from the rat *MLC2V* promoter in ventricular muscle of transgenic mice.[81] This is corroborated by in vitro data which demonstrates that each site enhances transcription when linked upstream of a non-homologous minimal promoter, and that mutations within the sites interfere with protein binding and are detrimental to transcriptional activation.[80] Two further elements, an E-box and a sequence called HF-3, lie upstream of HF-1 and function as negative regulatory elements (Fig. 4.8): these may contribute to tissue-specificity by binding ubiquitous factors in order to prevent expression of MLC2V in other muscle types. Notably, mutations in the E-box, which presumably binds a bHLH protein, result in low, but significant, levels of expression in soleus and uterine muscles from transgenic mice harboring the MLC2V promoter-reporter constructs (Fig. 4.8B). As tissue-specific bHLH factors are absent from heart, and the *MLC2V* gene is also expressed in skeletal muscle, probably through binding of myogenic bHLH proteins to E-boxes located outside the cardiac enhancer, these results suggest that bHLH factors may bind the *MLC2V* cardiac enhancer, but act as non-activators. An alternative hypothesis would be that an unidentified repressor binds the E-box, as has been suggested as a possible mechanism for the inability of MyoD to activate the IgH promoter.[29]

The *MLC2V* cardiac enhancer may serve as a simple model for cardiac-specific gene expression. HF-1 binds ubiquitous and tissue-restricted factors in order to specify and maintain high-level expression in ventricular muscle. Also, it is an example of an E-box independent pathway of cardiac gene expression in that bHLH proteins do not act as positive regulators, unlike in skeletal muscle. Nevertheless, *MLC2V* contains a canonical E-box sequence which appears essential from transgenic studies to prevent ectopic expression in fast or smooth muscles. In other words, ventricular muscle-specific expression of MLC2V may not be due simply to the presence of a positive regulatory factor expressed in ventricular myocytes, but results from the combined interactions of positive and negative regulators, perhaps through certain factors stabilizing transcription. This is an important concept as none of the factors binding to HF-1 are truly cardiac-specific in their expression. In this respect, it is interesting that expression from the *MLC2V* promoter in the hearts of transgenic mice containing the mutated E-box was significantly higher than that in mice containing the wild-type transgene. This suggests that in cardiac muscle negative regulators may partially antagonise high-level gene expression, whereas in skeletal and smooth muscles their presence would be sufficient to prevent low level MLC2V expression.

MEF-2 OR HF-1b ZFP: WHAT BINDS THE HF-1b MOTIF?

The precise physiological roles of MEF-2 factors and of the HF-1b zinc finger protein in the expression of MLC2V in vivo is unclear. The HF-1b zfp transcript is expressed in both cardiac and skeletal muscle, and also brain, where it is highly abundant. MEF-2 proteins are enriched in brain, cardiac and skeletal muscle compared to other tissues. It is apparent from band-shift assays using super-shifting antibodies specific for MEF-2 or HF-1b zfp that the HF-1b motif is bound principally by the HF-1b zfp in vitro when the assay is carried out with cardiac myocyte nuclear proteins. In contrast the MEF-2 site in the *MCK* enhancer binds MEF-2, but not HF-1b.[61] As described earlier, the HF-1b sequence is identical in 10 consecutive positions (out of 11), although in the reverse orientation to the *MCK* MEF-2 site (see Fig. 4.6). Interference assays demonstrate that the protein-DNA contact points of MEF-2 and HF-1b zfp to the *MCK* MEF-2 site and to the *MLC2V* HF-1b sequence respectively are identical. The contacts made to the center of the MEF-2 site and to the invariant TA dinucleotides, which are similar if not identical to contacts made by MEF-2 proteins in several other genes (see Fig. 4.6B), may govern docking of MEF-2 dimers to the half-sites. However, each protein makes unique additional contacts with DNA outside of the MEF-2 core sequence, implying that flanking sequences contribute to determining which factor binds to which gene. It will be important to address the role of each factor in the whole animal to see if results obtained in vivo correlate with those obtained by transient transfection into cultured myocytes. This could be done by inactivating the expression of the HF-1b zfp and individual MEF-2 genes by homologous recombination.

TROPONIN GENE REGULATION

The three components of the Troponin (Tn) complex, TnI, TnC and TnT, constitute a calcium-sensitive switch for contraction in striated muscle (see chapter 3). The isoforms expressed in cardiac muscle are encoded by separate genes and studies on the regulation of their expression have shed new light on how cardiac-specific gene expression is controlled. In particular, some studies have indicated the importance of two binding sites for transcription factors used commonly in muscle genes. One is a C-rich sequence known as a CCAC box, which binds at least two unrelated factors, and the other is a sequence called the M-CAT motif, the binding site for members of a novel family of TFs first described as proteins binding to viral gene promoters.

SLOW SKELETAL/CARDIAC TROPONIN C

The slow skeletal/cardiac isoform of Troponin C (TnC) is expressed in slow-twitch skeletal fibers and cardiac muscle.[83] Like several other genes expressed in skeletal and cardiac muscle (for example, MCK) this is achieved through the use of separable tissue-specific enhancers within the *TnC* gene. Each enhancer comprises several binding sites for transcription factors. The cardiac-specific enhancer is located immediately upstream of the promoter, whereas a region within the first intron directs expression in skeletal muscle (Fig. 4.9). The skeletal enhancer comprises three motifs: a CCAC box with the sequence CCCCACCCC, a MEF-2 site and a sequence called MEF-3. This enhancer has been linked to a minimal promoter in front of the CAT gene of *E. coli,* and the effects of mutations within each element upon

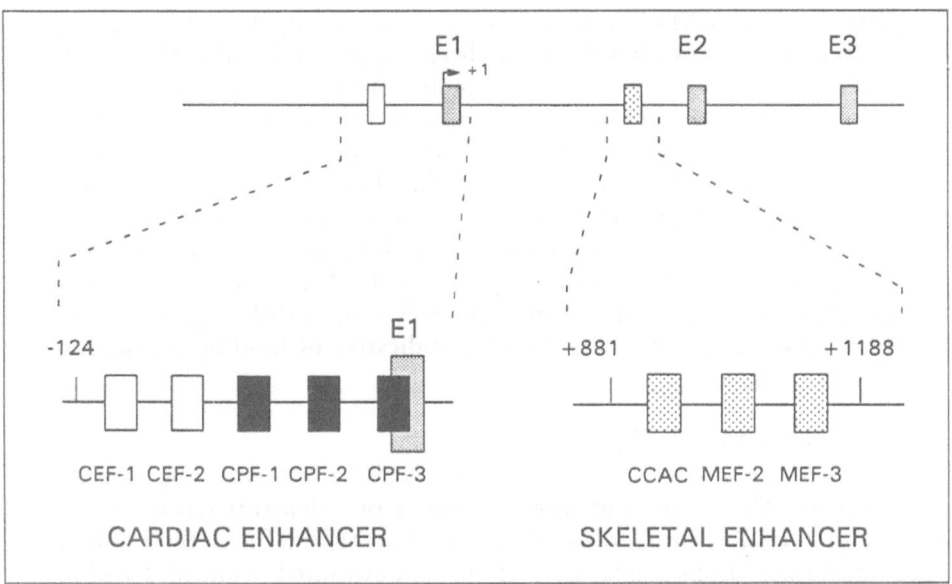

Fig. 4.9. The mouse slow skeletal/cardiac troponin C (TnC) gene contains separable tissue-specific enhancers. The cardiac enhancer overlaps with the TnC promoter and five binding sites have been identified by DNaseI footprinting: enhancer elements CEF-1 and CEF-2 (white boxes), which are necessary for high-level expression, and three elements CPF-1-3 (hatched boxes) overlapping the promoter and beginning of exon 1.[87] The skeletal enhancer, located at +881/+1188 within the first intron, contains a CCAC box and binding sites for MEF-2 and MEF-3.[86]

activity assessed by transfection into C2C12 myoblasts. Mutations in the CCAC box, MEF-2 or MEF-3 sites reduce wild-type levels of activity by approximately 99, 50 and 98%, respectively.[84] From bandshift assays with C2C12 nuclear extracts it appears that the CCAC box binds five different protein complexes. The protein binding the MEF-2 sequence was developmentally up-regulated during skeletal myocyte differentiation in culture, in agreement with published observations for the expression pattern of MEF-2 proteins. The identity of the protein(s) binding the MEF-3 site, which is conserved between the human and mouse slow skeletal/cardiac *TnC* genes, remains unclear at present. However, one of two enhancers in the rat aldolase A gene contains a sequence similar to the MEF-3 site and binds a novel nuclear protein.[85] This protein, originally termed MAF1 for <u>M</u>uscle <u>A</u>ldolase binding <u>F</u>actor <u>1</u>, but now renamed the MEF-3 protein on account of similarity to the factor binding the *TnC* skeletal enhancer, is expressed in a number of cell lines and tissues. These include cardiac muscle, and skeletal myoblasts and myotubes. The MEF-3 binding site in the aldolase A gene is nearly identical to the MEF-3 site of the *TnC* skeletal enhancer, suggesting that the sequences bind the same factor. Several other promoters, including that of the mouse myogenin

gene, contain MEF-3 sequences in close proximity to a MEF-2 site
and often a CCAC box.[86] It has been suggested for the aldolase A
gene that MEF-3 may stabilize the binding of MEF-2.[85] Although MEF-
2 genes are subject to alternative mRNA splicing and show a high
degree of specificity in the pattern of expression of the protein isoforms
these encode, the A/T-rich sequences they bind are rather similar (see
Fig. 4.6). MEF-3 protein(s) may help direct particular MEF-2 dimers
to specific target genes or stabilize their binding. A similar role has
been proposed for the ubiquitously expressed HF-1a factor which binds
the *MLC2V* gene.[80] Superficially, there is some DNA sequence similarity between the two sites, possibly indicative of binding a common
or related protein.

THE TROPONIN C CARDIAC ENHANCER

The *TnC* cardiac enhancer is less well-defined than the skeletal
enhancer. The cardiac enhancer comprises two elements termed CEF-
1 and CEF-2 (for Cardiac Enhancer Factor), defined by DNaseI
footprinting studies, upstream of three footprinted sequences within
the proximal promoter region termed CPF-1 (Cardiac Promoter Fac-
tor-1), CPF-2 and CPF-3[87] (Fig. 4.9). Furthermore, both CEF-1 and
CEF-2 bind factors which are present only (or enriched) in cardiac
myocytes. Mutations in either CEF-1 or CEF-2 reduce *trans*-activa-
tion to less than 10% of wild-type when expressed in neonatal cardiac
myocytes in culture. CEF-2 also contains a CCAC-like sequence,
CCACCCC, on the complementary (non-coding) strand.

Recently GATA-4, which is expressed solely in cardiac muscle and
some endodermally-derived tissues,[88,89] has been implicated in directing cardiac-specific expression from the *TnC* cardiac enhancer.[90] CEF-
1 and CEF-2 contain the sequences AGATTA and GGATAT, respectively, which resemble the consensus GATA binding sequence of
$(^A/_T)GATA(^A/_G)$ (see Table 4.1). GATA-4 binds to CEF-1 and, using
specific antisera, was shown to account for most of the endogenous
protein from cardiac myocyte nuclei that bound to the CEF-1 site in
band-shift assays.[90] Transient expression in cultured cardiac myocytes
revealed that mutation of the GATA-binding sequence of CEF-1 reduced transcriptional activation by 90%, suggesting GATA-4 plays a
major role in regulating the expression of the cardiac isoform of TnC
in the heart in particular and may be a major determinant of cardiac-
specific expression in general. Several other observations support this.[90]
First, TnC is not expressed in non-muscle cells and GATA-4 expression is highly tissue-restricted. Co-transfection of a GATA-4 expression plasmid and a reporter construct containing the *TnC* cardiac enhancer-promoter region into fibroblasts in culture results in a 45-fold
increase in cardiac-specific TnC expression over background,[90] demonstrating that GATA-4 is a potent *trans*-activator of the *TnC* cardiac
enhancer. Secondly, in situ hybridization with specific RNA probes

for *GATA-4* and cardiac *TnC* reveals strong similarities in their patterns of expression in cardiac and pre-cardiac tissues and shows that GATA-4 expression precedes that of cardiac TnC by 12-24 hours. Third, many other genes expressed in cardiac muscle contain potential GATA-4 binding sites. Taken as a whole, these findings indicate that expression of the slow/cardiac *TnC* gene in slow skeletal myotubes and cardiac muscle is controlled by two very different regulatory mechanisms operating through two separable complex enhancers.

CARDIAC TROPONIN T: REGULATION AT THE M-CAT SITE BY TEF-1

The M-CAT motif (sequence: CATTCC($^A/_T$)) was identified initially as the binding site for a nuclear protein from cardiac myocytes which bound two M-CAT sites in the chicken cardiac Troponin T (*cTnT*) promoter.[91] This protein, called M-CAT binding factor (MCBF), regulates expression of the *cTnT* gene in cardiac and embryonic skeletal muscle. Mutagenesis demonstrates that the two M-CAT sites in *cTnT* are essential, but not sufficient, for expression in cardiac myocytes.[92] An enhancer upstream of the M-CAT sites also contributes to cardiac-specific expression. This enhancer contains an A/T-rich sequence similar to a MEF-2 binding site. A nuclear protein called CEBF (for Cardiac-Element Binding Factor), present in cardiac and skeletal myocytes, liver and brain, has been shown to bind to the enhancer by band-shift assay, but relatively little is known of its nature and the gene encoding it has yet to be isolated.[92]

MCBF is present in several embryonic chicken tissues, including liver and gizzard, and is particularly enriched in skeletal and cardiac muscle.[93] Immunological studies show that MCBF is indistinguishable from the human factor TEF-1, which binds a similar sequence in several viral gene promoters.[93] TEF-1 is a member of the TEA-domain gene family (for TEF-1, TEC-1 (a yeast TF) and the *Aspergillus* ABAA regulatory protein), a group of TFs regulating developmental gene expression in eukaryotes and plants which are characterized by a novel three-helix DNA-binding domain.[94] Recently, several *TEF-1* cDNAs were cloned from chicken cardiac muscle which appear to arise from alternative splicing of a common primary transcript.[95] The cDNAs are expressed in several tissues and cell lines, but enriched in cardiac and skeletal muscle. When the cDNAs are transcribed and translated in vitro and their proteins tested for DNA-binding activity by band-shift assay, they bind M-CAT sequences in a manner indistinguishable from the MCBF activity detected in cardiac or skeletal nuclear extracts. Ubiquitous and muscle-specific isoforms of TEF-1 have also been reported to bind an M-CAT motif in the rabbit β-MHC promoter.[96] Together, these data strongly suggest that MCBF is a product of the *TEF-1* gene and, from its expression pattern, a major determinant of muscle gene expression. Interestingly, protein isoforms produced by alternative splicing of the *TEF-1* transcript differ in their ability to activate transcription, suggesting possible developmental differences in the regulation by TEF-1 proteins of M-CAT-containing promoters in different cells.[95] This might be functionally relevant as a TEF-1 binding site in the *β-MHC* gene has recently been demonstrated to be the site of responsiveness of this gene to protein kinase C and α_1-adrenergic agonists.[97,98]

KNOCKING OUT THE *TEF-1* GENE

Conclusive evidence of the importance of TEF-1/MCBP in the expression of cardiac contractile protein genes comes from analysis of mice in which the *TEF-1* gene had been disrupted using an enhancer-trap strategy[99] (Fig. 1.11). A line of homozygous mice were bred in which the *TEF-1* disruption was embryonic lethal, embryos dying midway through gestation. These embryos developed a four-chambered heart, but the ventricular wall was unusually thin compared to wild-type embryos at the same stage of development. Death appeared to result from an inability of the heart to pump blood out of the ventricles. The *cTnT* and cardiac *MHC* genes, which contain M-CAT sequences in their promoters, were expressed in mutant myocardium at levels similar to those seen in wild-type cardiac muscle. These results suggest that TEF-1 is not essential for cardiac-specific gene expression (consistent with results seen from mutating the *cTnT* M-CAT sites in vitro and assessing their effect upon expression in cultured cells),[92] nor for the initiation of cardiogenesis, but that it may be necessary for maintenance of cardiac-specific gene expression later during development or for responding to trophic stiluli. For example, TEF-1 and SRF binding sites are essential for TGF-β induction of skeletal α-actin[70,100] and expression of an MCBF-related protein binding the rat α-MHC promoter is up-regulated during hypertrophy.[101]

CARDIAC TROPONIN I

The cardiac Troponin I (*TnIc*) gene is unique among cardiac contractile protein genes in that it is only expressed in cardiac muscle. During development, the predominant isoform of TnI expressed in the developing heart is that encoded by the slow skeletal *TnI* gene (*TnIs*), whereas the *TnIc* gene is expressed at low levels. However, towards the time of birth in humans[102] and rodents,[103] *TnIc* gene expression is up-regulated dramatically, while that of TnIs declines such that soon after birth the only isoform expressed is TnIc. The molecular basis of this transition is unclear at present, but expression of human TnIc may occur at the transcriptional level as changes in protein accumulation mirror those seen in mRNA abundance.[102] The mouse[104] and human (P. Bhavsar, N. Brand and P. Barton, unpublished results) *TnIc* genes have been cloned, which will allow their regulatory regions to be defined. Both mouse and human proximal promoters and upstream flanking sequences contain possible binding sites for MEF-2, MCBF and GATA factors. In addition, the human gene contains several potential binding sites for CCAC-box binding factors and an intronic sequence similar to the *MLC2V* HF-1a site and a MEF-3 consensus sequence (PB, NB and PB, unpublished results).

MYOGLOBIN: IDENTIFICATION OF CCAC-BOX BINDING FACTORS

Myoglobin is a monomeric oxygen-binding protein found in the cytosol of cardiac and skeletal myocytes (see ref. 105). It is a major constituent of mitochondria and its expression is very sensitive to changes in demand for mitochondrial respiration. Expression of the myoglobin gene is developmentally regulated, with low levels of expression during fetal life, and a notable rise in expression which occurs soon after birth due to increased metabolic demand. Tissue-specific expression of the human myoglobin gene, which is limited to skeletal myocytes and ventricular myocardium, depends on the interactions of various TFs binding to a specific enhancer in the promoter.[106] In transgenic mice the region from -373 to +7 of the human myoglobin gene is sufficient to recapitulate the endogenous pattern of expression.[105]

A CCAC box with the sequence CCCCACCCC in the human myoglobin gene promoter has been shown to be essential for expression of the myoglobin promoter in vitro[107] and in vivo.[108] In skeletal myocytes the protein which binds the CCAC box acts co-operatively with other factors to direct skeletal muscle-specific expression, including a MEF-2-related protein binding to a downstream A/T-rich sequence which is flanked by two E-boxes. In cardiac muscle, mutations within the A/T-rich sequence and the CCAC box severely disrupt expression.[108] However, the CCAC box, whilst able to act as a classical enhancer when linked to a heterologous promoter, is not sufficient alone to direct muscle-specific expression, suggesting that appropriate expression from the myoglobin promoter requires the interactions of various factors. Similarly, expression of the tryptophan oxidase (*TO*) gene requires the co-operative interplay of glucocorticoid receptor, binding to its cognate response element within the *TO* promoter, with a ubiquitous protein binding to an adjacent CCAC box.[109] Two nuclear proteins have been identified which can bind to the myoglobin CCAC box. One is a 40 kDa protein called CPF40 present in the nuclei of differentiated rat sol8 myotubes, undifferentiated myoblasts and fibroblasts;[107] a cDNA for CPF40 has not been cloned as yet so its nature remains unknown. A gene encoding a 40 kDa protein which binds to the CCAC box in the human T-cell receptor (TCR) Vβ8.1 promoter and which may be the same as, or similar to, the CPF40 binding activity has been identified through cDNA cloning.[110] This factor, called htβ, contains four zinc fingers of the C_2H_2 class and is expressed as several transcripts in a variety of human cell lines, suggesting the possibility of different protein isoforms. A second CCAC-binding protein has been identified from mouse by screening a cDNA expression library (in which recombinant clones express the proteins encoded by the cDNAs they harbor) with a radio-labeled CCAC-box probe and isolating clones whose proteins bind the sequence.[105] DNA sequencing of a cDNA clone whose protein product bound the CCAC sequence revealed that it was a member of the winged-helix group of DNA-binding proteins which also includes the liver-specific factor HNF-3α and forkhead, a homeobox gene specifying terminal development in the *Drosophila* embryo. The clone was named MNF for Myocyte Nuclear Factor.[105] Two major MNF transcripts have been detected in skeletal and cardiac muscle by Northern blot analysis, although one is also present in mouse brain and kidney. Three distinct isoforms of MNF protein have been identified. The largest isoform can be phosphorylated, which may regulate its potency as a transcription factor. In this respect it has been noted that expression of the phosphorylated isoform is up-regulated in response to chronic motor nerve stimulation, a stimulus known to affect myoglobin expression in skeletal muscle.[105] As myoglobin expression responds quickly and dramatically according to physiological demand, it is possible that MNF is a direct nuclear target for signaling mediated through intracellular protein kinases.

CARDIAC MYOSIN HEAVY CHAIN GENES

In rat, α-MHC is expressed transiently in the forming embryonic heart tube and then not again until its re-expression is induced by a rise in circulating thyroid hormone (see chapter 3). At the same time, the *β-MHC* gene (the predominant fetal isoform expressed in the heart) is dramatically down-regulated, such that α-MHC accounts for approximately 95% of myosin heavy chain isoforms in the neonatal and adult rodent heart. Evidence suggests that this is regulated at the level of transcription.[111] The promoters and 5' flanking sequences of the α- and *β-MHC* genes show little similarity at the level of DNA sequence, suggesting that the two genes are regulated by different mechanisms.[112] Responsiveness to thyroid hormone (T_3) is mediated through the thyroid hormone receptor (TR), a member of the steroid hormone family of C_4 zinc twist transcription factors (see chapter 1) which, like all members of this family, only binds DNA when activated by its cognate hormone ligand. TR binds to a particular DNA-binding site, the thyroid hormone response element (TRE), which is present in the promoters of target genes. The regulation of *MHC* gene expression by T_3 has been intensively studied, in culture and in vivo, in both small and large animals (reviewed by Morkin[112]). Some of the differences seen between species in terms of MHC isoform expression and responsiveness to hormonal or pressure overload may reflect differences in how the cardiac *MHC* genes are regulated and the transcription factors involved. For example, the *α-MHC* gene of rat has one TRE, while that of human contains an additional weaker TRE.[113] The *β-MHC* gene has one TRE, located in the proximal promoter region, which may mediate down-regulation of the *β-MHC* gene in response to T_3 by interfering with the binding of basal transcription factors[112] (see chapter 3). The rat *β-MHC* TRE has not been well-characterized as yet, but the sequence GGGTGGG, identified through DNaseI footprinting, resembles non-consensus TRE half-sites found within the promoters of some other genes which are negatively regulated by T_3, such as those encoding rat growth hormone or the α-subunit of thyroid stimulating hormone.[114,115]

α-MHC: TOO MANY FACTORS FOR COMFORT?

The promoter and upstream sequences of rat *α-MHC* have been the subject of extensive studies and reveal the presence of multiple *cis*-acting elements which bind a variety of transcription factors. Data obtained from transgenic studies have shown that up to 3 kb of upstream sequence are required to recapitulate cardiac-specific α-MHC expression.[12,116] The role that several of these sequences play individually in regulating α-MHC expression has been determined (largely in culture) and it is clear that some enhance T_3-induced transcriptional activity.[117,118] Regulation of α-MHC is, therefore, a highly complex process.

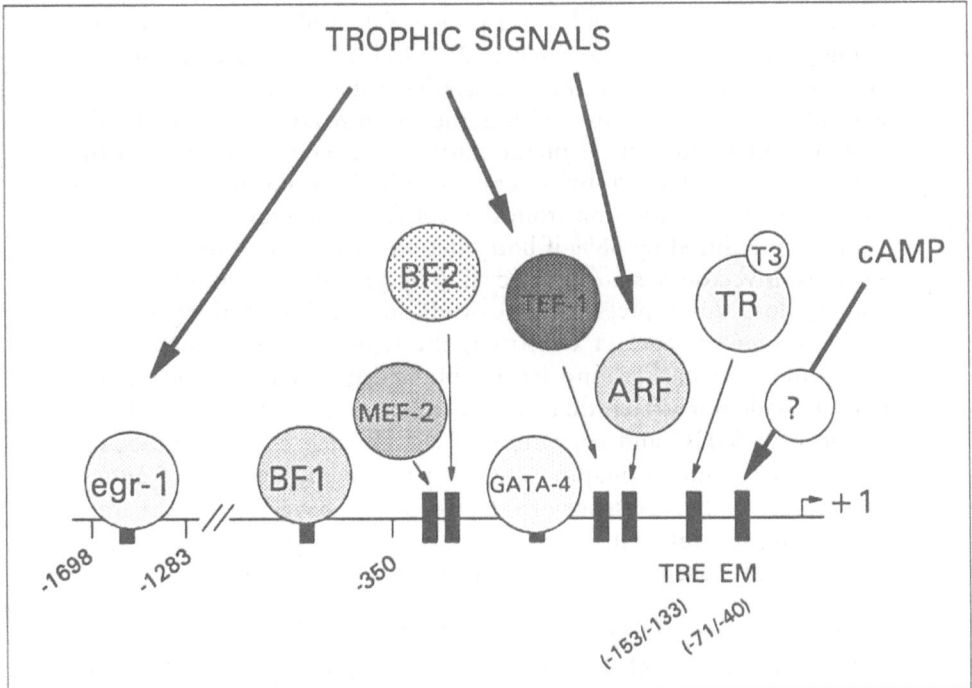

Fig. 4.10. Schematic representation of transcription factors binding the rat α-MHC promoter and 5' flanking sequence (not to scale)

There is considerable DNA sequence similarity between the promoters and 5' flanking sequences from rat and human *α-MHC* genes, suggesting they might be regulated in part by common factors, despite the differences in expression and cardiac development between the two species. For example, the human *α-MHC* promoter is active and inducible by T_3 when expressed in rat primary cardiac myocyte cultures.[119] This is mediated through a TRE located at -153/-133. Upstream of this in the rat gene lies a cluster of binding sites for TFs which have been shown to act as positive regulatory elements (Fig. 4.10). These include: an A/T-rich sequence at -229/-213 which binds a factor termed ARF(A-Rich binding Factor) which is antigenically related to MADS proteins of the MEF-2/RSRF class, though it clearly has different DNA-binding properties from MEF-2 proteins;[101] an M-CAT sequence, binding site for an MCBF/TEF-1 protein, at -247/-230;[101] two binding sites for GATA-4 at -270/-253[13] and a MEF-2 binding site adjacent to an E-box (BF-2, for binding factor 2) between -335 and -303.[14,120] A negative regulatory element called BF1 is at -599/-576.[117,119]

GATA-4

Recent work has focused on the role of GATA-4 in directing cardiac-specific expression. Molecular cloning has shown that GATA proteins bind DNA through a novel zinc-chelating domain. The expression of GATA-4 is developmentally regulated and is induced by the morphogen retinoic acid,[88] making GATA-4 a potential key player in interpreting morphogenetic signals (see chapter 2). GATA-4 is expressed in presumptive cardiac mesoderm prior to cardiac tube formation in mouse.[90] Two GATA-4 binding sites within the rat *α-MHC* promoter bind a factor present in the nuclei of rat cardiac myocytes which is indistinguishable from GATA-4 by several criteria.[13] Mutation of either of these sites reduces expression from the *α-MHC* promoter by about 50%; expression is reduced by 88% if both sites are disrupted. When a GATA-4 expression vector was co-injected directly into skeletal muscle (which normally does not express either α-MHC or GATA-4) along with an *α-MHC* promoter-reporter construct, the reporter was activated three to four-fold over background levels obtained by injecting the reporter alone. This demonstrates that GATA-4 can directly *trans*-activate expression of α-MHC and suggests that the absence of α-MHC expression in skeletal muscle may be due, in part, to the absence of endogenous GATA-4. The presence of GATA-4 in the developing heart infers it may be responsible, either wholly or partly, for the transient expression of α-MHC seen during early stages of cardiogenesis.[13]

THE MEF-2 SITE: IS IT ESSENTIAL IN VIVO?

The role of the MEF-2 site in *α-MHC* gene expression has been studied both in cultured cardiac myocytes and in transgenic mice. Surprisingly, there is some discrepancy between the results obtained when the MEF-2 site is mutated and examined in the two systems. However, these may also reflect species differences. A rat α-*MHC* promoter-reporter construct bearing a mutation in the MEF-2 site exhibits only 15-20% of the wild-type activity when assayed by direct injection into rat myocardium or by transient transfection into rat neonatal cardiac myocytes.[14] Similar results have been obtained with the rat *MLC2V* cardiac enhancer where mutating the HF-1b binding site results in reduced expression in culture[80] and in transgenic animals.[81] In contrast, transgenic mice in which the MEF-2 site of the mouse α-*MHC* gene has been mutated (TTAAAAATAA replaced by TgAccAcTcc) express higher levels of α-*MHC* mRNA in ventricular and atrial muscle compared to wild-type transgenic mice, as well as showing ectopic expression in aorta.[121] This suggests that the MEF-2 site is dispensable for cardiac-specific α-MHC expression in the mouse in vivo. The effect of mutating the site and assaying in culture to determine whether the same results are seen has not been addressed. These data are complicated by the fact that the MEF-2 mutation creates an E-box immediately upstream of the MEF-2 sequence,[121] which may have had some consequence on α-MHC expression in this transgenic mouse model. However, these and other observations underline the need to examine enhancer mutations carefully, preferably by backing up data obtained in culture with results from transgenic animals.[122]

ROLE OF THE E-BOX

The factor BF-2 that binds to the E-box at -313/-308 in the rat α-*MHC* gene remains un-cloned, but band-shift assays performed with various cell extracts show it is expressed widely in adult and neonatal rat tissues.[120] It has been proposed that this factor may be the same that recognizes the HF-1a motif in the rat *MLC2V* cardiac enhancer as each sequence competes for the other efficiently in band-shift assays.[120] The rat α-*MHC* E-box (CATGTG) has a five out of six match to the sequence CATGgG in the rat *MLC2V* HF-1a site. Both sequences are close or adjacent to MEF-2 sites. Mutating the binding sites for MEF-2 and BF-2 together is significantly more deleterious to α-MHC expression in cultured rat neonatal cardiac myocytes than mutating each site individually,[14] and expression of the factors binding the MEF-2 and BF-2 sites peaks during late fetal and early neonatal life, declining to lower levels in adult animals.[14,120] It should be noted that if there is co-operativity between the factors binding the MEF-2 and BF-2 sites in α-MHC expression, it is probably species-specific: the human[123] and mouse[121] genes lack the E-box.

MEDIATING THE RESPONSE TO TROPHIC STIMULI

Several observations suggest that the rat α-*MHC* gene is regulated in response to different stimuli by the interactions of multiple transcription factors. The expression of the genes encoding ARF and MCBF are developmentally regulated and increase in the immediate post-natal period. ARF and MCBF are also up-regulated during hypertrophy.[101] The A/T-rich sequence to which ARF binds has the sequence TAAAA-AAAGG: this sequence is also conserved in the mouse and human α-*MHC* gene.[121,123] Interference assays show that the contact points of ARF to this site are remarkably similar to those made by proteins binding the MEF-2 sites of *MLC2V* and α-*MHC* (Fig. 4.6B). Binding of ARF (present in neonatal rat cardiac myocyte nuclear extracts) to the A/T-rich site in band-shift assays showed that binding was not competed by the *MCK* MEF-2 site, nor by a CArG-like sequence CCAAATTTAG (see below) present at -70/-61 in the α-*MHC* promoter. This indicates that ARF is clearly distinct from other MEF-2/MADS factors characterized to date. The A/T-rich site was mutated to TGCAGGAAGG in α-MHC expression constructs which were subsequently injected directly into the myocardium of adult rats. Mutation reduced α-MHC expression by about 85%, demonstrating the necessity of this site in vivo.[101]

The M-CAT site is located just upstream of the ARF-binding site. Mutations in this site also reduced activation of the α-*MHC* promoter by approximately 85% when assayed by direct injection into rat heart. Band-shift assays using protein-DNA complexes which had been pre-incubated with an antibody raised against chicken TEF-1 gave a super-shifted complex. This indicated that the protein binding the α-*MHC* M-CAT site is TEF-1 or related to it antigenically.[101] The protein binding the M-CAT site is expressed in several tissues, as determined by band-shift assay, whereas ARF is enriched in brain and heart.[101] MCBF expression is detected in the fetal heart, unlike ARF. The expression of both factors is up-regulated 4- to 6-fold in rat models of left ventricular hypertrophy. M-CAT sites are found within several cardiac contractile protein genes, including β-*MHC*, cardiac *Troponin T* and α-skeletal actin, and TEF-1 has been proposed to mediate the response of the β-*MHC* gene to $α_1$-adrenergic stimulation in culture.[98,100] The atrial natriuretic peptide (*ANP*) gene, which is also induced in the hypertrophic response, contains two sites similar or identical to that in α-*MHC* which binds ARF.[101] Together, these data suggest that the response of different genes to hypertrophic stimuli may be channeled through specific transcription factors such as TEF-1/MCBF or ARF.

Another factor which binds the α-*MHC* gene and which is an early marker of the hypertrophic response is the early response gene *egr-1*, a C_2H_2 zinc finger gene (also called *krox24*) whose expression is induced by various growth signals. The level of *egr-1* mRNA, which is abundant in adult cardiac myocytes, is up-regulated by serum in cultured cardiac myocytes: this effect is not due to induction by thyroid hormone.[124] Deletion mutagenesis of the α-*MHC* 5' sequences revealed that a binding site for Egr-1 exists between -1698 and -1283: a sequence GTGGGGGTG within this region is similar to the consensus binding site for Egr-1 (GCGGGGGCG).[124]

Rat α-MHC is also inducible by cyclic AMP (cAMP).[125] This is mediated through a complex sequence from -71 to -40 which contains overlapping E-box and M-CAT sequences (called the EM motif[125]) flanked by a CArG-like element at the 5' side and the TATA box 3' to it. Deletion of this region, which does not resemble a consensus cAMP-responsive element,[126] reduces expression from the *α-MHC* promoter by about 80% and abolishes responsiveness to cAMP.[125] Inducibility by cAMP has been defined further to the EM motif by making point mutations within these sequences. In band-shift assays the *MLC2V* HF-1a sequence competes with the EM element for proteins present in cardiac myocyte. Interference studies reveal that three out of five contacts made to guanines around the E-box are identical to contacts made by proteins to HF-1a, BF-2 and E-box E_2 of α-cardiac actin.[125] Of these four sequences, only HF-1a does not agree with a consensus E-box sequence. Whether these sequences bind bHLH factors remains to be proven. The data obtained from analyses of the BF-2 site and the EM element reveal some similarities to HF-1a (competition for binding in gel-shift assays to the HF-1a sequence, interference assays). It is possible that BF-2 and EM bind similar or related proteins and that these may be members of the bHLH family.

GArC BOXES: A NOVEL MOTIF REGULATING CARDIAC MHC GENE EXPRESSION IN MAN?

Differences in cardiac MHC expression between species undoubtedly reflect subtle variations in the DNA sequences of the regulatory regions of the genes and the transcription factors that bind them. The human *α-MHC* gene contains two copies of a novel sequence termed the GArC motif that are not present in the rat gene.[118] The GArC sequences (GArC stands for G,AT-rich,C) were identified by DNaseI footprinting and are found at -904/-896 (GAAAAATCT) and -823/-816 (GAAAATCT) in the rat *α-MHC* 5' flanking sequences. Despite the difference in the number of A nucleotides at the center of the two motifs, they efficiently compete with each other for binding to nuclear proteins, whereas unrelated A/T-rich sequences such as the SRE/CArG box do not compete. One of these regions functions as a positive regulatory element, the other as a negative element.[118] Both human and rat nuclear extracts make similar DNaseI footprints over the two elements and they bind several proteins of different molecular weights which are present in a human fibroblast cell line (HeLa) and several rat tissues. The distribution of some of these proteins appears to vary between tissues. Together, these results suggest that GArC motifs bind proteins, perhaps belonging to a family of TFs, that are conserved between species and widely expressed. Differences in expression between rat and human *α-MHC* genes may, in part, be due to the presence in the human gene of binding sites for GArC proteins. As the -924/-851 and -851/-762 regions have opposite effects functionally, it remains to be established whether the GArC elements within them bind different factors, some functioning as activators, others as repressors. Alternatively, GArC proteins may work synergistically or competitively with other TFs to effect positive and negative regulation through the two sequences.

The rat *β-MHC* gene contains a GArC sequence at -1162/-1154 on the non-coding strand, but exhibits poor binding of nuclear protein binding from footprinting and band-shift assays. Differences in sequences flanking GArC motifs may be important in determining the type or specificity of proteins binding these elements.[118]

CONCLUDING REMARKS

Tissue-specific expression in skeletal and cardiac muscle is regulated by the co-operative interactions of multiple transcription factors binding to specific DNA sequences within genes. Expression of factors involved in cardiac gene regulation ranges from ubiquitous to tissue-specific and some factors are phosphorylated in response to extracellular cues or trophic stimuli which alters their function. Many transcription factors belong to large gene families, allowing, in theory, the formation of different combinations of functional heterodimeric molecules. In skeletal muscle, the best-studied system to date, tissue-specific gene regulation is largely achieved through the formation of functional dimers between tissue-restricted and ubiquitously expressed members of the bHLH family.[9,22,35] There is also evidence that some *cis*-acting sequences can bind more than one type of transcription factor. This may allow competition for binding of different factors to a particular site on a gene, which may be modulated by extracellular cues (e.g., competition between SRF and MyoD for overlapping sites on the *c-fos* promoter in culture as a result of growth factor stimulation[32] or binding of MEF-2 and TATA-box binding protein to the *Xenopus MyoDa* TATA element[73]).

The rat *MLC2V* and *α-MHC* genes contain binding sites for a variety of transcription factors and the function of some of these sites has been assessed both in culture and in transgenic animal models. The dissection of promoters of genes expressed in cardiac muscle in these ways is a major part of understanding gene regulation in the cardiac myocyte. For the most part such studies have, for technical reasons, focused on ablating one binding site at a time and assessing the impact mutation has upon cardiac expression. However, results obtained in culture can sometimes differ from those obtained in transgenic animals, often due to positional effects on the DNA in vivo (see ref. 122 and references within). These include the presence of remote enhancers necessary for correct spatial and temporal expression in vivo, which may not be present in the transgenic construct, the site of insertion of the transgene into the genome and the number of copies inserted.

In transgenic mice in which a gene at a particular locus is inactivated by homologous recombination (knock-out mouse), ablation results in disruption of that gene in all tissues in which it would normally be expressed and throughout development. The inter-breeding of two lines of animals in which the expression of two separate transcription factor genes has been inactivated may reveal something of the interactions occurring between factors in vivo. A recent advance in knock-out technology is the development of strategies for knocking out a gene in a particular cell type or lineage. Called the *Cre-loxP* system,[127,128] it involves inter-breeding two distinct lines of transgenic mice. One line harbors a transgene for the P1 bacteriophage enzyme Cre (a protein which excises any DNA sequence located between DNA sequences called *loxP* recognition sites) in which *Cre* expression is under the transcriptional control of a particular tissue-specific promoter. Though every cell of the animal carries the transgene, the *cre* enzyme will only be produced in those cells in which its targeting promoter is active. The second transgenic line contains the gene to be inactivated, flanked by *loxP* sequences. The progeny resulting from a cross between two such lines will thus lack the targeted gene product only in those tissues in which the Cre enzyme is produced. This approach will prove useful in studying the effect of inactivation of certain widely expressed transcription factors by targeting knock-outs to the cardiac myocyte. With appropriate targeting constructs (for example, inducible by metal ions, pharmacologic agents or hormones) this may be extended to perturbing specific time-points during cardiac development. Such radical new tools, together with the crossing of established lines of transgenic mice in which particular transcription factor genes have been inactivated, should allow the interplay between different factors binding to a gene in vivo to be evaluated more fully.

REFERENCES

1. Gaul U, Jäckle H. How to fill a gap in the *Drosophila* embryo. Trends Genet 1987; 3:127-31.
2. Kenyon C. If birds can fly, why can't we? Homeotic genes and evolution. Cell 1994; 78:175-80.
3. Krumlauf R. Hox genes in vertebrate development. Cell 1994; 78:191-201.
4. Davis RL, Weintraub H, Lassar HB. Expression of a single transfected cDNA converts fibroblasts to myoblasts. Cell 1987; 51:987-1000.
5. Edmondson DG, Olson EN. A gene with homology to the *myc* similarity region of MyoD1 is expressed during myogenesis and is sufficient to activate the muscle differentiation program. Genes Devel 1989; 3:628-40.
6. Wright WE, Sassoon DA, Lin VK. Myogenin, a factor regulating myogenesis, has a domain homologous to MyoD. Cell 1989; 56:607-17.
7. Braun T, Buschhausen-Denker G, Bober E et al. A novel human muscle factor related to but distinct from MyoD1 induces myogenic conversion in 10t1/2 fibroblasts. EMBO J 1989; 8:701-9.

8. Rhodes SJ, Konieczny SF. Identification of MRF4: a new member of the muscle regulatory factor gene family. Genes Devel 1989; 3:2050-61.

9. Buckingham M. Molecular biology of muscle development. Cell 1994; 78:15-21.

10. Chambers AE, Kotecha S, Towers N et al. Muscle-specific expression of SRF-related genes in the early embryo of Xenopus laevis. EMBO J 1992; 11:4981-91.

11. Yu YT, Breitbart RE, Smoot LB et al. Human myocyte-specific enhancer factor 2 comprises a group of tissue-restricted MADS box transcription factors. Genes Devel 1992; 6:1783-98.

12. Subramaniam A, Gulick J, Neumann J et al. Transgenic analysis of the thyroid-responsive elements in the α-cardiac myosin heavy chain gene promoter. J Biol Chem 1993; 268:4331-6.

13. Molkentin JD, Kalvakolanu DV, Markham BE. Transcription factor GATA-4 regulates cardiac muscle-specific expression of the α-myosin heavy-chain gene. Mol Cell Biol 1994; 14:4947-57.

14. Molkentin JD, Markham BE. Myocyte-specific enhancer-binding factor (MEF-2) regulates α-cardiac myosin heavy chain gene expression *in vitro* and *in vivo*. J Biol Chem 1993; 268:19512-20.

15. Tapscott SJ, Davis RL, Thayer MJ et al. MyoD1: a nuclear phosphoprotein requiring a myc homology region to convert fibroblasts to myoblasts. Science 1988; 242:405-11.

16. Braun T, Bober E, Winter B et al. *Myf-6*, a new member of the human gene family of myogenic determination factors: evidence for a gene cluster on chromosome 12. EMBO J 1990; 9:821-31.

17. Miner JH, Wold B. Herculin, a fourth member of the *MyoD* family of myogenic regulatory genes. Proc Natl Acad Sci USA 1990; 87:1089-93.

18. Johnson JE, Birren SJ, Anderson DJ. Two rat homologues of *Drosophila achaete-scute* specifically expressed in neuronal precursors. Nature 1990; 346:858-60.

19. Murre C, McCaw PS, Baltimore D. A new DNA binding and dimerization motif in immunoglobulin enhancer binding, daughterless, MyoD, and myc proteins. Cell 1989; 56:777-83.

20. Nelson C, Shen L, Meister A et al. Pan: a transcriptional regulator that binds chymotrypsin, insulin, and AP-4 enhancer motifs. Genes Devel 1990; 4:1035-43.

21. Sun XH, Baltimore D. An inhibitory domain of E12 transcription factor prevents DNA binding in E12 homodimers but not in E12 heterodimers. Cell 1991; 64:459-70.

22. Weintraub H, Davis R, Tapscott S et al. The *MyoD* gene family: nodal point during specification of the muscle cell lineage. Science 1991; 251:761-6.

23. Bengal E, Flores O, Rangarajan PN et al. Positive control mutations in the MyoD basic region fail to show cooperative DNA binding and transcriptional activation *in vitro*. Proc Natl Acad Sci USA 1994; 91:6221-5.

24. Johnson JE, Birren SJ, Saito T et al. DNA binding and transcriptional regulatory activity of mammalian *achaete-scute* homologous (MASH) pro-

teins revealed by interaction with a muscle-specific enhancer. Proc Natl Acad Sci USA 1992; 89:3596-600.

25. Ma PCM, Rould MA, Weintraub H et al. Crystal structure of MyoD bHLH domain-DNA complex: perspectives on DNA recognition and implications for transcriptional activation. Cell 1994; 77:451-9.

26. Ellenberger T, Fass D, Arnaud M et al. Crystal structure of transcription factor E47: E-box recognition by a basic region helix-loop-helix dimer. Genes Devel 1994; 8:970-80.

27. Ferré-D'Amaré AR, Prendergast GC, Ziff EB et al. Recognition by Max of its cognate DNA through a dimeric b/HLH/Z domain. Nature 1993; 363:1003-9.

28. Brennan TJ, Chakraborty T, Olson EN. Mutagenesis of the myogenin basic region identifies an ancient protein motif critical for activation of myogenesis. Proc Natl Acad Sci USA 1991; 88:5675-9.

29. Weintraub H, Genetta T, Kadesch T. Tissue-specific gene activation by MyoD: determination of specificity by cis-acting repression elements. Genes Devel 1994; 8:2203-11.

30. Wright WE, Funk WD. CASTing for multicomponent DNA-binding complexes. Trends Biochem Sciences 1993; 18:77-80.

31. Treisman R. The serum response element. Trends Biochem Sciences 1992; 17:423-6.

32. Trouche D, Grigoriev M, Lenormand J et al. Repression of *c-fos* promoter by MyoD on muscle cell differentiation. Nature 1993; 363:79-82.

33. Buckingham ME. Muscle: the regulation of myogenesis. Curr Op Genet Develop 1994; 4:745-51.

34. Yaffe D, Saxel O. Serial passaging and differentiation of myogenic cells isolated from dystrophic mouse muscle. Nature 1977; 270:725-7.

35. Olson EN, Klein, WH. bHLH factors in muscle development: dead lines and commitments, what to leave in and what to leave out. Genes Devel 1994; 8:1-8.

36. Hasty P, Bradley A, Morris JH et al. Muscle deficiency and neonatal death in mice with a targetted mutation in the *myogenin* gene. Nature 1993; 364:501-6.

37. Nabeshima Y, Hanaoka K, Hayasaka M et al. *Myogenin* gene disruption results in perinatal lethality because of severe muscle defect. Nature 1993; 364:532-5.

38. Hughes SM. Running out of control. Nature 1993; 354:485

39. Rudnicki MA, Schnegelsberg PN, Stead RH et al. MyoD or Myf-5 is required for the formation of skeletal muscle. Cell 1993; 75:1351-9.

40. Olson EN. Regulation of muscle transcription by the MyoD family. The heart of the matter. Circ Res 1993; 72:1-6.

41. Weintraub H. The MyoD family and myogenesis: redundancy, networks, and thresholds. Cell 1993; 75:1241-4.

42. Li L, Zhou J, James G et al. FGF inactivates myogenic helix-loop-helix proteins through phosphorylation of a conserved protein kinase C site in their DNA-binding domains. Cell 1992; 71:1181-94.

43. Miner JH, Miller JB, Wold BJ. Skeletal muscle phenotypes initiated by ectopic MyoD in transgenic mouse heart. Development 1992; 114:853-60.
44. Springhorn JP, Ellingsen O, Berger HJ et al. Transcriptional regulation in cardiac muscle. Coordinate expression of Id with a neonatal phenotype during development and following a hypertrophic stimulus in adult rat ventricular myocytes in vitro. J Biol Chem 1992; 267:14360-5.
45. Sartorelli V, Webster KA, Kedes L. Muscle-specific expression of the cardiac alpha-actin gene requires MyoD1, CArG-box binding factor, and Sp1. Genes Devel 1990; 4:1811-22.
46. Sartorelli V, Hong NA, Bishopric NH et al. Myocardial activation of the human cardiac α-actin promoter by helix-loop-helix proteins. Proc Natl Acad Sci USA 1992; 89:4047-51.
47. Skerjanc IS, McBurney MW. The E box is essential for activity of the cardiac actin promoter in skeletal but not in cardiac muscle. Dev Biol 1994; 163:125-32.
48. Brand NJ. Principles and applications of the polymerase chain reaction. In:Latchman DS (ed): PCR Applications in Pathology. Oxford: Oxford University Press, 1995:1-16.
49. Benezra R, Davis RL, Lockshon D et al. The protein Id: a negative regulator of helix-loop-helix DNA binding proteins. Cell 1990; 61:49-59.
50. Wang Y, Benezra R, Sassoon DA. Id expression during mouse development: a role in morphogenesis. Devel Dynam 1992; 194:222-30.
51. Evans SM, Walsh BA, Newton CB et al. Potential role of helix-loop-helix proteins in cardiac gene expression. Circ Res 1993; 73:569-78.
52. Barone MV, Pepperkok R, Peverali FA et al. Id proteins control growth induction in mammalian cells. Proc Natl Acad Sci USA 1994; 91:4985-8.
53. Kurabayashi M, Jeyaseelan R, Kedes L. Doxorubicin represses the function of the myogenic helix-loop-helix transcription factor MyoD. J Biol Chem 1994; 269:6031-9.
54. Breitbart RE, Liang C, Smoot LB et al. A fourth human MEF2 transcription factor, hMEF2D, is an early marker for the myogenic lineage. Development 1993; 118:1095-106.
55. Martin JF, Miano JM, Hustad CM et al. A *Mef2* gene that generates a muscle-specific isoform via alternative mRNA splicing. Mol Cell Biol 1994; 14:1647-56.
56. Lilly B, Zhao B, Ranganayakulu G et al. Requirement of MADS domain transcription factor D-MEF2 for muscle formation in *Drosophila*. Science 1995; 267:688-93.
57. Gossett LA, Kelvin DJ, Sternberg EA et al. A new myocyte-specific enhancer-binding factor that recognizes a conserved element associated with multiple muscle-specific genes. Mol Cell Biol 1989; 9:5022-33.
58. Pollock R, Treisman R. Human SRF-related proteins: DNA-binding properties and potential regulatory targets. Genes Devel 1991; 5:2327-41.
59. Cserjesi P, Olson EN. Myogenin induces the myocyte-specific enhancer binding factor MEF-2 independently of other muscle-specific gene products. Mol Cell Biol 1991; 11:4854-62.

60. Edmondson DG, Lyons GE, Martin JF et al. Mef2 gene expression marks the cardiac and skeletal muscle lineages during mouse embryogenesis. Development 1994; 120:1251-63.

61. Zhu H, Nguyen VT, Brown AB et al. A novel, tissue-restricted zinc finger protein (HF-1b) binds to the cardiac regulatory element (HF-1b/MEF-2) in the rat myosin light-chain 2 gene. Mol Cell Biol 1993; 13:4432-44.

62. Jaynes JB, Johnson JE, Buskin JN et al. The muscle creatine kinase gene is regulated by multiple elements, including a muscle-specific enhancer. Mol Cell Biol 1988; 8:62-71.

63. Horlick RA, Benfield PA. The upstream muscle-specific enhancer of the rat muscle creatine kinase gene is composed of multiple elements. Mol Cell Biol 1989; 9:2396-413.

64. Cserjesi P, Lilly B, Bryson L et al. MHox: a mesodermally restricted homeodomain protein that binds an essential site in the muscle creatine kinase enhancer. Development 1992; 115:1087-101.

65. Cserjesi P, Lilly B, Hinkley C et al. Homeodomain protein MHox and MADS protein myocyte enhancer-binding factor-2 converge on a common element in the muscle creatine kinase enhancer. J Biol Chem 1994; 269:16740-5.

66. Grueneberg DA, Natesan S, Alexandre C et al. Human and Drosophila homeodomain proteins that enhance the DNA-binding activity of serum response factor. Science 1992; 257:1089-95.

67. Lints TJ, Parsons LM, Hartley L et al. *Nkx-2.5*: a novel murine homeobox gene expressed in early heart progenitor cells and their myogenic descendants. Development 1993; 119:419-31.

68. Komuro I, Izumo S. *Csx*: A murine homeobox-containing gene specifically expressed in the developing heart. Proc Natl Acad Sci USA 1993; 90:8145-9.

69. Lee T, Zhang Y, Schwartz RJ. Bifunctional transcriptional properties of YY1 in regulating muscle actin and c-*myc* gene expression during myogenesis. Oncogene 1994; 9:1047-52.

70. MacLellan WR, Lee T, Schwartz RJ et al. Transforming growth factor-β response elements of the skeletal α-actin gene. J Biol Chem 1994; 269:16754-60.

71. Seto E, Shi Y, Shenk T. YY1 is an initiator sequence-binding protein that directs and activates transcription *in vitro*. Nature 1991; 354:241-5.

72. Usheva A, Shenk T. TATA-binding protein-independent initiation: YY1, TFIIB, and RNA polymerase II direct basal transcription on supercoiled template DNA. Cell 1994; 76:1115-21.

73. Leibham D, Wong MW, Cheng TC et al. Binding of TFIID and MEF2 to the TATA element activates transcription of the Xenopus MyoDa promoter. Mol Cell Biol 1994; 14:686-99.

74. Yee SP, Rigby PW. The regulation of myogenin gene expression during the embryonic development of the mouse. Genes Dev 1993; 7:1277-89.

75. Cheng T, Wallace MC, Merlie JP et al. Separable regulatory transcription elements in mouse embryogenesis. Science 1993; 261:215-8.

76. Kaushai S, Schneider JW, Nadal-Ginard B et al. Activation of the myogenic lineage by MEF2A, a factor that induces and cooperates with MyoD. Science 1994; 266:1236-40.

77. O'Brien TX, Lee KJ, Chien KR. Positional specification of ventricular myosin light chain 2 expression in the primitive murine heart tube. Proc Natl Acad Sci USA 1993; 90:5157-61.

78. Lee KJ, Ross RS, Rockman HA et al. Myosin light chain-2 luciferase transgenic mice reveal distinct regulatory programs for cardiac and skeletal muscle-specific expression of a single contractile protein gene. J Biol Chem 1992; 267:15875-85.

79. Zhu H, Garcia AV, Ross RS et al. A conserved 28-base-pair element (HF-1) within the rat cardiac myosin light-chain-2 gene confers cardiac-specific and α-adrenergic-inducible expression in cultured neonatal rat myocardial cells. Mol Cell Biol 1991; 11:2273-81.

80. Navankasattusas S, Zhu H, Garcia AV et al. A ubiquitous factor (HF-1a) and a distinct muscle factor (HF-1b/MEF-2) form an E-box-independent pathway for cardiac muscle gene expression. Mol Cell Biol 1992; 12:1469-79.

81. Lee KJ, Hickey R, Zhu H et al. Positive regulatory elements (HF-1a and HF-1b) and a novel negative regulatory element (HF-3) mediate ventricular muscle-specific expression of myosin light-chain 2-luciferase fusion genes in transgenic mice. Mol Cell Biol 1994; 14:1220-9.

82. Goswami S, Qasba P, Ghatpande S et al. Differential expression of the myocyte enhancer factor 2 family of transcription factors in development: the cardiac factor BBF-1 is an early marker for cardiogenesis. Mol Cell Biol 1994; 14:5130-8.

83. Gahlmann R, Kedes L. Tissue-specific restriction of skeletal muscle troponin C gene expression. Gene Expr 1993; 3:11-25.

84. Parmacek MS, Leiden JM. Structure, function, and regulation of troponin C. Circulation 1991; 84:991-1003.

85. Hidaka K, Yamamoto I, Arai Y et al. The MEF-3 motif is required for MEF-2 mediated skeletal muscle-specific induction of the rat aldolase A gene. Mol Cell Biol 1993; 13:6469-78.

86. Parmacek MS, Ip HS, Jung F et al. A novel myogenic regulatory circuit controls slow/cardiac troponin C gene transcription in skeletal muscle. Mol Cell Biol 1994; 14:1870-85.

87. Parmacek MS, Vora AJ, Shen T et al. Identification and characterization of a cardiac-specific transcriptional regulatory element in the slow/cardiac troponin C gene. Mol Cell Biol 1992; 12:1967-76.

88. Arceci RJ, King AA, Simon MC et al. Mouse GATA-4: a retinoic acid-inducible GATA-binding transcription factor expressed in endodermally derived tissues and heart. Mol Cell Biol 1993; 13:2235-46.

89. Heikinheimo M, Scandrett JM, Wilson DB. Localization of transcription factor GATA-4 to regions of the mouse embryo involved in cardiac development. Dev Biol 1994; 164:361-73.

90. Ip HS, Wilson DB, Heikinheimo M et al. The GATA-4 transcription factor transactivates the cardiac muscle-specific Troponin C promoter-enhancer in nonmuscle cells. Mol Cell Biol 1994; 14:7517-26.

91. Mar JH, Ordahl CP. A conserved CATTCCT motif is required for skeletal muscle-specific activity of the cardiac troponin T gene promoter. Proc Natl Acad Sci USA 1988; 85:6404-8.

92. Iannello RC, Mar JH, Ordahl CP. Characterization of a promoter element required for transcription in myocardial cells. J Biol Chem 1991; 266:3309-16.

93. Farrance IK, Mar JH, Ordahl CP. M-CAT binding factor is related to the SV40 enhancer binding factor, TEF-1. J Biol Chem 1992; 267: 17234-40.

94. Bürglin TR. The TEA domain: A novel, highly conserved DNA-binding motif. Cell 1991; 66:11-2.

95. Stewart AFR, Larkin SB, Farrance IKG et al. Muscle-enriched TEF-1 isoforms bind M-CAT elements from muscle-specific promoters and differentially activate transcription. J Biol Chem 1994; 269:3147-50.

96. Shimizu N, Smith G, Izumo S. Both a ubiquitous factor mTEF-1 and a distinct muscle-specific factor bind to the M-CAT motif of the myosin heavy chain β gene. Nuc Acids Res 1993; 21:4103-10.

97. Kariya K, Karns LR, Simpson PC. Expression of a constitutively activated mutant of the β-isozyme of protein kinase C in cardiac myocytes stimulates the promoter of the β-myosin heavy chain isogene. J Biol Chem 1991; 266:10023-6.

98. Kariya K, Karns LR, Simpson PC. An enhancer core element mediates stimulation of the rat β-myosin heavy chain promoter by an α1-adrenergic agonist and activated β-protein kinase C in hypertrophy of cardiac myocytes. J Biol Chem 1994; 269:3775-82.

99. Chen Z, Friedrich GA, Soriano P. Transcriptional enhancer factor 1 disruption by a retroviral gene trap leads to heart defects and embryonic lethality in mice. Genes Devel 1994; 8:2293-301.

100. Karns LR, Kariya K, Simpson PC. M-CAT, CArG, and Sp1 elements are required for α_1-adrenergic induction of the skeletal α-actin promoter during cardiac myocyte hypertrophy. J Biol Chem 1995; 270:410-7.

101. Molkentin JD, Markham BE. An M-CAT binding factor and an RSRF-related A-rich binding factor positively regulate expression of the α-cardiac myosin heavy-chain gene in vivo. Mol Cell Biol 1994; 14:5056-65.

102. Sasse S, Brand NJ, Kyprianou P et al. Troponin I gene expression during human cardiac development and in end-stage heart failure. Circ Res 1993; 72:932-8.

103. Gorza L, Ausoni S, Merciai N et al. Regional differences in troponin I isoform switching during rat heart development. Dev Biol 1993; 156:253-64.

104. Ausoni S, Campione M, Picard A et al. Structure and regulation of the mouse cardiac troponin I gene. J Biol Chem 1994; 269:339-46.

105. Basel-Duby R, Hernandez MD, Yang Q et al. Myocyte Nuclear Factor, a novel winged-helix transcription factor under both developmental and neural regulation in striated myocytes. Mol Cell Biol 1994; 14:4596-605.

106. Devlin BH, Wefald FC, Kraus WE et al. Identification of a muscle-specific enhancer within the 5' flanking region of the human myoglobin gene. J Biol Chem 1989; 264:13896-900.

107. Bassel-Duby R, Hernandez MD, Gonzalez MA et al. A 40-kilodalton protein binds specifically to an upstream sequence element essential for muscle-specific transcription of the human myoglobin promoter. Mol Cell Biol 1992; 12:5024-32.

108. Bassel-Duby R, Grohe CM, Jessen ME et al. Sequence elements required for transcriptional activity of the human myoglobin promoter in intact myocardium. Circ Res 1993; 73:360-6.

109. Schule R, Muller M, Otsuka Murakami H et al. Cooperativity of the glucocorticoid receptor and the CACCC-box binding factor. Nature 1988; 332:87-90.

110. Wang Y, Kobori JA, Hood L. The htβ gene encodes a novel CACCC box-binding protein that regulates T-cell receptor gene expression. Mol Cell Biol 1993; 13:5691-701.

111. Boheler KR, Chassagne C, Martin X et al. Cardiac expressions of α- and β-myosin heavy chains and sarcomeric α-actins are regulated through transcriptional mechanisms. J Biol Chem 1992; 267:12979-85.

112. Morkin E. Regulation of myosin heavy chain genes in the heart. Circulation 1993; 87:1451-60.

113. Flink IL, Morkin E. Interaction of thyroid hormone receptors with strong and weak *cis*-acting elements in the human α-myosin heavy chain gene promoter. J Biol Chem 1990; 265:11233-7.

114. Crone DE, Kim H, Spindler SR. α and β thyroid hormone receptors bind immediately adjacent to the rat growth hormone gene TATA in a negatively hormone-responsive region. J Biol Chem 1990; 265:10851-6.

115. Chatterjee V, Lee J, Rentoumis A et al. Negative regulation of the thyroid stimulating hormone α gene by thyroid hormone: Receptor interaction adjacent to the TATA box. Proc Natl Acad Sci USA 1989; 86:9114-8.

116. Subramaniam A, Jones WK, Gulick J et al. Tissue-specific regulation of the α-myosin heavy chain gene promoter in transgenic mice. J Biol Chem 1991; 179:24613-20.

117. Markham BE, Bahl JJ, Gustafson TA et al. Interaction of a protein factor within a thyroid hormone sensitive region of rat α-myosin heavy chain gene. J Biol Chem 1987; 262:12856-62.

118. Mably JD, Sole MJ, Liew C. Characterization of the GArC motif. A novel *cis*-acting element of the human cardiac myosin heavy chain genes. J Biol Chem 1993; 268:476-82.

119. Tsika RW, Bahl JJ, Leinwand LA et al. Thyroid hormone regulates expression of a transfected human α-myosin heavy-chain fusion gene in fetal rat heart cells. Proc Natl Acad Sci USA 1990; 87:379-83.

120. Molkentin JD, Brogan RS, Jobe SM et al. Expression of the α-myosin heavy chain gene in the heart is regulated in part by an E-box-dependent mechanism. J Biol Chem 1993; 268:2602-9.

121. Adolph EA, Subramaniam A, Cserjesi P et al. Role of myocyte-specific enhancer-binding factor (MEF-2) in transcriptional regulation of the α-cardiac myosin heavy chain gene. J Biol Chem 1993; 268:5349-52.

122. McGrew MJ, Rosenthal N. Transgenic analysis of cardiac and skeletal myogenesis. Trends Cardiovasc Med 1994; 6:251-6.

123. Yamauchi-Takihara K, Sole MJ, Liew J et al. Characterization of human cardiac myosin heavy chain genes. Proc Natl Acad Sci USA 1989; 86:3504-8.

124. Gupta MP, Gupta M, Zak R et al. Egr-1, a serum inducible zinc finger protein regulates transcription of the rat cardiac α-myosin heavy chain gene. J Biol Chem 1991; 266:12813-6.

125. Gupta MP, Gupta M, Zak R. An E-box/M-CAT hybrid motif and cognate binding protein(s) regulate the basal muscle-specific and cAMP-inducible expression of the rat cardiac α-myosin heavy chain gene. J Biol Chem 1994; 269:29677-87.

126. Karin M. Complexities of gene regulation by cAMP. Trends Genet 1989; 5:65-7.

127. Gu H, Marth JD, Orban PC et al. Deletion of a DNA polymerase β gene segment in T cells using cell type-specific gene targeting. Science 1994; 265:103-6.

128. Barinaga M. Knockout mice: Round two. Science 1994; 265:26-8.

129. Parmacek MS, Bengur AR, Vora AJ et al. The structure and regulation of expression of the murine fast skeletal troponin C gene. Identification of a developmentally regulated, muscle-specific transcriptional enhancer. J Biol Chem 1990; 265:15970-6.

GENE EXPRESSION IN CARDIAC HYPERTROPHY

The molecular basis for the long-term adaptation to cellular stress remains one of the fundamental problems in molecular biology and molecular genetics. This is particularly true for the cardiovascular system.[1-4] Fifteen years ago, the molecular features of cardiac myocytes were described in terms of a fixed phenotype typical of terminally differentiated cells. More recently, the use of recombinant DNA technologies in conjunction with cellular biological techniques has led to the consensus that the phenotype of these differentiated cells changes with ontogeny, senescence and hypertrophy.[5-10] Cardiac hypertrophy usually results from an increase in workload imposed on the heart, frequently resulting from either an increased afterload in patients with hypertension or after a myocardial infarction where the remaining muscle must assume the workload of the segment that was lost due to injury. One of the most important properties of the heart is its ability to adapt to an altered hemodynamic load. In the short-term this is accomplished by the Frank-Starling mechanism, but in the long-term or after repeated increases in load, the heart adapts through an increase in myocardial mass (Fig. 5.1) frequently involving changes in cardiac gene expression.[8,9,11] The resulting hypertrophy helps compensate for excessive loading through an increase in wall thickness and a reduction in wall stress per unit area of muscle.[12-14] Although it is thought to be an adaptive response, the development of left ventricular hypertrophy in man is one of the best criteria for predicting the occurrence of cardiovascular disease.[15]

To understand the adaptive process and the limitations of cardiac hypertrophy, which is best described in terms of an increase in the size of pre-existing myocytes beyond that normally seen for a given age, weight and sex,[16] we must first attempt to understand what options cardiac cells have available to adapt to external stimuli, such as mechanical overload. Fetal and, to some extent, neonatal cardiocytes

Molecular Biology of Cardiac Development and Growth, by Paul J.R. Barton, Kenneth R. Boheler, Nigel J. Brand, Penny S. Thomas. © 1995 R.G. Landes Company.

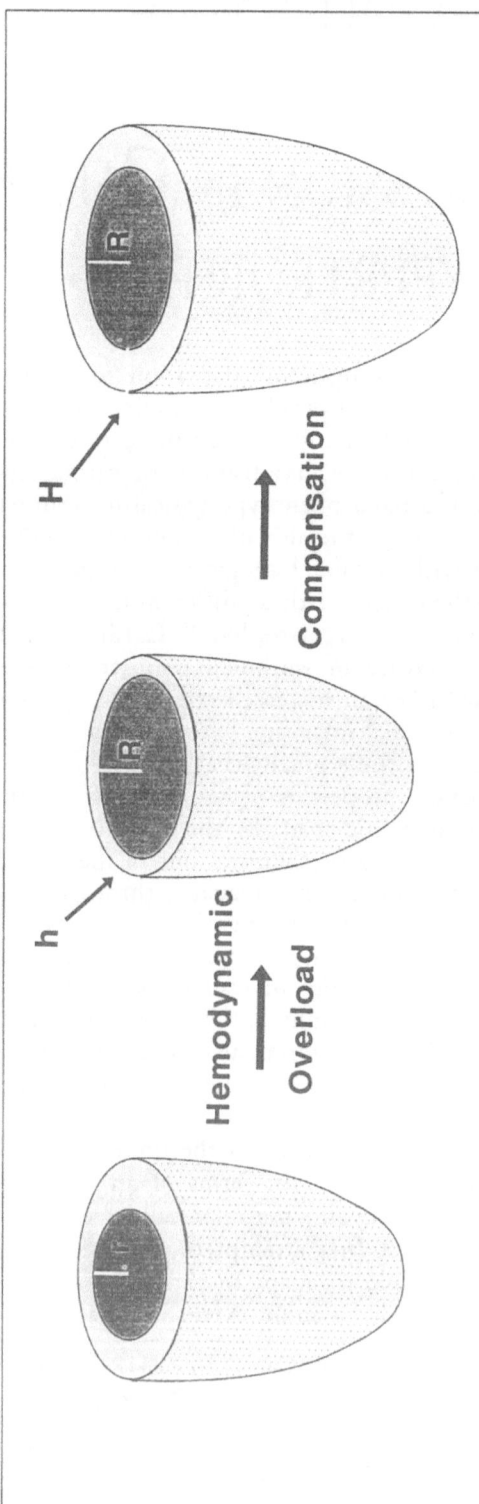

Fig. 5.1. In response to a hemodynamic overload, the increase in ventricular volume results in an increase in the radius of the cavity at end-diastole and a decrease in the free-wall thickness. These changes are normally compensated for by the Frank-Starling mechanism; however, with sustained pressure overload, the free-wall thickness will increase through enhanced protein synthesis and the addition of new sarcomeres, until the ventricular wall stress has returned to 'normal'. Molecularly, these changes are brought about by an increase in total RNA expression and the induction of a number of gene products not always seen in the normal myocardium, including a number of molecular markers implicated in the immediate-early response. (Adapted from Boheler In: Yacoub M, Pepper J, eds. Annual of Cardiac Surgery. Philadelphia: Current Science Ltd, 1994:41-47.)

retain the capacity to divide while in the adult, cardiac myocytes do not normally divide;[16-19] therefore, both in response to aging and to hemodynamic stimuli such as hypertension, adult myocardium increases its mass through hypertrophy of a pre-existing and fixed population of myocytes. Non-myocytes, in contrast, can proliferate.[16-19] Their contribution to the increased heart mass, although relatively small by comparison to that of the myocytes, is achieved by both hypertrophy and hyperplasia.[17] Since hypertrophy is not associated with a change in myocyte cell number, except perhaps in the senescent heart,[20] examination of those changes within myocytes that contribute to this adaptation is essential.

In an early attempt to characterize some of the events of hemodynamic-overload induced cardiac hypertrophy, the cardiac growth process had been divided into three phases that were based on morphological and functional criteria.[16,21] During phase I, heart mass increased rapidly to compensate for an increase in wall stress; during phase II, the increase in heart mass attained an apparent equilibrium between the increased afterload and the growth processes responsible for "normalizing" the stress across the ventricular wall;[12] and during phase III, due to a progressive decrease in force generation and dilatation, growth processes were no longer able to compensate for the elevated workload, resulting in a heart that was incapable of meeting its functional demands.[16,21]

In rat models of cardiac hypertrophy, the process of rapid cell growth seen in phase I resulted in an increase in the number of sarcomeres[22] and an increase in the efficiency of contraction.[4,5] A limited number of cellular and biochemical events were thought to be responsible for these changes. For example, an enhanced RNA polymerase activity, detectable as early as one hour after acute pressure overload,[23,24] should and did result in an increase in the rate of total RNA synthesis;[25-27] consequently, the increase in RNA concentration and content[27] accompanied by an increase in protein synthesis[28,29] and a decrease in protein degradation[30] were thought sufficient to explain the increase in cardiac mass. Additionally, the overall increase in RNA was conjectured to be through an increase in the pre-existing mRNA population.[31] Support for this model came from studies showing that inhibition of RNA synthesis by actinomycin D prevented the development of hypertrophy,[32] and that disruption of RNA and protein synthesis ultimately led to cardiac failure.[33] The biochemical data, therefore, suggested that cardiac hypertrophy was primarily the result of increased RNA and protein synthesis; however, the discovery of myosin isoforms, whose expression were regulated with hypertrophic stimuli, called into question some of these early conclusions.[5,34-36]

Since then molecular analyses have shown that pressure and volume overload are accompanied by numerous changes in the expression of individual genes. Included in the molecular adaptation is the rapid induction of a number of genes associated with the immediate-early response[2,8,9,37-42] that involves primarily protooncogene and growth promoting gene products. "Recapitulation of a fetal gene program" involving the expression of genes that are normally present during fetal development, but which are poorly expressed in the neonate or adult,[8,9,35,43,44] and the altered rate of synthesis of some cellular components, resulting in phenotypic changes and modification of the contractile activities of the heart, have also been described.

The details of the exact changes in the composition of the heart gained from the molecular analyses and the regulation of these fundamental processes are now becoming available due to recent advances in molecular biology and their application to cardiovascular research. These discoveries have enhanced our basic understanding of the complex problems of cardiovascular physiology and medicine and are beginning to have a direct impact on the development of strategies for the treatment of cardiovascular disease. The aim of this chapter is to describe some of the techniques and some of the results that have led to our current understanding of the mechanisms involved in the development and maintenance of cardiac hypertrophy. We will emphasize results at a molecular, cellular and functional level, most of which have been best described in animal or cellular models, but also to some extent in humans. After an initial overview, we will describe in greater detail how changes in gene expression for the proteins responsible for contraction and relaxation play an essential role in altering the function of the hypertrophied myocardium and how application of molecular biological techniques to the study of heart disease should enable us to elucidate some of the underlying mechanisms responsible for altered gene expression and function in normal and hypertrophied myocardium.

CARDIAC MOLECULAR BIOLOGY

How do molecular biologists study cardiac hypertrophy? Cardiac hypertrophy almost always occurs following an increase in workload. In most instances, an increase in pressure or volume loading of the left ventricle is also associated with an increase in release of neurotransmitters or in plasma concentrations of hormones that may have direct effects on cardiac myocyte growth, including activation of the sympathetic nervous system and/or development of renal hypertension.[7] A fundamental question related to this growth process has, therefore, been "Is cardiac hypertrophy the result of mechanical and/or hormonal contributions?" To dissociate between these contributing factors, experiments must be designed to test specific hypotheses that can be performed in vitro and/or in vivo. Classically, the aim of the experiments performed by the cardiac physiologist has been to describe the functions of the heart, including the physical and chemical factors and processes involved in regulating these functions, while the role of the molecular biologist has been to describe the chemical and physical composition and activities of the molecules making up the component systems. For the cardiac molecular biologist, it is necessary to combine both roles; consequently, it is necessary to study simpler systems where one or several variables may no longer be present in the model system. The results then need to be used to make testable predictions about the effects that any one molecule or component of the simpler system has in the presence of a more complete model system, and finally in vivo.

One experimental approach has been to study in vivo the expression of molecular markers associated with the development of cardiac hypertrophy in vivo. The assumption behind this experimental approach is that examination of the mechanisms that regulate the expression of genes whose expression in vivo is altered with cardiac hypertrophy may lead to the identification of molecular switches or molecules that regulate the cardiac phenotype. A stable cardiac muscle cell line would be ideal for such a study since it would enable the effects of different perturbations to be studied over long periods of time. As yet, there are only a limited number of cardiac-like cell lines or cell clones available, many of which have lost some of the phenotypic characteristics typical of cardiac cells.[45-50] For this reason, primary cardiocytes prepared from freshly isolated tissue have become the model system of choice. The increasing availability of well characterized cardiac genes and of cDNAs is also proving tremendously informative, particularly following their introduction into primary neonatal or adult cardiocytes in culture through the molecular techniques of transfection, microinjection, and infection. Finally, with the advent of transgenics, it is now possible to alter a specific gene in an animal model and examine the functional consequences in a physiological setting.

GENETIC MARKERS FROM IN VIVO STUDIES ON CARDIAC HYPERTROPHY

INDUCTION OF A FETAL-TYPE GENE PROGRAM?

Myosin heavy chains (MHC) in mammalian hearts encoded by two separate genes exist as two isoforms, α-MHC and β-MHC.[51] Transcripts for α-MHC are almost exclusively expressed in cardiac muscle[52] while those for β-MHC are found abundantly in both cardiac and slow twitch skeletal muscle. The relative abundance of these isoforms in tissues depends on the mammalian species and its developmental stage. The differential expression of these isoforms can play an important role in the dynamics of contraction and adaptation in cardiac muscle,[5,34,36,53-63] and their expression in cardiac tissue is regulated developmentally,[64-66] hormonally[64,65,67-69] and with hemodynamic loading.[35,70-75] The gene products of the *β-MHC* gene, mRNA transcripts and proteins, predominate in fetal and neonatal rat heart.[64] With cardiac pressure overload, the *β-MHC* gene is activated and the *α-MHC* gene is repressed,[11,35,76] leading to the idea that cardiomyocytes may adapt to pressure overload by synthesis of 'fetal proteins'. The potential for an increase in β-MHC, however, depends upon the initial phenotype: it is high in rat ventricles and in human atria which normally contain about 90% α-MHC and is small in human ventricles which contain about 90% β-MHC.[9] Among species then, the abundance of β-MHC gene transcripts in ventricles of adult animals varies considerably.

In mammalian hearts, the sarcomeric α-actins consist of two isoforms,[77-82] cardiac α-actin and skeletal α-actin, encoded by different genes.[79,81] In heart, the expression of the two α-actin isoforms are differentially regulated in response to developmental[77,83-85] hormonal[86,87] and hemodynamic stimuli.[38,44,75] In rat heart, both isoforms are expressed during fetal and neonatal development although in adult mammalian hearts, cardiac α-actin predominates.[83,84,88] An exception to this pattern of expression is in adult human ventricles where both cardiac and skeletal α-actins are abundantly expressed.[88,89] Skeletal α-actin expression can be rapidly and/or transiently increased in models of rodent ventricular hypertrophy following either constriction of the aorta (a model for pathological hypertrophy) or injection of thyroid hormone (a model for physiological hypertrophy).[38,44,86] The differences between physiologic and pathologic forms of cardiac hypertrophy are based primarily on functional data and have been described elsewhere.[12,16,90] Skeletal α-actin transcripts, like those for β-MHC, are expressed abundantly in normal fetal but not adult rodent heart, and since these transcripts are induced in the adult rodent heart following pressure overload, the idea of a 'fetal program' of gene expression induced with hemodynamic overload was further developed.[38,44]

The exact mechanisms leading to induction of β-MHC and skeletal α-actin expression are only partially understood, but the physiological consequences are hypothesized to be beneficial. Support for this hypothesis has come from experimental data correlating myosin heavy chain gene expression with changes in the maximum velocity of contraction, heat production, and the speed of contraction of the myofilaments (see Functional Consequences of Cardiac Hypertrophy).[5,62,63] Correlative data linking sarcomeric α-actin gene expression with alterations in the contractile parameters of the myofilaments has generally been lacking,[91,92] although recent results suggest a possible physiological role for altered sarcomeric α-actin gene expression in cardiac contractility.[93] It should be emphasized however, that ventricular hypertrophy in man or other large mammals is not associated with large increases in the amounts of either the β-MHC or skeletal α-actin isoforms as these isoforms are already constitutively expressed. The induction of a 'fetal type gene program' with cardiac hypertrophy, although a useful term in initially describing the changes in gene expression associated with hemodynamic overload in rat, is not a ubiquitous feature of cardiac hypertrophy. As such and from an embryological point of view, expression of these genetic markers in cardiac hypertrophy do not, in fact, represent the re-induction of a fetal-type gene program.

Other useful markers induced prior to or during hypertrophy have been sought that could be used ubiquitously in the study of mammalian cardiac hypertrophy. One of the best examples of an increase in the levels of a marker mRNA following pressure overload is the ventricular expression of atrial natriuretic factor (ANF).[94-100] During embryonic development, the *ANF* gene is expressed in both the ventricles and the atria, but shortly after birth its expression is largely limited to atria.[101,102] With cardiac hypertrophy, this gene product is re-expressed in the ventricles of mice, rats and humans, implying that its induction is a generalized response to pressure overload.[8,103-107] In fact, among species and between models, its expression now serves as one of the best markers for the development of cardiac hypertrophy.[8,9] While the exact pathways that lead to its induction remain to be clearly defined, its regional specificity and increased expression associated with cardiac hypertrophy can be easily assessed by molecular means e.g., Northern blot, RNase protection assays, PCR, etc. (see chapter 1). As the ANF peptide plays an important role in regulating blood pressure and natriuresis, its presence may contribute to a reduction in cardiac preload and afterload, and so functionally it may also serve as a useful adaptation to the hypertrophic stimuli.

The availability of these molecular markers and others (*c-fos, c-jun, c-myc,* Egr-1) has proven very informative in the study of the mechanisms responsible for the changes in gene expression associated with cardiac hypertrophy. Much of the data have been generated through analyses of their expression in tissue culture models, the subject of the next section of this chapter.

I. TECHNIQUES FOR STUDYING CARDIAC HYPERTROPHY IN VIVO

TISSUE CULTURE SYSTEMS

Primary cultures of both neonatal and adult cardiocytes from a variety of species and man have been developed over the past decade. One of the most commonly used culture systems is prepared from neonatal rat heart (Fig. 5.2).[108-112] Although these cells can not be passaged, they can be maintained in culture for at least two weeks and experimentally manipulated through molecular means, e.g., transfections, injections, etc.[10,113] For the most part, neonatal cardiocytes do not undergo cytokinesis.[10,114] Only fetal heart cells isolated from early stages seem to retain the capacity to divide.[10] A small percentage of the neonatal cardiocytes in culture do, however, synthesize DNA and become multinucleated[114] a typical characteristic of adult cardiocytes.[115,116] During the preparation of the neonatal rat cardiocytes, the cells go through a number of profound morphological changes, including round-

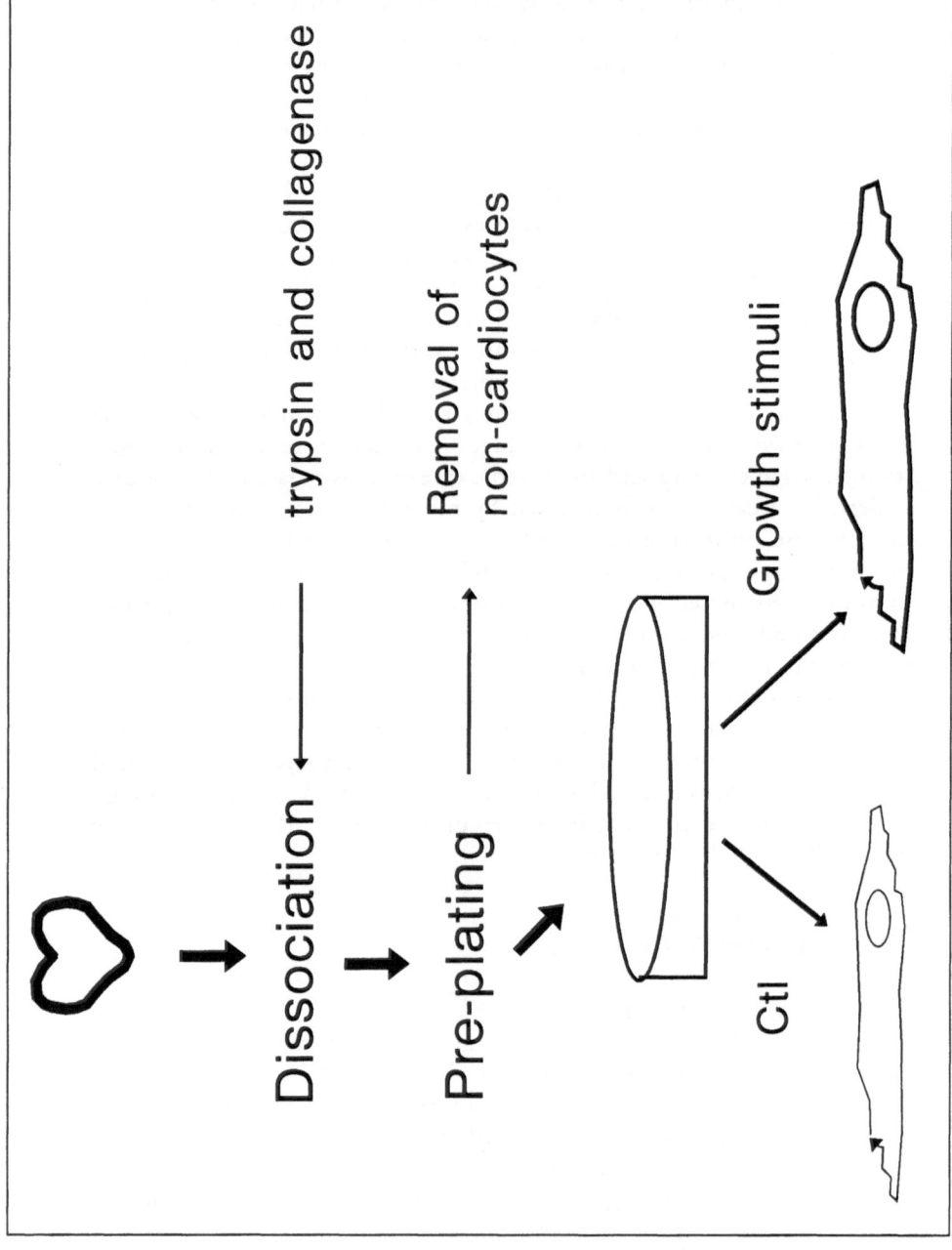

Fig. 5.2. Schematic diagram showing some of the major steps in the isolation and exploitation of an isolated cardiocyte preparation. In such a preparation, individual cells are isolated from whole heart through enzymatic digestion with trypsin and/or collagenase. Cardiomyocytes are then selected for by preplating to allow non-cardiocytes to adhere to the petri dish. The enriched cardiocytes are then plated and allowed to attach over a period of about 18-24 hours. At this time, the cardiocytes can be manipulated through the addition of a number of pharmacological agents which may cause the cells to grow. By taking advantage of the cells ability to "hypertrophy" under these conditions, it has been possible to exploit this system to analyze some of the pathways that may be involved in the hypertrophic response.

ing up and degeneration of sarcomeric cross striations.[108,117] Once the cells, plated at low densities, have attached to the tissue culture plates, they flatten, appear quiescent, and for the most part, do not re-initiate beating. When plated in the presence of serum, attached cells begin to grow (hypertrophy) and spread. Depending on the density of cell plating, the cells may form cell-cell contacts and re-initiate sarcomerogenesis and contractile activity. Growth factors in serum seem to be important for this process, because if the cells are plated in a defined medium lacking growth factors, the cells remain viable but often do not regain either their typical pattern of striation or contract.[108] Exploitation of the characteristic hypertrophic response seen in cardiocytes after addition of either growth factors or serum has led to an increasing body of data concerning cardiac growth (hypertrophy) in vivo.

TRANSCRIPTIONAL OR POST-TRANSCRIPTIONAL REGULATION

The expression of an individual protein might be regulated at any one of a number of steps during its synthesis, i.e., at the transcriptional, post-transcriptional or pretranslational, translational or post-translational stages.[16] How might the regulatory mechanisms be delineated? An ideal model would be a cell-free system consisting of known components that is capable of normal gene expression and that can be examined in clearly defined ways. Establishing such a system to distinguish between the regulatory stages is not easy, although the use of cell nuclei or cell culture systems is common.

An essential prerequisite for studying the mechanism of specific cardiac gene transcription is the establishment of an in vivo transcription system. It is possible to look indirectly at transcriptional regulation of cardiac genes, using essentially a DNA-dependent system with putative regulatory DNA promoter regions linked to reporter genes (Fig. 5.3A), such as chloramphenicol acetyltransferase (CAT) or luciferase, as templates to assay transcriptional activity (see chapter 1). Such an experiment is designed to determine which DNA sequences and protein factors may be required for "accurate" expression of cloned genes after transfection (introduction of foreign DNA into a cell). From these sorts of analyses, both *cis*-acting regions (regions of DNA important for transcription) and *trans*-activating factors (proteins that bind to regions of DNA to activate or inhibit DNA transcription) have been described, including positive and negative DNA elements and protein factors required for accurate expression of a cloned gene. Such studies have allowed a classification of conserved DNA regions necessary for gene expression both ubiquitously and/or in a tissue or species specific manner (see chapters 1 and 4). In the cardiac system, they have also been useful in showing how a number of mechanical or trophic stimuli can affect the expression of markers of cardiac hypertrophy, including those for β-MHC, skeletal α-actin, *c-fos* and others (see below).

An alternative to transfection into whole cells is isolation of intact, functioning nuclei and subsequent continuation of transcription

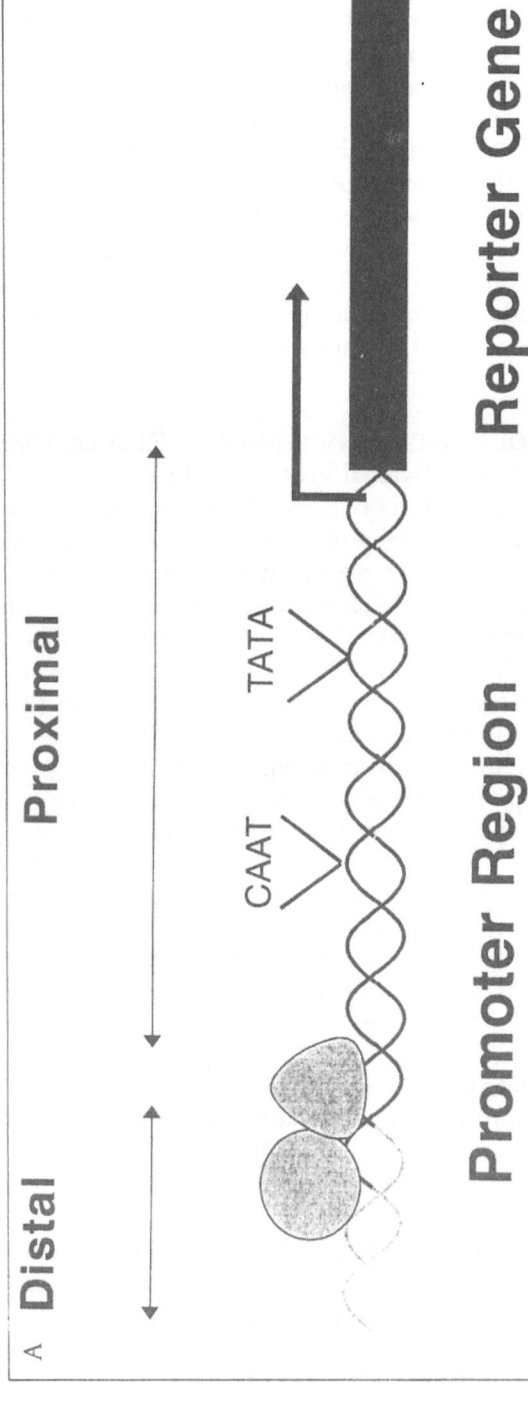

Fig. 5.3A. Schematic diagram showing two methods for analyzing transcriptional regulation of genes in vivo. (A) The use of promoter constructs linked to reporter genes allows determination of potential cis-elements which may be important in regulating the expression of individual genes, including some involved in tissue-specificity and activity (positive or negative regulation) of transcription. Use of such a system is also extremely useful in determining what trans-activating factors may be responsible for transcriptional regulation of an individual gene. As with any in vivo system, however, the results from these analyses are not always identical to the in vivo situation. An alternative method has therefore been used. (Continued on facing page.)

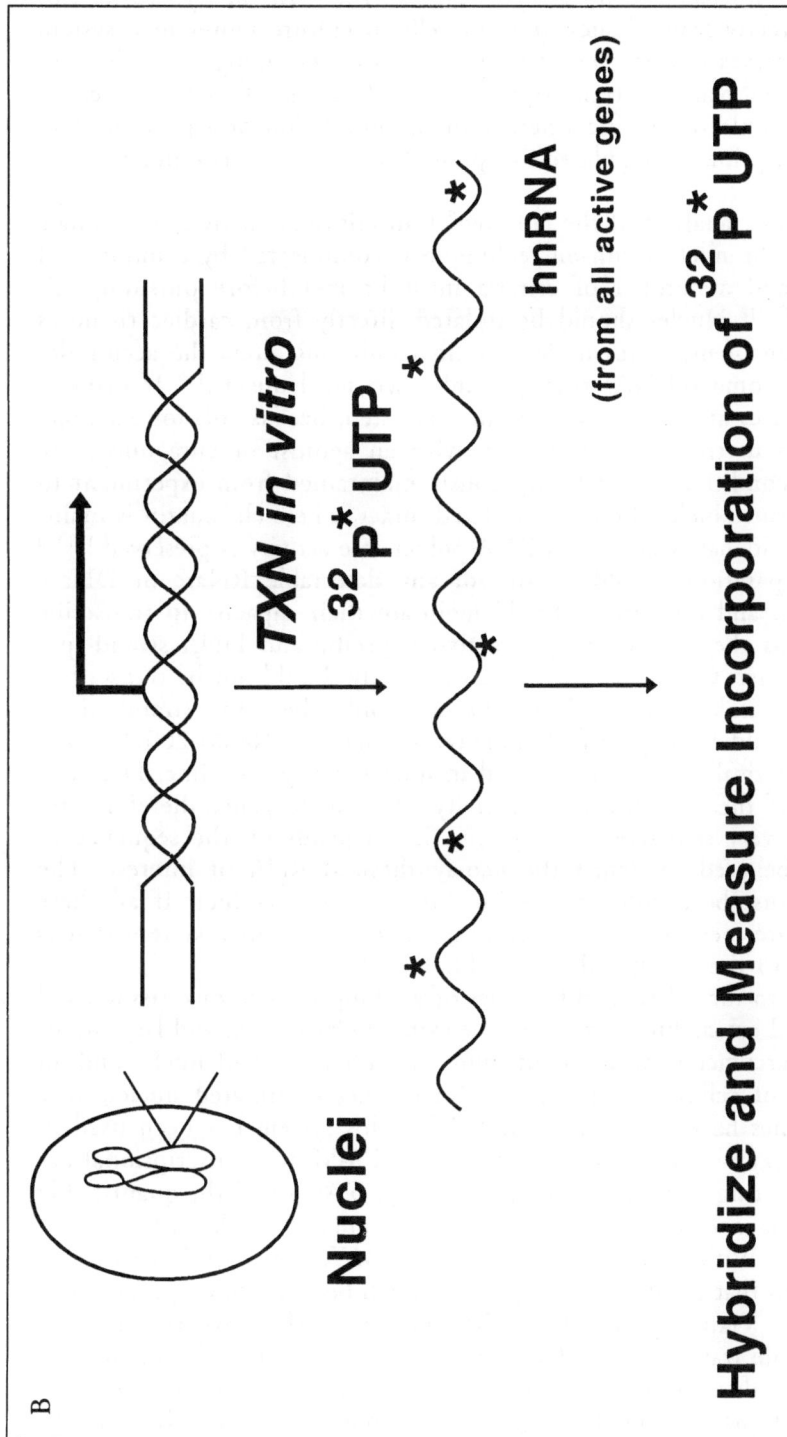

Fig. 5.3B. (Continued) In this system, nuclei are isolated directly from whole tissue and incubated in vivo in the presence of a radioactive label like ^{32}P-UTP which should only be incorporated into RNA. Incorporation of the radiolabel into endogenous transcripts through continuation of DNA transcription (TXN) permits examination of the activity of an individual gene, and as such, should reflect the transcriptional activity of that gene in vivo.

in vivo ("nuclear run-on assay") (Fig. 5.3B).[118] In theory, nuclei isolated directly from tissues or from cells in culture represent a system that preserves nuclear activities in vivo and uses endogenous chromatin as a template for transcription. The advantage of such a system is that it should determine whether the accumulation of a gene product is transcriptionally regulated as opposed to post-transcriptionally regulated.

Indirect analysis of the "in vivo" transcriptional activity of a single gene by the nuclear run-on technique is complicated by a number of factors and a number of criteria must be met before pursuing this approach.[118] Nuclei should be isolated directly from cardiac tissue as preplating of myocytes under certain conditions alters the accumulations of some mRNAs through unknown mechanisms.[119] If primary cell cultures are used as an intermediate step, because of, for example, the need to treat the cardiocytes with an agonist or antagonist, the culture conditions must be rigorously maintained from experiment to experiment. Nuclei should be isolated intact, where chromatin is maintained in its native state, and RNA polymerase activity is preserved.[118,120] The preparation should be free of any detectable RNase or DNase activities, and the nuclei should maintain their capacity to transcribe in vivo as they do in vivo, in a tissue-specific and DNA strand-specific manner. Cardiac specific genes normally should not be transcribed in nuclei isolated from liver cells, and only the sense strand of an RNA should be transcribed. Finally, as many as 10000-20000 genes are theoretically being transcribed in nuclei at any one time. Determination of the transcriptional activity of a single gene, therefore, requires a very sensitive assay system that depends on the sequence of the probe used to detect the neo-synthesized RNA of interest. The probe must be absolutely specific for the gene product. If all these criteria are met, the run-on assays probably represent a system that is as close to the intact cell as possible.[118,120-122]

Due to the inherent difficulties of working with cardiac tissue (well developed cytoarchitecture, extensive extracellular matrix, and large number of sarcomeres) which contribute to a poor yield of nuclei and an increase of cellular debris in the final pellet of isolated nuclei, very few studies have been reported.[120,123-126] This system has been used in rat to assay sarcomeric α-actin and α- and β-MHC transcriptional activities during neonatal development (Fig. 5.4). At birth, quantifiable levels of *α-* and *β-MHC* gene transcription are detectable with the run-on assays. With postnatal development, *α-MHC* transcription increases relative to that for *β-MHC*—a pattern which becomes more pronounced with age. Transcription of *α-MHC* increases well above that seen at birth while transcription of *β-MHC* decreases to barely detectable levels 20-25 days after birth. In contrast, the levels of accumulation of *β-MHC* transcripts predominate at birth but are replaced by *α-MHC* transcripts over a period of several weeks.[64] The accumulations of the

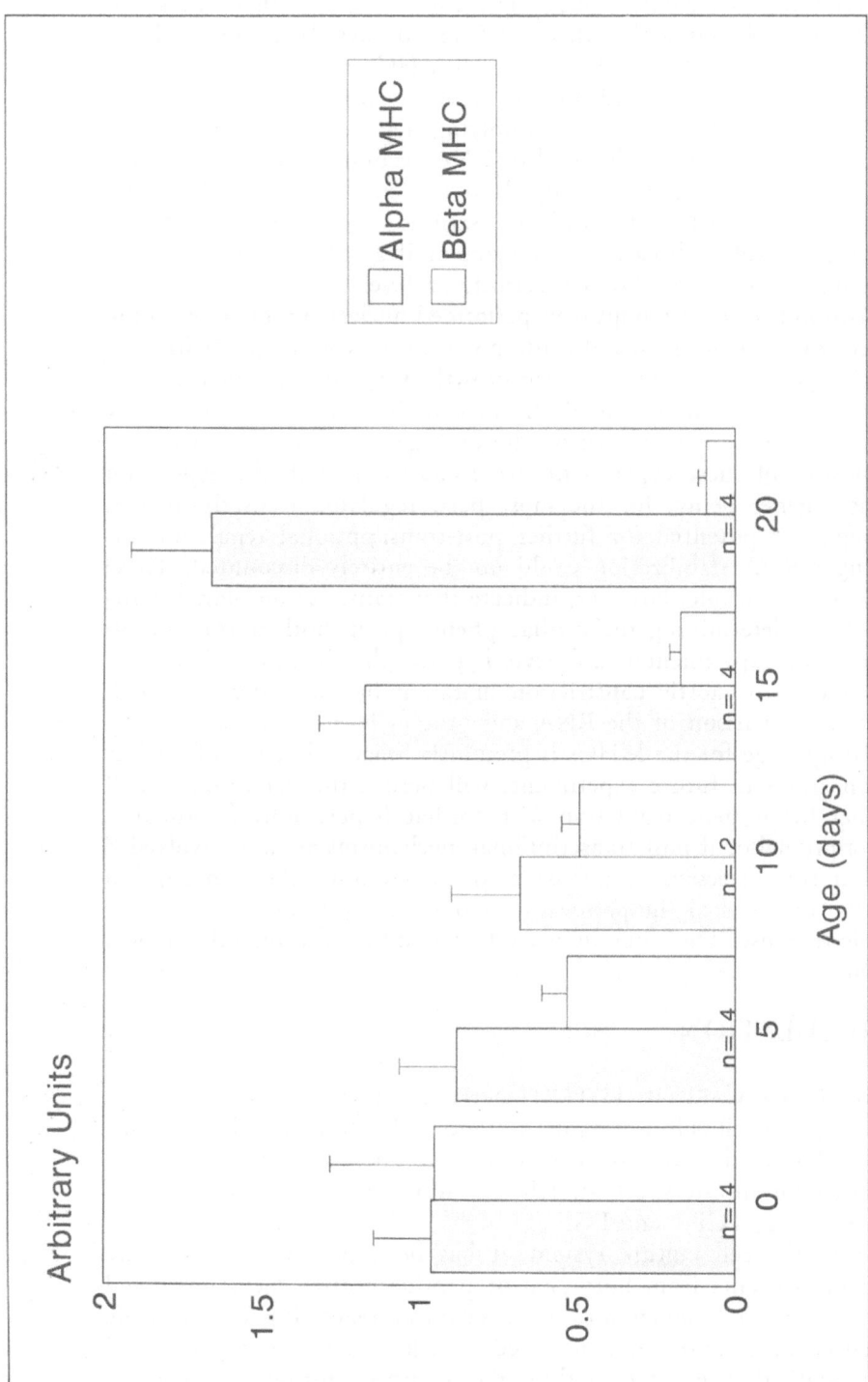

Fig. 5.4. Results from nuclear run-on assays designed to examine the expression of α- and β-MHC gene activity during post-natal development. The data indicate that β-MHC gene activity decreases after birth, and that for the most part, α-MHC gene activity increases. As the relative accumulation of the mRNAs from these genes follows the transcriptional activity of these genes, it can be concluded that α- and β-MHC gene expression is regulated primarily through transcriptional mechanisms.

transcripts for the MHCs thus mimic the changes in their transcriptional activities, but with a delay of 5 to 10 days. In 24 day old rat, the mRNA accumulations for α- and β-MHC as well as cardiac and skeletal α-actins correspond directly to the relative transcriptional activity for each respective gene,[120] implying that a steady state situation has been attained where the level of RNA accumulation is maintained by a basal transcriptional activity. It is not clear, however, if the decreased levels of sustained *β-MHC* transcription seen between 5-25 days after birth involve all cardiac myocytes or if a subpopulation of cells maintain this relatively low transcriptional level.

Two independent groups have performed nuclear run-on assays using nuclei isolated from neonatal cardiomyocytes in vivo treated with adrenergic agonists to induce cardiac growth. One focused on the transcription and accumulation of the myosin light chain 2 single copy gene[127] and the other focused on the multigene sarcomeric α-actins.[128] The results of these experiments were consistent with the expression of these genes being, for the most part, regulated transcriptionally, although the potential for further post-transcriptional regulation including mRNA stabilization could not be entirely discounted. These results on the whole, however, indicate that transcription plays a critical role in determining the cardiac phenotype in both normal development and experimental cardiocyte hypertrophy. In cardiac hypertrophy induced by aortic constriction in rat, it has been demonstrated, through comparison of the RNA and protein levels, that the primary regulatory stage for the MHCs is pretranslational.[35] The use of nuclear run-on assays in future experiments will permit the determination of whether MHC gene regulation with cardiac hypertrophy is primarily transcriptional or if post-transcriptional mechanisms are also involved.[52] Nuclear run-on assays cannot be used to determine the transcription factor(s) involved or the pathways responsible for inducing the hypertrophic response. These questions are better addressed using other model systems.

II. REGULATION

INDUCERS OF CARDIAC HYPERTROPHY IN VIVO

Using the cell culture system, a number of factors have been shown to stimulate cardiac cell growth in culture. These include neurotransmitters, phorbol esters, cell stretch, and numerous peptide growth factors, including TGF-β and FGF.[2,7,8,37,39,109,117,129,130-137] Through manipulation of the cell culture system, it has been possible to dissociate mechanical events from hormonal or peptide growth factor-mediated events. A general model for the mechanism regulating cardiac gene expression involves receptor-mediated cascades. Such receptor-mediated events work through at least three mechanisms. Hormones, when introduced to a cell, may bind to a specific receptor located either on the cell surface or in the cytoplasm. Once bound, this hormone receptor-complex may be translocated into the cytoplasm to initiate a response, or to the nucleus where the interaction of the hormone-receptor complex with DNA-binding sites has the potential to modulate gene expression. For example, cardiac growth induced by thyroid hormone leads to an increase in α-MHC gene expression with a concomitant decrease in β-MHC gene expression.[52] This action is mediated through binding of thyroid hormone to its receptor, translocation to the nucleus, and binding to thyroid hormone responsive elements located in the promoter regions of both α-MHC and β-MHC.[138-141] In the case of α-MHC, binding acts to stimulate transcription from this gene; whereas, for β-MHC, binding acts to repress transcription (see chapter 4).

The second general mechanism whereby receptor-mediated events can modulate gene expression is through coupling proteins (Fig. 5.5), like guanine nucleotide-binding protein-linked receptors (G-protein).[136,142-149] These receptors are transmembrane proteins whose amino acid sequences are usually characterized by seven hydrophobic segments containing distinctive sequence patterns.[150] The receptors from this large family of transmembrane proteins are thought to be involved in transmission of extracellular signals to the cytoplasm, i.e., ligand binding to the extracellular side of the receptor leads to activation and binding to a coupling protein, that in turn, may modify the amounts or activities of second messengers, e.g., cAMP, cGMP, inositol phosphates, etc., or may modify the properties of ion channels. The second messengers or change in ion flux then may act through other proteins to initiate a "cascade" capable of amplifying the initiating signal. The classic example, in liver and muscle cells, involves the break-down of intracellular stores of glycogen upon exposure to epinephrine. Epinephrine, by direct coupling of its receptor with a G-protein, induces adenylate cyclase activity and increases the levels of cAMP, which in turn, activates cAMP-dependent protein kinase (PKA). Once activated, this kinase acts to phosphorylate the catalytic subunit of glycogen phosphorylase, and the net result is an increase in glycogen catabolism.

The third mechanism involves catalytic receptors that are capable of functioning as an enzyme. These receptors are almost always transmembrane proteins, that when activated by their ligand, function as tyrosine-specific protein kinases.[151-153] When activated, the tyrosine-specific protein kinases act to phosphorylate tyrosine residues of selected proteins in the cell, frequently including tyrosine residues on the receptor itself. Phosphorylation of an enzyme may either activate or inhibit the activity of a protein. Included in this family of receptors are those for insulin and a number of growth factors, including insulin-like growth factors I and II, acidic and basic fibroblast growth factors.[144] The classic example is synthesis of glycogen in the liver. In this case, insulin binds to its receptor, triggering what is believed to be phosphoryla-

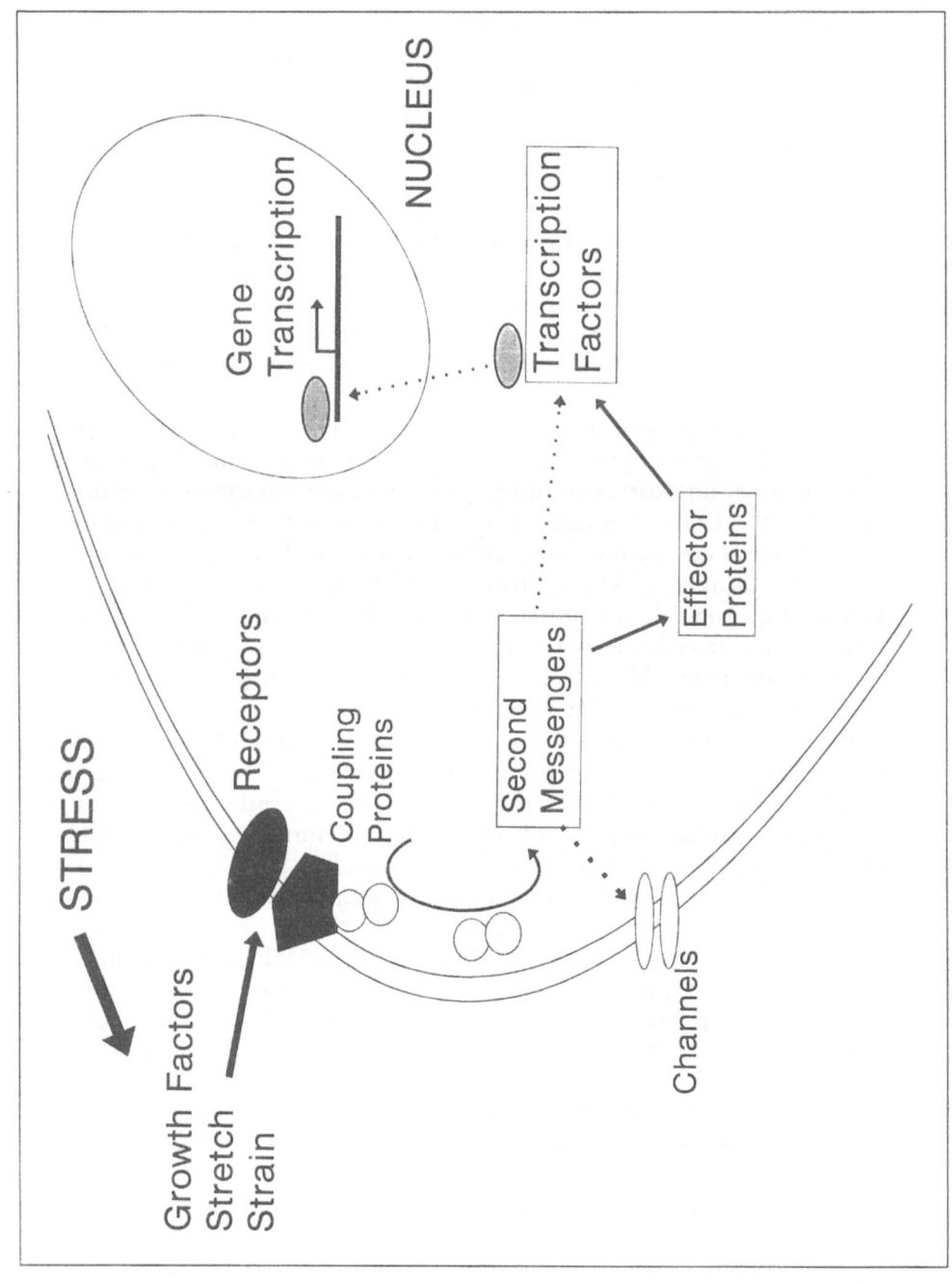

Fig. 5.5. Schematic diagram showing how signals from the cell surface like trophic factors, adrenergic agonists, and/or stretch may be transduced to the nucleus to increase gene expression. The idea behind such a system is that coupling proteins and/or second and third messengers would be needed to transmit cell surface signals through the cytoplasm to the nucleus.

tion of a protein phosphatase, which in turn, dephosphorylates glycogen synthase. The net result is an increase in glycogen anabolism, that is achieved through enhanced glycogen synthase activity. In heart, all three receptor-mediated mechanisms have been identified.

ADRENERGIC AGONISTS

α-Adrenergic receptors[154] are present in the heart of various mammalian species including man.[155-162] Adrenoceptor (α- and β-) stimulation is capable of eliciting a number of physiological responses, but the mechanisms whereby the adrenoceptor subtypes mediate their response requires different sets of receptors and coupling proteins. In human, they are thought to be coupled via a number of G-proteins to activate numerous intracellular responses. α-Adrenoceptor stimulation leads to positive inotropic effects that are not accompanied by increases in the second messenger concentrations of cAMP or cGMP. In fact, cAMP levels may decrease due to increased phosphodiesterase activity.[163] The degree of stimulation, however, is very minor relative to those invoked by β-adrenoceptor stimulation. From data obtained with nonhydrolyzable analogs of GTP on agonist binding, it does appear that the α1-adrenergic receptor subtype is coupled via a pertussis toxin-sensitive guanine nucleotide binding protein to mediate intracellular events through activation of phospholipase C and protein kinase C.[143,158,162,164]

β-Adrenoceptors have been shown to exist as at least two isoforms in the heart of several mammalian species,[165] and there is considerable evidence that these proteins are regulated by a wide variety of drugs, hormones, and physiological conditions.[166-171] β1- and β2-adrenoceptors (AR) are structurally distinct molecules[172] that are coupled via the stimulatory G-protein (Gs) to the adenylate cyclase and to the dihydropyridine sensitive L-type channel in cardiomyocytes,[173] both of which are intimately involved in maintaining calcium homeostasis to regulate beat-to-beat control of contractility.[174] In human ventricle, increased contractility is usually thought to be mediated by the β1-adrenoceptors (β1AR),[175] although β2AR can also have important effects.[176] In human heart failure, β1AR are markedly down-regulated with a 60% loss of receptor number in the ventricle,[177,178] whereas β2AR numbers appear unchanged. The net result is a shift of β1AR to β2AR ratio from 80:20 in normals to about 60:40 in human heart failure. β2AR numbers may remain relatively constant but their response may be decreased by 30% as a consequence of uncoupling to adenylate cyclase.[177,179] Recent data suggest that an increase in the β-adrenergic receptor kinase is associated with β-adrenoceptor uncoupling.[180]

The study of the regulation of contractile protein gene expression by adrenergic agonists in vivo has provided many clues towards the mechanisms of how cardiac gene expression is regulated with hypertrophy.[119,128,181] The in vivo work has primarily depended upon the use of a cell culture system first described by Simpson,[108,109] where cells are plated at a density of 100-150 cells/mm^2 in the presence of bromodeoxyuridine, a substance that inhibits cell proliferation. At this plating density, the cardiomyocytes are, for the most part, quiescent, isolated, and unable to form cell-cell contacts, but addition of the non-specific catecholamine agonist, norepinephrine (NE), produces myocyte enlargement without hyperplasia via an α1-adrenoceptor-mediated pathway. The contractile activity of these cells is mediated through a β-adrenoceptor-mediated pathway.[110]

Many of the studies reported in the literature are based on this system, although frequently with modifications.[131,182,183] Using these culture conditions, α1-adrenergic agonist-stimulated hypertrophy has been shown to increase total transcription by 2- to 3-fold and selectively upregulate transcription of the genes for myosin light-chain 2 gene,[127] skeletal α-actin[119] and β-MHC.[181] In the case of MHC gene expression, α1-adrenergic agonists selectively increased the cellular content of β-MHC mRNAs and proteins during hypertrophy; however, no changes in the levels of α-MHC mRNAs or proteins could be demonstrated. These data and others suggest that α1-adrenergic and not β-adrenergic receptors are somehow involved in the pretranslational regulation of β-MHC expression.

Sarcomeric α-actin expressions have proven particularly informative as a model system for gene activation.[119,129,135,184] Using the conditions of the tissue culture system just described, skeletal α-actin gene expression is activated through an α1-adrenoceptor-mediated pathway, independent of β-adrenoceptor-mediated events. However, results from the laboratory of N. Bishopric have clearly demonstrated a role for β-adrenergic agonists in the regulation of the contractile proteins in cultures of neonatal cardiocytes plated at densities greater than those first described.[117,131] Under these plating conditions, the cardiomyocytes formed extensive intercellular contacts and contracted synchronously at >300 beats/min, similar to the normal beating rate of the intact rat heart. In these experiments, where the promoter regions of the human skeletal α-actin gene were linked to the CAT reporter vector, increased CAT activity was coupled to calcium channel activation and compartmentalized calcium release. Furthermore, sarcoplasmic reticulum calcium release was found to be essential and sufficient for the activation of skeletal α-actin gene promoter constructs[131] under high density culture conditions. It is difficult to know if these results are a function of the reporter gene constructs used or representative of actual changes in endogenous skeletal α-actin expression, even if indicative of a potential molecular pathway. The apparent plasticity of the cells to regulate skeletal α-actin expression by adrenergic agonists in low density versus high density cultures may indicate potential complexity in the interpretation of how gene expression in cultured neonatal cardiocytes is regulated and cautions against an oversimplistic interpretation of experiments using specific gene products as markers.

Can endogenous skeletal α-actin expression be regulated by both α- and β-adrenergic agonists? By modifying the cell culture system first

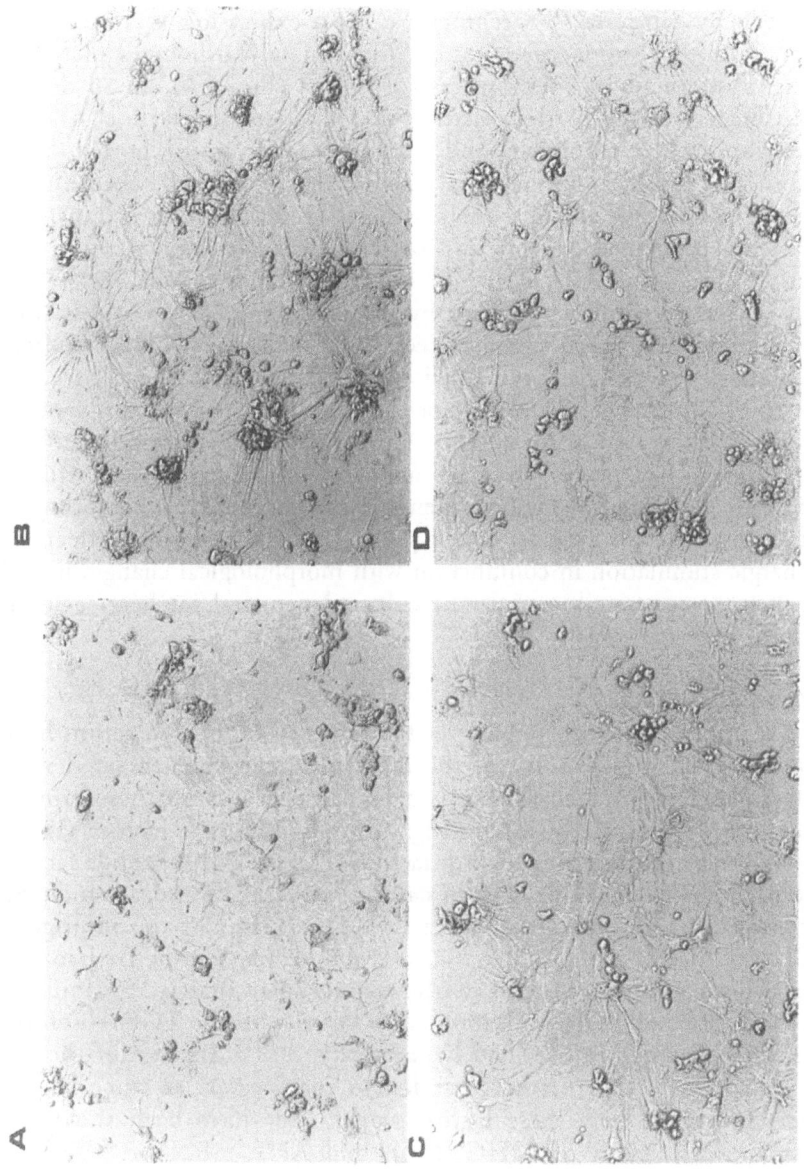

Fig. 5.6. Photographs of primary cultures of neonatal cardiocytes in culture that have been treated with (A) vehicle, (B) norepinephrine (NE), (C) phenylephrine (PE), or (D) isoproterenol (Iso) for 70 hours after incubation for 18 to 24 hours in serum-containing medium to allow cell attachment. These cells were not washed after the initial plating so that rounded up dead or dying cells could be seen. Before preplating, all cardiocytes have lost their characteristic striated pattern and are rounded up. The viability or percentage of living cells in the preparation must be determined by trypan blue exclusion so that cells can be plated at a pre-determined density. Following plating, viable cells will attach, elongate and form striations that can be readily dectected through immunofluorescent techniques with antibodies specific for proteins of the contractile apparatus. With the addition of adrenergic agonists, the viable cells undergo a number of morphological changes, cellular hypertrophy (NE and PE) and cytoplasmic extensions (NE, PE, Iso), and physiological changes, increased speed and frequency of spontaneous contractions.

described by Simpson,[109] sarcomeric α-actin expression was examined after addition of adrenergic agonists to neonatal cardiocytes plated at densities intermediate to those just described (Fig. 5.6). In so doing, it has been possible to demonstrate that endogenous skeletal α-actin expression can be regulated, under some conditions, by both α- and β-adrenergic agonists in neonatal cardiocytes plated at subconfluent densities (Fig. 5.7 - middle panel), suggesting that there is a transition in adrenergic-mediated regulation between low and high density cultures. If this is the case, it would imply that endogenous skeletal α-actin gene expression is regulated by the combined effects of cell contact and morphological changes seen in association with adrenergic stimulation. Since it has been previously demonstrated that contractile activity of cultured neonatal cardiocytes modulates the protein synthesis and turnover rate of total actin[185] and since skeletal α-actin mRNA expression is regulated by both α- and/or β-adrenergic agonist stimulation in cells plated at medium densities, it is reasonable to speculate that skeletal α-actin expression is regulated by the combined effects of adrenergic stimulation in conjunction with morphological changes linked to enhanced contractile activity of cultured neonatal rat heart cells in culture.

Trophic Factors

A number of growth peptide factors normally found in serum have been shown to induce some of the molecular markers associated with cardiac hypertrophy in cardiac myocytes. Two groups of these growth factors seem particularly interesting: fibroblast growth factors (FGF) and the β-type transforming growth factors (TGFβ). These peptide factors have been shown to be present in cardiac tissues[186,187] and are induced in some forms of myocardial disease.[134] Two forms of the multigene family for FGFs, acidic and basic, are abundantly synthesized in the myocardium and accumulate in the extracellular matrix.[188] Similarly several isoforms of TGFβ are expressed in heart, with the TGFβ1 isoform induced with pressure overload hypertrophy.[2,137,189] Analysis of the expression of molecular markers has led to the greatest advance in our understanding of how these factors work. They have been shown to affect the expression of MHC, actin and ANF mRNAs.[2,43,135,137,190] Sarcomeric α-actin expression is a good example. When added to neonatal cardiocytes in culture, basic FGF and TGFβ induce the expression of skeletal α-actin mRNA with little or no change in cardiac α-actin mRNA expression.[135,190] Acidic FGF, in contrast, leads to a decrease in the levels of mRNAs encoding both skeletal and cardiac α-actins. The response of sarcomeric α-actins to these growth factors differs fundamentally from that seen in striated muscle, because in skeletal muscle their expression is characterized by suppression.[135,190] These results demonstrate that acidic and basic FGF have a different effect on

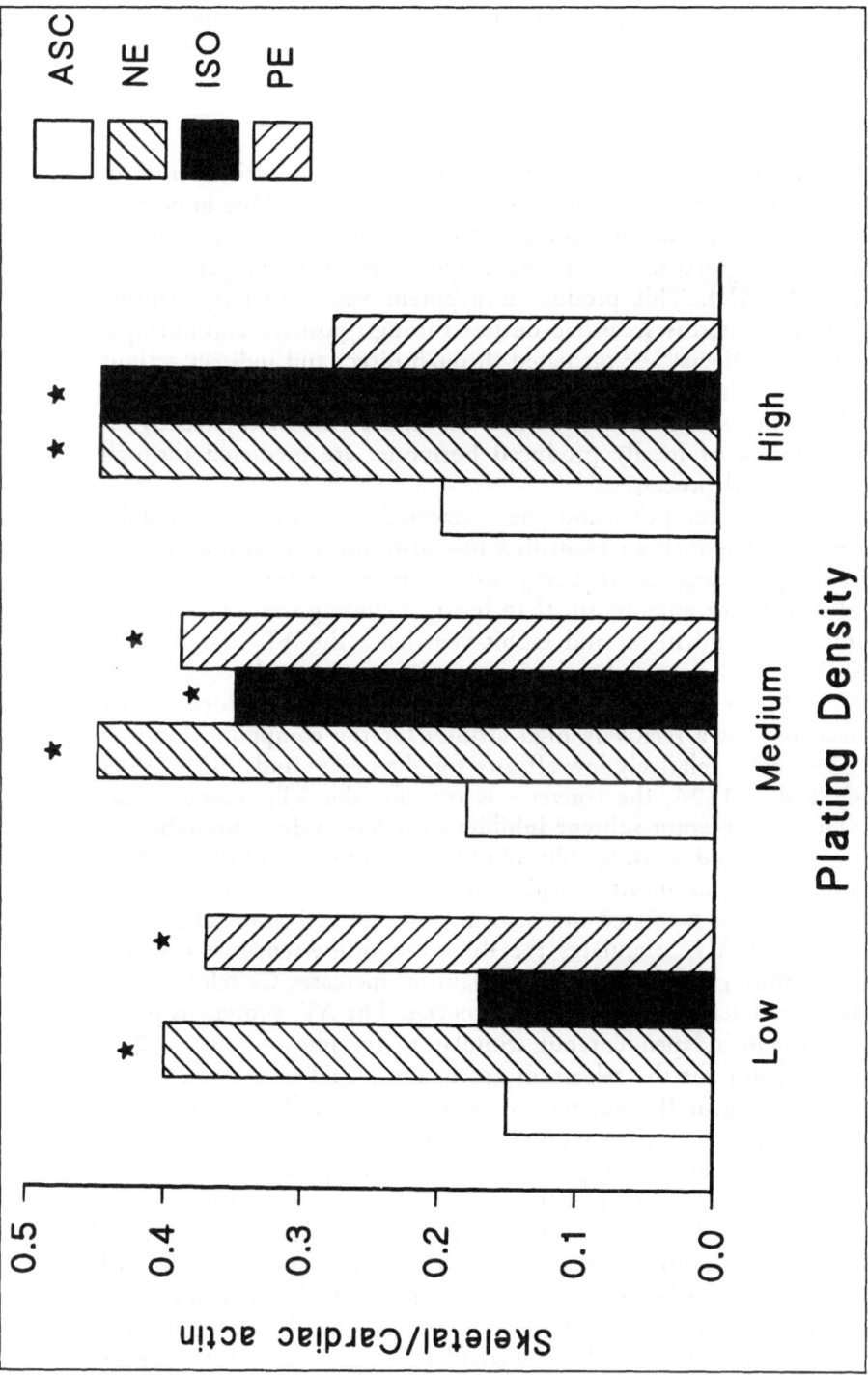

Fig. 5.7. Analysis of skeletal α-actin to cardiac α-actin mRNA ratios in cultures of primary neonatal cardiocytes plated at different densities. The induction of skeletal α-actin depends on the plating density: at low densities (100-150 cells/mm2), its expression is 'regulated primarily by α1-adrenergic agonists, at medium densities (400-500 cells/mm²), by α1- and β-adrenergic agonists, and at high densities (1000-1400 cells/mm²), apparently by β-adrenergic agonists. Data are adapted from Bishopric et al. J Clin Invest 1987;80:1194-1199, Bishopric and Kedes. Proc Natl Acad Sci USA 1991;88:2132-2136 and unpublished).

the regulation of skeletal α-actin gene expression, in skeletal versus cardiac muscle. This implies that the factors or signal transduction mechanism(s) that ultimately regulate the effects of these growth factors differ between muscle types.

ANGIOTENSIN II

The renin-angiotensin system in mammals plays a critical role in the regulation of cardiovascular and renal homeostasis. One important component of this system, produced by the proteolytic actions of renin and the angiotensin-converting enzyme (ACE), is the peptide angiotensin II (AII). This product is a potent vasoconstrictor peptide hormone that increases cardiac output through positive chronotropic and inotropic effects[191,192] mediated through direct and indirect actions on heart.[7,193] The indirect effects involve both pressure changes and central nervous system actions, while the direct effects, elicited by either circulating or locally produced hormone, are mediated through receptor-coupled processes.[194]

Distinct AII-receptors and the intracellular responses elicited by these receptors have been identified primarily through ligand-binding studies using nonpeptide AII-antagonists.[193] Two receptor-subtypes have been described for angiotensin II in heart, including the AT_1 and AT_2 receptors,[195-197] and in rat and rabbit ventricles, the AT_1 and AT_2 receptors have been reported to be present in approximately equal numbers.[195,198] In rat heart, there are two types of AT_1 receptors.[199] The AT_1 receptor has a relatively high affinity for the compound losartan (DuP753) and a relatively low affinity for the compounds PD 123319 and CGP 4222112A; the converse is true for the AT_2 receptor subtype. The AT_1 receptor-subtype inhibits adenylate cyclase through coupling with Gi, and activates phospholipase C which hydrolyzes phospholipids to release inositol triphosphates and diacylglycerol (DAG). The AT_1 receptor-mediated mechanism, through activation of protein kinase C by DAG, stimulates the dihydropyridine-sensitive Ca currents, and through an undefined mechanism, increases Ca release from sarcoplasmic reticulum in quiescent myocytes. The AT_2-subtype is probably G protein-independent but stimulatory for phospholipase A2, a known mediator for the release of arachadonic acid as a second messenger. Angiotensin II-mediated events in cardiac cells, therefore, act through at least three transducing/coupling systems.[193,194]

Recent evidence has suggested that autocrine, paracrine and circulating biologically active mediators initiate events that result in the development of CH.[194,200-203] Of particular interest has been the role of AII and ACE inhibition on the development and regression of CH.[204-208] AII is capable of increasing protein synthesis and heart mass. In the hypertrophied left ventricle of spontaneously hypertensive rats AII content correlates with left ventricular weight,[209] and AII receptor density increases 2-fold as cardiac hypertrophy develops.[210] Thus, rats when administered AII, even at subpressor doses, develop cardiac hypertrophy. Addition of the relatively specific angiotensin type I (AT_1) receptor blocker, losartan, inhibits this response, suggesting that cardiac growth mediated by AII is independent of pressure changes. For example, treatment of rats that have had their abdominal aortas constricted, with angiotensin converting enzyme inhibitors (ACE), prevents the development of cardiac hypertrophy.[206,207] Blockade of angiotensin-mediated responses has also been described as one of the best mechanisms available to induce regression of cardiac hypertrophy[211] From these experiments and others, AII has been proposed as a mediator of cardiac hypertrophy.

MECHANICAL STRESS AS AN INDUCER OF CARDIAC GROWTH IN VIVO

How can a mechanical stress be perceived by the heart or myocyte and how does this act as a stimulus to cardiac growth? It has been well documented that hemodynamic overload after adrenoceptor blockade or sympathectomy in vivo leads to cardiac hypertrophy and that increased load leads to an increase in protein synthesis.[212,213] Many systems have been used to address this question in vivo, including cultured myocytes isolated from rat neonatal and feline adult hearts, isolated papillary muscles, skinned fibers and beating or arrested isolated perfused hearts subjected to various pressure or volume loading conditions.[13,214-216] For example, an increase in protein synthesis of a Langendorf heart preparation correlates with stretching (increase in the aortic perfusion pressure) of the ventricular wall.[213] Myocardial stretch also increases inositol monophosphate, biphosphate and triphosphate content in isolated perfused rat hearts[217] probably through activation of phospholipase C, and it leads to small increases in the levels of 1,2-diacylglycerol, a known activator of PKC. In the in vivo system, the mechanisms responsible for the growth response remain allusive, again necessitating the use of a simpler system to examine this problem.

The best model system to study load-dependent cardiocyte growth has come from work with cultured neonatal rat cardiocytes that have been plated on a deformable plate which permits stretch of the cells in vivo. Stretching of cultured cardiomyocytes in vivo, where there are no neural or humoral factors, stimulates protein synthesis, activates specific gene expression, and induces myocyte hypertrophy.[39,129,216] Specifically, stretch-activated cardiac myocyte hypertrophy causes an increase in the immediate-early gene molecular markers *c-fos*, *c-jun*, c-*myc* and Egr-1 and in the late hypertrophy markers skeletal α-actin and ANF.[218] In fact, the use of this isolated neonatal cardiocyte system has made it possible to show that passive stretch alone is able to activate multiple second messenger systems in cardiocytes, including phospholipases C, D, and A2, tyrosine kinases, Ras, mitogen-activated protein kinases, and protein kinase C, all of which might be involved in regulating gene expression[200,219] These data suggest that this in vivo model is suitable for the study of the molecular mechanisms associated with load-induced cardiac hypertrophy.

The use of *c-fos* transcripts as a molecular marker for hypertrophy has been instrumental in delineating some of the pathways involved in hypertrophy associated modifications of gene expression.[37] *c-Fos* is rapidly induced following stretch. Cardiocytes subjected to stretch and transfected with a CAT construct containing promoter regions of the *c-fos* gene produce an increase in CAT activity, demonstrating that the *c-fos* promoter is activated in this model system. Support for transcriptional activation of its promoter comes from nuclear run-on assays using nuclei isolated from the stretched cells. In these experiments, *c-fos* DNA transcription increases.[39] Subsequently, one sequence in the promoter region that contained a serum-responsive element (SRE) was found to be necessary and sufficient for transcription in this isolated system.[129] Mutation analysis of the promoter region indicated that the SRE-binding proteins ($p67^{SRF}$ and $p62^{TCF}$) were important in *c-fos* induction by mechanical stretch[200,219,220] and involved DNA elements present within 356 base pairs of the 5'-flanking region. Subsequent pharmacological studies have supported these data and enabled development of a model for a "phosphorylation cascade" necessary for the transduction of signals from the outside of the cell to the nucleus. These studies have shown that increased *c-fos* gene expression following stretch is not inhibited by streptomycin and gadolinium (stretch-activated ion channel blockers), propranolol (β-adrenergic antagonist), and sterazosin and yohimbine (α1- and α2-adrenergic antagonists, respectively).[37,221] The mechanism of stretch-stimulated changes in gene expression, therefore, does not obviously involve any of the adrenergic receptors commonly found in myocardium. Inhibitors of protein kinase activity (staurosporine) and, more specifically, down-regulation of protein kinase C (PKC), effectively block stretch-activated *c-fos* upregulation; however, activation of PKC with phorbol esters up-regulated *c-fos* gene expression. These data suggest that PKC-mediated events may be essential for stretch-mediated activation of gene expression (see below, protein kinase C).

Although very informative, these studies failed initially to identify the activator molecule or mechanism that might be responsible for the stretch-mediated events. As already described, AII and a number of components of the renin-angiotensin system may be implicated in the development and regression of cardiac hypertrophy, and therefore, AII was suggested as an important mediator of the hypertrophic response. To test the possibility that AII mediates some of the growth associated responses seen with stretch, Sadoshima and Izumo examined the stretch-induced hypertrophy response of neonatal cardiocytes in culture with relation to angiotensin production and release.[37,200,218,220-223] A number of important findings came from these studies. It was demonstrated that cardiocytes do, in fact, contain secretory granules that are immunoreactive to AII.[223] AII addition to neonatal rat cardiocytes leads to an increase in myocyte protein synthesis in the absence of any change in DNA synthesis; non-myocytes, in contrast, increased their protein synthesis, were mitotically active, and underwent cell division. The hypertrophic effects of AII, like stretch, were shown to lead to the rapid induction of the immediate-early gene products *c-fos, c-jun, junB*, Egr-1 and *c-myc*, and the "late" molecular markers for cardiac hypertrophy, skeletal α-actin and ANF.[211,218,222] This response was blocked by addition of AT_1 receptor antagonists, but not AT_2 antagonists; therefore, the hypertrophic effects were primarily mediated through the actions of the AT_1 receptor.[195] Finally, as the phenotypic molecular changes of cardiac cells following AII administration were very similar to those seen with pressure overload induced hypertrophy in vivo, it was suggested that the endogenous renin-angiotensin system in cardiocytes must play a critical role in the hypertrophic response associated with stretch.

From these and other experiments performed in simple model system, we can conclude that once a receptor has been activated, signals from the cell surface can be transduced to the nucleus. The initiating stimuli certainly include adrenergic agonists, angiotensin II, trophic growth factors, and stretch. This list of stimuli, however, should not be considered comprehensive, as other stimuli have been[185,224-227] and will certainly be described in the future. With the vast array of known initiating stimuli, it is logical to assume that a plethora of coupling and effector proteins will also be involved in the transduction of signals from the cell surface to the nucleus—some of which have already been alluded to.

COUPLING PROTEINS

In the heart, hormone and neurotransmitter signals are propagated by heterotrimer G-proteins, monomeric G-proteins, and non-G-protein-linked receptors, which are in turn, coupled to many effectors e.g., adenylate cyclase, phospholipases, ion channels, protein kinases and phosphatases (Fig. 5.8). In this section, how signals from the cell surface can be transmitted to the nucleus will be described. The focus will be on (1) heterotrimeric G-protein-mediated events and, to a lesser extent, on (2) monomeric G-protein-mediated events. Establishing the existence of individual components of a signal transduction pathway is relatively easy to demonstrate in an isolated well-defined model system, but the experiments often do not reveal the complexity of any single pathway, particularly since signal transduction pathways are subject to multiple levels of regulation i.e., feedback regulation or cross-talk among various pathways.[144,228]

HETEROTRIMER G-PROTEINS

G-proteins are guanine nucleotide-binding proteins that were first discovered in the late 1970s.[229,230] They are composed of a family of

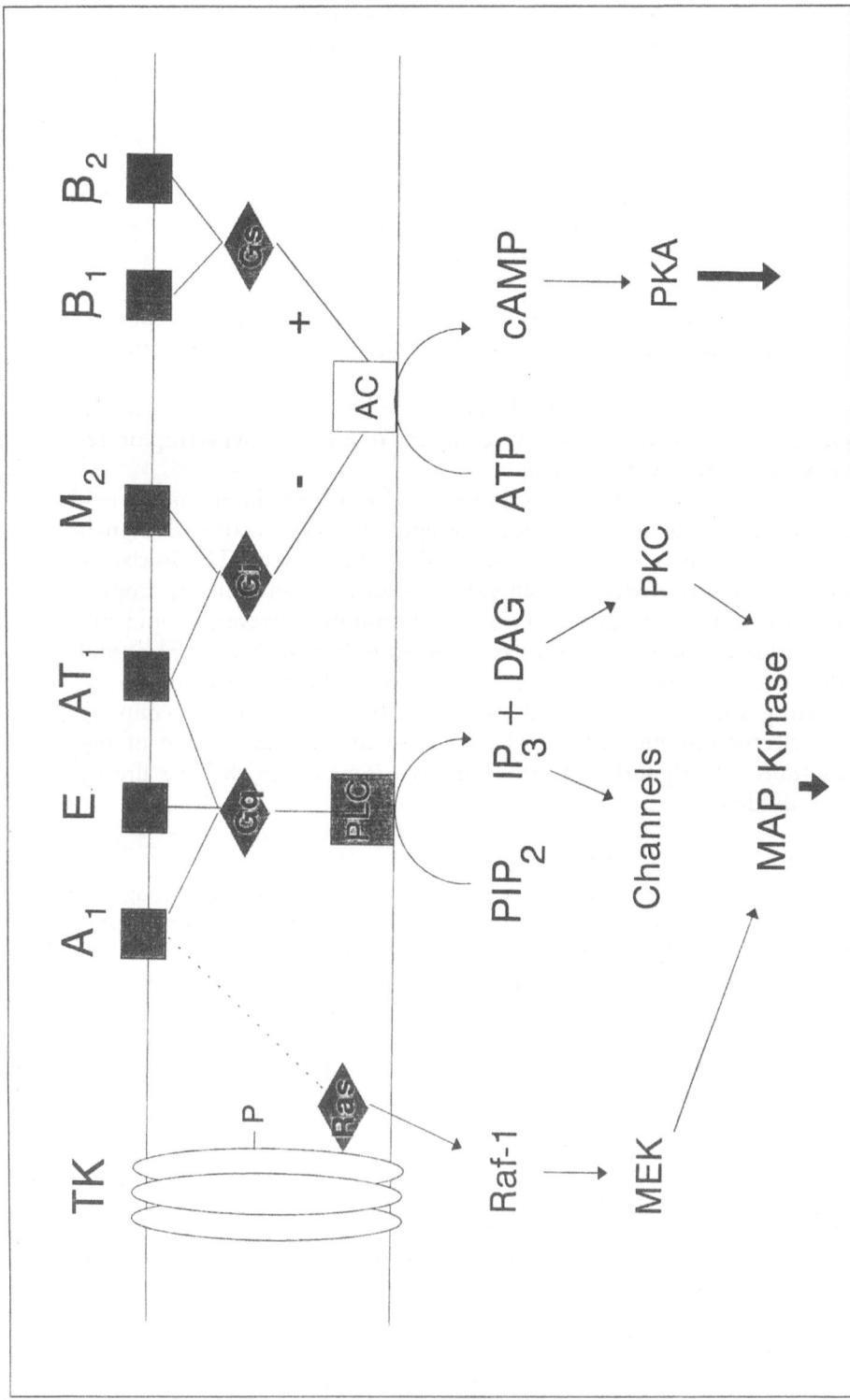

Fig. 5.8. Schematic diagram showing some potential G-protein (Gi, Gs, Gq, Ras) linked coupling mechanisms for signal transduction from cell-surface membranes to coupling proteins in the cytoplasm. (TK, tyrosine kinase-associated receptors; A, α1-adrenergic; E, endothelin; AT_1, angiotensin receptor type-1; M2, muscarinic receptors; β1 and β2, β-adrenergic receptortypes 1 and 2; PLC, phospholipase C; AC, adenylate cyclase.

highly homologous intracellular proteins that couple a large family of transmembrane receptors (>100), characterized by seven hydrophobic segments containing distinctive sequence patterns, with protein effector enzymes.[146-148] Ligand binding to the receptor elicits a signal generating process that activates the receptor to bind to an intracellular heterotrimeric G-protein, composed of α, β and γ subunits.[150,231] Interaction of activated transmembrane receptors with G-proteins promotes the exchange of GDP, bound to the α subunit, for GTP and subsequent dissociation of the α subunit from the βγ heterodimer. The α-GTP complex and the βγ heterodimer are then free to initiate a second messenger system of intracellular signals or couple with different ion channels. Termination of the signal occurs when the GTP bound to the α-subunit is hydrolyzed to GDP, permitting the reassociation of the heterotrimeric G-protein complex. The time needed to terminate the signal depends on the intrinsic GTPase activity found on the α-subunit.

Effector proteins include enzymes that generate second messengers or regulatory molecules. Some effectors include phosphodiesterases, phospholipase C, phospholipase A2, adenylate cyclase, and ion channels.[136,231] In heart, G-protein linked receptors are involved in mediating inotropic and chronotropic responses by both direct and indirect coupling to ion channels, including those for Ca^{2+} and K^+ ions.[145] GTP-binding proteins also play an important role in signal transduction of other cytoplasmic events, including transcriptional regulation.[232,233] At least four sets of G proteins are found in heart: (a) receptors coupled to cholera toxin-sensitive G proteins (Gs); (b) receptors coupled to pertussus toxin-sensitive G proteins (Go and Gi); (c) receptors coupled to phospholipase C activation (Gq and others); and (d) receptors coupled to tyrosine kinase activity.[144,145,231,234] The stimulatory G protein (Gs) and the inhibitory G protein (Gi) which are coupled with adenylate cyclase, and the Gq proteins that stimulate PLC signal transduction are prominently represented in mammalian atrial and ventricular myocardium.[235] These coupling enzymes are responsible for mediating a wide variety of physiological and biochemical responses.

Activation of Gs occurs predominantly through β-adrenoceptor mediated events and leads to an increase in adenylate cyclase activity and the production of cAMP.[136,145,236] cAMP in turn activates protein kinase A (PKA), an enzyme capable of phosphorylating a host of proteins involved in regulating the contractility of the heart. One of the best described events mediated by PKA is phosphorylation of phospholamban at Serine 16.[237,238] Phospholamban is an inhibitor of the sarcoplasmic reticulum CaATPase (SERCA), the enzyme with the primary responsibility of removing Ca^{2+} from the cytoplasm during end-systole.[237,239-241] In the dephosphorylated state, phospholamban inhibits SERCA activity and prolongs the time that Ca^{2+} is present in the cytoplasm. Upon phosphorylation, phospholamban is no longer able to interact with SERCA, enabling a greater and a more rapid uptake of cytosolic Ca^{2+} into the sarcoplasmic reticulum (SR); consequently, subsequent Ca^{2+}-activated Ca^{2+} release from the SR is greater. This β-adrenergic receptor-mediated change in uptake and release by the SR results both in a shorter time of contraction and a stronger force of contraction in the next cardiac cycle.[237]

Other substrates for PKA include protein kinases, phosphatases, receptors, G-proteins, the L-type voltage sensitive Ca^{2+} channels, transcription factors, DNA-binding proteins and protooncogenes.[144] Increases in cAMP[7,242,243] and PKA activity are important in accelerating protein synthesis and ribosome formation after an increase in rat heart perfusion pressure, and the production of cAMP has been implicated in the induction of cardiac hypertrophy.[244-246] β-adrenergic agonists, in addition to the effects on contractility and protein synthesis mediated by G-proteins, are capable of modifying the expression and function of G-proteins in heart.[125,179,247] This regulation can best be described as a feedback loop. When β-adrenergic receptors are chronically stimulated, the receptors themselves are down-regulated through internalization, receptor synthesis is decreased, and to some extent, effector proteins are 'uncoupled' from the receptors.[177,248-250] Uncoupling occurs via a number of mechanisms, some of which involve phosphorylation events mediated by PKA, PKC, the β-adrenergic receptor kinase (βARK)[249] and by tyrosine kinase.[250] Concomitant with chronic activation and uncoupling of the β-adrenergic receptor from the adenylate cyclase, the expression and function of inhibitory G proteins are increased.[247,251,252] Gi stimulation, in turn, acts to antagonize the effects of Gs on adenylate cyclase and decrease further the levels of cAMP. The increase in Gi has been demonstrated in cell culture and animal models systems, and in human, the failing myocardium has a diminished capacity to produce cAMP in response to adrenergic stimulation. This change, in human, is once again accompanied by an apparent increased functional activity of Giα and a decreased functional activity of Gsα subunits.[143,252-254] The mechanism responsible for the increase in Gi, in animal models, appears to depend on cAMP and its activation by PKA, since the increase in Gi is observed experimentally with both β-adrenergic agonists and the addition of forskolin, an agent that activates adenylate cyclase directly.[143]

As chronic β-adrenergic stimulation increases Gi protein content without any demonstrable change in Gs content, it was hypothesized that the increase was due to an increase in the gene expression of Gi. The effects of isoproterenol (β-adrenergic specific agonist) infusions in the rat were therefore analyzed.[125,255] From total RNA and from isolated nuclei, it was possible to determine the relative levels of expression of Gs and Gi, and the transcriptional activity of these genes, respectively. The results were then compared to determine if the expressions of Gs and Gi were regulated at a transcriptional level. The data from these experiments showed that infusions of isoproterenol for 96 hours increased the concentrations of the adenylate cyclase inhibitory sub-unit Giα-2 transcripts without affecting the concentrations of the stimu-latory subunit Gsα. Results from the nuclear run-on assays were more difficult to interpret. They indicated a biphasic regulation of Giα-2 gene transcriptional activity. Forty-eight hours after initiation of isoproterenol infusions, Giα-2 transcription decreased by 37%, but after another 48 hours of infusion, the transcriptional activity had increased by 45% relative to controls. Gsα transcription was unchanged. From these results, it was speculated that the molecular mechanisms respon-sible for the difference in activation might reside in the promoter re-gions of the two genes, where Giα-2, unlike Gsα, was found to con-tain a *cis*-element in its promoter region that could be activated by the nuclear transcription factor commonly known as the cAMP re-sponsive element-binding protein (CREB).[256,257] By analogy with the positive/negative regulation of *c-jun* by CREB in a non-cardiac model system,[233] it was postulated that the biphasic response in Giα-2 tran-scriptional activity could be due wholly to CREB binding to the pro-moter sequence of this gene. In the case of *c-jun* regulation, CREB acted as a repressor of transcription when dephosphorylated and as an enhancer of transcription when phosphorylated.[233] Although enticing, proof of CREB-mediated biphasic regulation of Giα-2 is lacking.

Protein Kinase C

A role for a G-protein subunit in coupling α1-adrenergic receptor-mediated hypertrophy[99,127,258] has been reported.[136,144] The responsible G-protein probably involves Gq. Numerous studies have shown that GTP and its nonhydrolyzable analogs can stimulate phosphatidylinositol turnover,[259] and these results suggest that a guanine nucleotide bind-ing protein is actually involved in coupling receptors, like α-adrenoceptors, to PLC activation. In fact, activation of Gq proteins, which are coupled with phospholipase C (PLC), has been shown to be particularly im-portant in the development of cardiac hypertrophy through activation of PKC.[136,144,155,260-262] Hydrolysis of phosphatidylinositols into DAG and inositol phosphates is accomplished by a number of PLC isoenzymes, of which, there are at least 9 different isoforms.[263,264] The products of this hydrolysis are the second messengers diacylglycerol (DAG) and inositol mono, di, and triphosphates (IP3).[225,265] In vivo, DAG is thought to be the major activator of the multifunctional Ser-Thr protein ki-nase, protein kinase C (PKC), while inositol triphosphate has been shown to promote Ca^{2+} release after binding to IP3 sensitive receptors that are found in endoplasmic/sarcoplasmic reticulum.[266] DAG and IP3 are thus major physiological regulators of intracellular events.

PKC has been implicated in the action of numerous hormones and growth factors in various tissues. There are at least 8 related en-zymes of PKC that can be divided into two sub-families. The first subfamily is dependent on the presence of Ca^{2+} for activity and is composed of α, β1, β2, and γ isotypes.[267-269] The second subfamily lacks the Ca^{2+}-binding domain of the first group, is independent of Ca^{2+} for activation and is comprised of the δ, ε, ξ and η isotypes.[267,270-272] This second subfamily may be important in muscle, since members of this family would be able to regulate cellular events independently of the Ca^{2+} oscillations associated with excitation-contraction (E-C) coupling.[272] It is now known, from studies using antisera specific for the PKC isotypes, that PKC-ε, -ξ, and -η are present in adult rat ventricle and that the -ε isotype is predominant. PKC-α, -β1, -γ and -δ were not detected in adult rat ventricles.[273] In neonatal cardiocytes, however, PKC-ε, -α, -β and a non-PKC-α/non-PKC-β isotype of 70 kDa are detectable, suggesting that during maturation, transitions in the PKC isoforms occur.[274]

In the model system of isolated rat cardiocytes in culture, α1-adrenergic agonist and endothelin stimulation lead to increases in inositol phosphate hydrolysis and to myocyte hypertrophy. Several lines of evidence suggest that PKC might be involved in the effects of α1-adrenergic agonist-mediated cardiac hypertrophy, including α1-adrenergic agonist activated translocation of PKC from the cytoplasm to the nucleus[260,275] and increased PKC immunoreactivity[274] in cardiac myocytes. Additionally, α1-adrenergic stimulation of β-MHC gene expression is thought to be associated with PKC-mediated events. To determine if PKC isoforms might be involved in the gene changes associated with the hypertrophic process, co-transfections of neonatal cardiocytes with DNA containing β-MHC promoter regions linked to a reporter gene (as marker) and constitutively active forms of PKC-α or PKC-β were performed.[275] The results generated in this model system demonstrated that transcription of the β-MHC gene promoter is preferentially stimulated by β-PKC in cardiocytes co-transfected with these plasmid constructs. These results support the hypothesis that α1-adrenergic stimulation couples with Gq to activate phospholipase C and hydrolyze inositol phosphates, which in turn, activate PKC. Activation of β-PKC thus may lead to phosphorylation of enzymes and/or transcription factors that mediate some of the transcriptional processes associated with cardiac hypertrophy.

To examine the possibility that β-PKC was directly involved in α1-adrenergic agonist stimulated transcription of β-MHC and not through an indirect effect, it was postulated by Kariya et al that conserved sequences in the DNA promoter regions of the β-MHC gene would have to be responsive to both α1-adrenergic agonists and activated PKC. If a different *cis*-element were involved, then an alternative mechanism would need to be postulated. It was important therefore to identify a *cis*-element in the β-MHC promoter that was responsive to α1-adrenergic stimulation. Through the application of transfection techniques, the authors were able to identify a 20 base pair sequence in the β-MHC promoter (-215/-196) that was required for CAT activity.[276] Next, through co-transfections of β-MHC-CAT constructs and plasmids expressing constitutively active forms of β-PKC, the authors were able to demonstrate that the same *cis*-element activated by α1-adrenergic agonists was stimulated by β-PKC mediated events. Activation of transcription was prevented by either stimulus when the *cis*-element contained a 3-base pair mutation. These data supported the hypothesis that PKC-β is actually involved in α1-adrenergic regulation of gene expression with cardiac hypertrophy (see Nuclear Factors). Perhaps most importantly, the results from these studies identified a M-CAT/ enhancer core motif in the promoter region of the rat β-MHC gene that could be used to identify actual transcription factor(s) implicated in α1-adrenergic agonist-mediated cardiac hypertrophy.[276,277] Finally, since the β-PKC isoform is poorly abundant or absent in adult rat heart,[273] it is reasonable to speculate that an alternative mechanism(s) may function to mediate the changes in β-MHC gene expressions seen with adult rat cardiac hypertrophy e.g., induction of β-PKC expression, an alternative PKC-mediated pathway or a PKC-independent mediated pathway.

SMALL MONOMERIC GUANINE NUCLEOTIDE BINDING PROTEINS—RAS

In many cell systems, a common point of convergence for a number of growth factors is activation of Ras proteins. Ras is a product of three protooncogenes[278] and has been implicated in transformation of a number of cell lines. Ras proteins are associated with cell membranes and are small guanine nucleotide binding proteins with intrinsic GTPase activity. They do not, however, associate with any βγ subunits to form a G-protein complex. Activation of Ras occurs when GDP on inactive Ras is replaced by GTP, and once active, it mediates the effects of many growth factors or differentiation factors, including epidermal growth factor, insulin, and perhaps AII, most of which are either tyrosine kinases themselves, or are associated with tyrosine kinases.[151,279,280]

Ras has been suggested as an important signal transducer in heart[281] between the cytoplasmic membrane and cytoplasmic kinases. Ras does not regulate adenylate cyclase activity in mammalian cells, and in fact, cAMP can in some cell lines inhibit the signal transduction pathway from Ras,[282] suggesting the potential for cross-talk between the two systems. In models of pressure-overload induced rat cardiac hypertrophy, the mRNA levels of Ras increase.[42] Hyperthyroidism, in contrast, leads to a reduction in Ras mRNA levels.[283] In cultured neonatal cardiocytes, stretch increases Ras expression.[200] To examine the potential role of Ras, as a coupling protein, on gene expression in cardiac cell hypertrophy, Thorburn et al[281] have used the approach of direct cytoplasmic injection of activated Ras proteins and dominant interfering mutant Ras proteins into isolated neonatal cardiocytes. In this system, mutant Ras proteins would be expected to inhibit the down-stream effects of endogenous Ras proteins through competition. Likewise, injection of activated Ras i.e., a form of Ras that acts as if it were always in the GTP-bound state, would be expected to trigger intracellular events associated with endogenous Ras activation. Injection of the activated form of Ras into neonatal rat cardiocytes induced the expression of both *c-fos* and ANF, two molecular markers for cardiac hypertrophy. Conversely, expression of mutant variants of Ras inhibit the hypertrophic stimulation of the ANF promoter by phenylephrine, a known activator of the hypertrophic response in neonatal cardiocytes. Ras proteins therefore appear to have a role as a modulator of α1-adrenergic agonist-mediated hypertrophy.

Outside of the cardiovascular system, Ras has been shown to be critical for three signal transducing protein kinases including MAP kinase, Raf-1 and the S6 kinase.[280,284-287] The evidence suggests that the Ser/Thr kinase Raf-1 lies downstream of Ras in the phosphorylation cascade. As an example, the effects of microinjecting Ras neutralizing antibodies into cells can be overcome by expression of the constitutively activated Raf-1 allele.[278] Expression of dominant negative Ras, a mutant form of Ras that has lost its normal function and dominates the effects of any endogenous protein, blocks activation of Raf-1 by Tyr-kinase-mediated signaling events.[288,289] Likewise, antisense-mediated down regulation of Raf-1 blocks mitogenic stimulation of constitutively active Ras.[290] Finally there is direct evidence that Ras and Raf-1 are capable of forming a complex in mammalian cells.[280] Together these data from other experimental systems suggest that Ras directly interacts with Raf-1 in a signal transduction pathway.

G-protein and Ras-coupled pathways have been shown to regulate signals to mitogen activated protein kinases (MAP).[287,291] As such, the MAP kinases may serve as a common point of convergence in the mediation of effector proteins. This cascade system, therefore, should be viewed as a mechanism for amplifying/integrating signals from a diverse variety of receptors and effector proteins. The MAP kinases are ser/thr kinases that require phosphorylation of both the tyr and thr for activation. This is accomplished by the MAP kinase kinase (MEK). MEK, in turn, is regulated through phosphorylation by the product of the *c-raf* protooncogene (Raf-1), which has been shown in some model systems to be physically linked with activated Ras. How might MAP kinases be involved in altered gene expression with cardiac hypertrophy? As an example, MAP kinases can phosphorylate the protooncogene protein-product Jun,[292] and in certain instances after phosphorylation, Jun is better able to form a heterodimer with Fos (see Nuclear Factors).[256] Heterodimers of Fos and Jun or homodimers of Jun act cooperatively as transcription factors to induce flexure at consensus AP-1 sites in transcriptional promoter regions for a number of genes,[293] including the 9 base pair motifs found in the promoters for ANF and skeletal α-actin, two common molecular markers for cardiac hypertrophy. Since these gene products have each been shown to increase with several stimuli for hypertrophy involving unique receptors, it is reasonable to assume that there will be some convergence in the hypertrophic stimulation of certain gene products. For example, AII in cardiocytes can activate MAP kinase.[200] In other cell systems, cAMP is able to block this pathway, showing a potential convergence in signaling pathways.[282] The extent of "cross talk" between these pathways is incompletely understood for the moment, but it is anticipated that further investigation into these systems will further our understanding of the processes that lead to regulation and/or eventual dysregulation of cardiac gene expression. This leads us to the model system shown in Figure 5.9.

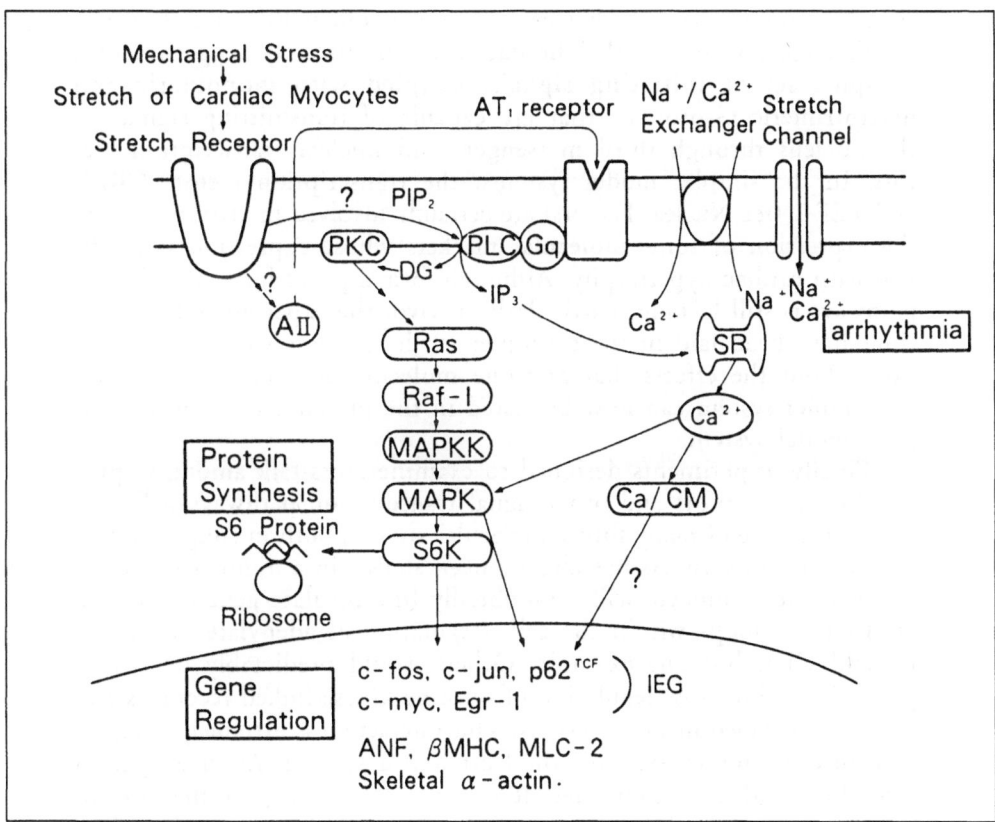

Fig. 5.9. Diagram depicting how passive stretch of neonatal cardiocytes in culture and activation of a phosphorylation cascade may regulate transcriptional activation of c-fos, skeletal α-actin and other genes. Of potential importance is the possibility of cross-talk among the various signal-transduction pathways found in the cardiocytes. AII, angiotensin II; ANF, atrial natriuretic factor; βMHC, β-myosin heavy chain; Ca^{2+}/CM, calcium-calmodulin kinase; DG, 1,2-diacylglycerol; IEG, immediate early genes; IP_3, inositol 1,4,5-trisphosphate; MAPK, MAP kinase; MAPKK, MAP kinase kinase (MEK); MLC-2, myosin light chain 2; PKC, protein kinase C; PLC, phospholipase C; S6K, S6 kinase; and SR, sarcoplasmic reticulum. (Reprinted by permission of the publisher from Intracellular signaling pathways in cardiac myocytes induced by mechanical stress. Komuro and Yazaki, Trends Cardiovasc Med Vol. No.4, Copyright 1994 by Elsevier Science Inc.)

In summary, three broad categories of hypertrophic stimuli thus act on neonatal cardiocytes. First, angiotensin II, endothelin-1, α-adrenergic agonists and other initiating signals all couple through their receptors (characterized by seven membrane-spanning helices) with specific G-proteins that lead to activation of the inositol phosphate/diacylglycerol pathway and protein kinase C activation; β-adrenergic agonists couple with the stimulatory G-protein to activate the adenylate cyclase, increase levels of cAMP, and activate PKA. Second, a number of growth factors (insulin, insulin-like growth factors I and II, and acidic and basic FGF) that bind to receptors with endogenous tyrosine kinase activity are capable of initiating a signaling cascade that may include the Ras and Raf-1 protooncogenes. Third, myocyte stretching works through a mechanism that does not involve any of the known stretch-activated channels that have been experimentally tested, but may act through release of AII. The data are thus strongly suggestive of a complex set of initiating signals, coupled with monomeric and heterotrimeric G-proteins that are capable of transmitting signals to the nucleus through third messengers and nuclear transcription factors. In the simpler model systems, the transcription factors CREB and TEF-1 (see Nuclear Factors) are certainly involved in vivo to regulate the expression of some molecular markers whose expression is modified with cardiac hypertrophy. Although formal proof for some of these pathways is still lacking, particularly in vivo, the experimental foundations have been laid in these simpler model systems whereby, predictions about the effects that any one molecule or component has on the simpler system can now be tested in the presence of a more complete model system.

Finally, experiments designed to examine cross-talk among G-protein-linked receptors and other signal transduction pathways will be a common theme of many future molecular studies, both in simpler model systems and in vivo. As has already been shown in a number of model systems, the cardiocyte will undoubtedly fine regulate gene expression via feedback loops within the cell. Regulation of adenylate cyclase by Gs and Gi is but one example. Others would predictably be anticipated, including cross-regulation of tyrosine-kinase-linked receptors and G protein-coupled mediated events. This interaction is extremely complex and, in other model systems, the least well-described. As investigation into these pathways continues, it is reasonable to expect that the in vivo system will maintain some of the regulatory pathways found in simpler model systems, but equally, due to the enhanced complexity of the system, individual regulatory pathways may be but one component of a remarkably integrated regulatory system capable of regulating cardiac growth and gene expression.

NUCLEAR FACTORS

Once a signaling-cascade is active, the activators must now be taken to the nucleus. If phosphorylation is, for example, the primary acti-

vating step, then it would be reasonable to assume that post-translational regulation of nuclear factors mediate many of the cardiac hypertrophy associated responses of cardiocytes. Although, we have already discussed to some degree the actions of Fos and Jun, further
discussion is necessary, and as we will see, the molecular mechanisms
associated with the cardiac hypertrophic response probably involve transcriptional, post-transcriptional and post-translation events.

IMMEDIATE-EARLY EVENTS

One of the first detectable changes in cardiac gene expression induced by pressure overload, is the activation of a set of genes characterized by their rapid and transient induction after mitogen stimulation[2,8,9] which can occur even in the absence of new protein synthesis.
This property makes it very likely that coupling proteins and/or phosphorylation events are responsible for their activation. A central question associated with this response is what role, if any, do these genes
play in the regulation of cardiac gene expression after pressure overload? The transient induction of a set of growth-related genes, including several protooncogenes (*c-fos, c-jun, junB, nur77, c-myc*) and transcription factors (Egr-1), is termed the immediate-early response. The
role of the immediate-early gene program in cardiac hypertrophy is at
present mostly speculative, although recent results in experimental systems
suggest an important role in the transcriptional activation of cardiac
genes. If the data are confirmed, it will show that after an acute increase in pressure, the heart is capable of "priming" itself to undergo
a phenotypic change in gene expression to meet the demands placed
upon it.

Two prototype genes for the immediate-early program in heart
include *c-fos* and *c-jun* which are activated not only by pressure overload but also by such diverse growth stimuli as stretch,[39,218] endothelin,[225]
angiotensin II,[222] α- and β-adrenergic agonists, aFGF, bFGF and hy-
poxia.[2,294,295] There are at least three homologues for *c-fos* (*fra-1, fosB*
and *fra2*),[2,296-298] and two for *c-jun* (*junB* and *junD*).[299] The protein
products of *c-fos* (Fos) and *c-jun* (Jun) are able to interact to form a
transcription factor complex.[293] In fact, transactivation by Fos requires
the formation of a dimer, preferentially with Jun (leucine zipper). This
heterodimer can bind to a short DNA sequence (TGACTCA) known
as a TRE/AP-1 binding domain. Since *c-fos* and *c-jun* are coactivated
together following a number of hypertrophic stimuli, it is logical to
postulate that their binding to an AP1 site on a target gene might act
as a positive or negative trans-activating factor. In its simplest form,
this supposition may be true; however, it is more complex, because
Fos and Jun have been shown to function differently despite both having
identical target sequences. These differences may partially be due to
the various isoforms, differences in their expression, and the potential
for phosphorylation influences.[256] For example, Jun can be phosphorylated on five major sites. The C-terminal phosphorylation sites of Jun

are dephosphorylated by phorbol ester activation of PKC and by the actions of several protooncogenes, including Ras. Dephosphorylation is correlated with an increase in AP-1 binding activity and may be one of the mechanisms for enhanced AP-1-mediated transcriptional activity in stimulated cells. This suggests that Jun in the cell is normally phosphorylated, but after dephosphorylation of its C-terminus, its DNA binding activity increases. Phosphorylation of N-terminal residues in Jun, however, can modify this activity.[256] For *c-fos*, the increased activity of protein phosphatases in the cell can abolish serum response factor (SRF) binding to its promoter region.[300] Conversely, *c-fos* induction by cAMP requires phosphorylation of the cAMP responsive element binding protein (CREB). Unphosphorylated CREB, however, represses the induction of the *c-jun* promoter by serum and hence may diminish the transcriptional activation of the AP-1 *cis*-binding element in other promoter regions by Fos/Jun heterodimers.[233] Hence, the induction of *c-fos* and *c-jun* transcription and the binding activity of the mature protein complex to an AP-1 site in a target gene are subject to a complex regulation involving multiple phosphorylation events in vivo. Activation of AP-1 sites by Fos/Jun heterodimers, therefore, depend on post-translational events and cross-talk between cellular regulatory events.

The most conclusive indirect evidence that Fos and Jun may play a role in regulating cardiac gene expression in cardiac hypertrophy has been shown using the molecular marker skeletal α-actin. First, the expression of *c-fos* and *c-jun* is activated early in response to isoproterenol treatment of high density cultures of neonatal cardiocytes.[117] Second, a reporter gene construct using human skeletal α-actin promoter regions when co-transfected with *c-fos* and *c-jun* constructs leads to an upregulation in reporter gene activity.[301] Although an exact copy of the consensus sequence for an AP-1 site (see chapter 4) is not present in the human skeletal α-actin promoter construct, a close homologue is present in the promoter region. Finally, purified Fos and Jun specifically bind to this promoter sequence. Neither α-MHC nor cardiac α-actin gene expression, which are not normally induced with cardiac hypertrophy, are responsive to transient *c-fos* and *c-jun* expression. These in vivo experiments show that *c-fos* and *c-jun* expression are associated with human skeletal α-actin promoter activity with isoproterenol treatment of neonatal cardiac myocytes,[301] and strongly support the signal transduction role of AP-1 site activation through the immediate-early response mechanism.

The inducible zinc-finger gene Egr-1 is also a member of the immediate-early response family of proteins and has been found to be co-regulated with *c-fos* in some systems.[258] In neonatal cardiomyocytes stimulated with adrenergic agonists, Egr-1 expression is primarily α1-adrenergic mediated, whereas *c-fos* expression is both α- and β-adrenoceptor mediated, providing the first evidence for independent transcription factor regulation in the immediate-early gene program of cardiac hypertrophy. This additionally demonstrates that induction of these protooncogenes is not necessarily sufficient to provoke a transcriptional program of cardiac hypertrophy. Egr-1 has also been shown to be important in the regulation of α-MHC but not β-MHC gene expression through the presence of a *cis*-acting element present in one promoter but absent in the other, again giving an example as to how different transcription factors can regulate individual gene expressions.[302]

Finally, the protooncogene *c-myc* (Myc) belongs to a family of transcription factors which are important in cell proliferation and differentiation. Its expression has been shown to be increased in various models of cardiac hypertrophy.[8,9] Although its role as a transcription factor in cardiac hypertrophy is unclear, its expression is probably more important in early cardiac development and cell division than with cardiac hypertrophy. In normal fetal hearts, *c-myc* is highly expressed, but it becomes almost undetectable shortly after birth. Cardiomyocytes of transgenic fetal mice that overexpress *c-myc* transcripts proliferate dramatically and have an increased cardiac mass.[303] The increase is primarily the result of a hyperplastic, not a hypertrophic response, supporting the idea that Myc plays an important role in stimulating cardiocyte cell division. Its enhanced expression in adult transgenic mice also seems to be of importance in the development of cardiomyocyte hypertrophy.[304]

The immediate-early gene program is thus characterized by the rapid induction of a number of protooncogenes and other transcription factors. In the case of Fos and Jun, their activation and enhanced expression may be involved in the altered gene expression of hypertrophic molecular markers (skeletal α-actin); whereas, Myc activation may only be involved in regulation of cardiocyte cell division during embryonic and fetal development, periods when myocytes maintain their potential for division (it may however be involved in regulating non-myocyte division).[303,304] Egr-1 can be induced with hypertrophic stimuli, but at present, it has only been associated with induction of the α-MHC gene, a molecular marker for a "physiological" form of hypertrophy induced by thyroid hormone. It is not known if Egr-1 is involved in activation of this molecular marker after stimulation by thyroid hormone. The immediate-early gene program, thus, seems to be a mechanism whereby cardiac cells prepare themselves to adapt to further stimuli or stress. Once activated, it is probable that their potential to activate specific gene promoter regions will be a function and a consequence of potential cross-talk between signaling pathways.

TRANSCRIPTIONAL ENHANCER FACTOR 1 (TEF-1)

Transcription factors, including many not normally associated with the immediate-early gene program, are important in the hypertrophic response. One of these has been implicated in regulating β-MHC gene expression in the model of α1-adrenergic stimulated hypertrophy of neonatal rat cardiocytes. As described earlier, the rat β-MHC promoter region contains a 20-base pair sequence (-215/-196) that is regulated by α1-adrenergic agonist stimulation and the expression of activated PKC-β.[275-277] Contained within this 20-base pair sequence is a M-CAT/ enhancer core sequence that is similar to a consensus *cis*-element originally identified in several viral enhancers. The study of these viral enhancers subsequently led to the identification of a transcription factor that could bind to variants of this site, and it has become known as transcriptional enhancer factor-1 (TEF-1). Through the use of a number of molecular techniques, including gel mobility shift assays and isolation of nuclear extracts from rat cardiocytes, it has been possible to show that a rat homologue of TEF-1 is capable of binding to the rat β-MHC promoter in the -215 to -196 region. Furthermore, the DNA binding activity of the β-MHC enhancer core region was found to closely correlate with TEF-1 immunoreactivity, and mutations in the core sequence eliminate TEF-1 binding in this promoter. These results suggest that TEF-1, in fact, may be one of the transcription factors responsible for β-MHC gene activation in cardiac hypertrophy in vivo, and that it is a target for transcriptional regulation by PKC. Similarly, TEF-1 has been shown to be involved in skeletal α-actin gene activation by α-adrenergic agonists in vivo.[305]

FUNCTIONAL CONSEQUENCES OF CARDIAC HYPERTROPHY

Thus far, we have described how the techniques of molecular and cellular biology can be used to delineate some of the mechanisms responsible for the altered gene expression associated with cardiac hypertrophy. In this final section, we will show how changes in gene expression can lead to altered cardiac function. Some of the molecular changes are probably of physiological benefit, some may be of little or no physiological importance, and some may be implicated in the long-term pathogenesis of cardiac muscle. We will therefore focus primarily on proteins of contraction and relaxation, since they are involved with the primary hemodynamic function of pumping oxygenated blood to all cells of the body.

CONTRACTILE PROTEINS

In heart as in skeletal muscle, the basic unit of contraction is the sarcomere which is composed of a diverse set of proteins working together to generate force and contraction. The two major components of the sarcomere are the thick and thin filament (see Fig. 3.1). The thick filaments are arranged in an hexagonal array, are located in the center of the sarcomere, and are composed primarily of myosin molecules. The thick filaments are birefringent, and as such, form the A bands. At rest, the central portion of the A band, called the H zone, is devoid of thin filaments, and the H zone is bisected by a dark line known as the M line. The thin filaments are composed primarily of actin, tropomyosin and troponin molecules. It extends from the Z line towards the center of the sarcomere through the I band and into the A band, and the thin filaments are anchored into the transverse filaments which form the Z line. Projections from the thick filaments touch adjacent thin filaments to form crossbridges, and it is the interactions between the thick and thin filaments which are responsible for "sliding" along the sarcomere. During shortening, the length of the thick filaments remains the same, but the thin filaments move towards the center of the sarcomere. Because of the primordial role played by the proteins of the thick and thin filaments in shortening and in the generation of force and contraction, it is reasonable to assume that any regulatory diversity or adaptational potential of these abundant proteins could play an important role in the process of cardiac hypertrophy.

The major protein constituent of the thick filament is myosin, a hexameric molecular consisting of two heavy chains (MHC), two myosin alkali light chains (MLC1), and two regulatory (phosphorylatable) myosin light chains (MLC2).[1,4] The general structure of myosin is similar in all types of muscle, and the 3-dimensional structure of the S1 fragment, as determined by X-ray crystallography, has been very informative in explaining how differences in heavy chain structure may play a role in determining their functional characteristics.[306,307] In the myosin dimer, each heavy chain molecule forms the bulk of one myosin head, and together, the carboxy-terminus of the two heavy chains form a coiled tail-region. The "head" region, encoded by the amino terminus, contains an actin binding domain (amino acids 626 to 647) and substrate pocket located on opposite sides of the head, and it has intrinsic actin-activated adenosine triphosphatase (ATPase) activity. The myosin molecule is unique by having an α helix formed by the carboxy terminus that runs down and ends at the base of the molecule. The light chains (see chapter 3) are associated with this helical structure and seem to stabilize it.

Isolation and analysis of the genes coding for α- and β-MHC have permitted a detailed analysis of the primary structures of these molecules.[51,308-311] In the rat, comparison of the amino acid sequences of these two isoforms shows only 131 amino acid differences out of a total of 1938 amino acids.[309] It is the position of these differences that explains the functional differences seen between the two isoforms. Many of the amino acid differences are located in subfragment-1 (myosin head), the region responsible for ATP hydrolysis, binding to actin, and force production.[309] In fact, the different binding affinities of α- and β-MHC isoforms with actin can be explained by the mere 52% mRNA sequence identity coding for amino acids 618 to 640.[312]

Functionally the two MHC isoforms are distinguishable by different ATPase activities. The V1 isoform (composed of two α-MHC subunits) can hydrolyze ATP more rapidly than the V3 isoform (composed of two β-MHC subunits). It is this intrinsic property of the heavy chain isoforms that was thought to have an important influence on the speed and efficiency of contraction in cardiac muscle. This is functionally very important since force and shortening are generated through myosin crossbridge attachment to actin, subsequent rotation of the myosin head and sliding of the thin filaments along the thick filaments. Since it had previously been shown[56] that differences in myosin isozyme composition between species were correlated with alterations in Vmax, it was assumed that the main determinant of the kinetics of crossbridge cycling lay with the myosin isoforms. Proof to this effect came from Alpert and his colleagues who have broken down into their component parts mechanical and myothermal data obtained from cardiac papillary muscle or muscle strips.[62,63] They were able to demonstrate in rabbit a number of changes in mechanical parameters following pressure overload (V3 myosin rich), including a reduced maximum shortening velocity (Vmax), decreased maximum rate of tension development, and an increased time to peak isometric tension without a significant change in peak tension. Additionally, both tension independent heat (heat associated with calcium cycling) and tension dependent heat (heat associated with crossbridge cycling) were significantly reduced relative to controls, all of which together suggested that cardiac muscle, following a hemodynamic overload, had an improved economy of isometric force. In terms of crossbridge behavior, these data were interpreted as meaning that there was prolonged crossbridge attachment time with a reduced crossbridge cycling rate, demonstrating changes in both the activation and contractile kinetics of hypertrophied myocardium. With hormonally induced cardiac hypertrophy (thyrotoxocosis) the contractile capability was also changed, albeit quite differently from that induced by hemodynamic overload,[61] as would be expected in a myocardium containing almost exclusively α-MHC (V1 myosin rich). In general, fibers with a high ATPase activity MHC shorten faster than fibers with a lower enzymatic activity; therefore, the type of MHC present in sarcomeres is of physiological importance.

The thin filaments are composed primarily of actin, tropomyosin, and three subunits of troponin. Actin is a ubiquitous protein in eukaryotic cells with a critical role in muscular movements through interaction with myosin. The three-dimensional structure of the actin monomer has been determined by X-ray crystallography and the structure of filamentous actin by electron microscopy.[313,314] The actin filament is composed of two monomeric strands composed of 13 actin molecules per 6 (left-handed) turns. Morphologically, this appears as two right-handed helices that rotate around each other. Actin filaments are associated with the TnT complex each 38.5 nm. Actin is a symmetric molecule composed of two domains that can be subdivided into two subdomains. The stability of this 3D-structure is due to salt bridges formed between the four domains. Interactions of actin with tropomyosin appear to be purely electrostatic.

Actin is encoded by a multigene family in mammals and exists in cardiac muscle as one of two sarcomeric isoforms, α-cardiac and α-skeletal actin—this nomenclature arose from the tissues where the isoforms were first isolated.[77-81,88,315] Cardiac α-actin diverges from skeletal α-actin at amino acid positions 2 and 3 (Asp to Glu and Glu to Asp, respectively) and at positions 298 and 357 (Leu to Met, Ala to Thr, respectively).[82] Structurally, actin molecules are highly conserved with an 8% divergence between yeast and mammalian skeletal muscle actins. As there are only four highly conserved amino acid differences between mammalian cardiac and skeletal α-actin proteins—two of which are located at the amino terminus of the molecule where the protein binds myosin—it is reasonable to assume that any changes in contractile function derived from a switch in isoform expression would be due to subtle differences in their ability to interact with other contractile proteins.[316] The structure of the six actin isoforms is so well conserved that sarcomeric actin isoforms can associate with cytoskeletal structures and smooth and skeletal muscle actins are mechanically indistinguishable from each other in in vivo motility assays.[91,92] Mutations in the molecule can however lead to a loss of function.[317] The functional significance of the two sarcomeric isoforms was therefore unknown.

Recently, however, Hewett et al presented data suggesting that overexpression of skeletal α-actin isoforms in hearts of BALB/c mice may lead to an increased contractile function.[93] BALB/c mice express abnormally high levels of skeletal α-actin in the heart, probably due to a spontaneous mutation and duplication in the promoter region of the cardiac actin gene. These mice normally express between 26 and 54% skeletal α-actin in their hearts with no apparent cardiac dysfunction.[318,319] When myocardial contractility determined from the maximum rate of contraction (+dP/dt) and the time to peak pressure, and myocardial relaxation from -dP/dt and the time to half relaxation of the left intraventricular pressure were measured in these mice,[93] a correlation was found between skeletal α-actin mRNA expression and +dP/dt (r = 0.80) and the time to peak pressure (r = -0.80). A weaker correlation was seen between the relaxation parameters and the skeletal α-actin content. Although it had previously been shown that some actin isoforms could activate the ATPase of specific myosin isoforms,[316,320,321] the functional data described here was the first to indicate that a correlation exists between the expression of the sarcomeric actin isoforms and the contractile behavior of the myocardium.

Tropomyosin (TM), in striated muscle, forms an elongated dimer which binds along the major groove of actin. Alone it prevents actin-myosin interactions, but upon binding of calcium to troponin T (and a conformational change), the tropomyosin dimer displaces along the groove permitting myosin heads to interact with actin molecules, thus initiating contraction. Tropomyosin is a member of a multigene family that is regulated both developmentally and in a tissue specific manner.[322,323] In the heart, two major isoforms, α-TM and β-TM, are found which differ in their binding sites to troponin I and T. With cardiac hypertrophy, the β-TM isoform is transiently activated at the onset of pressure-overload in a manner similar to that seen for α-skeletal actin.[38] The functional significance for this activation remains to be determined. With the troponin (Tn) complex, the Tn-TM complex acts to inhibit actin-myosin interactions at low intracellular free Ca^{2+} but with an increase in cytoplasmic Ca^{2+} the inhibition is released. The troponin (Tn) complex is comprised of three different troponin molecules: TnT (tropomyosin binding subunit), TnC (calcium binding subunit), and TnI (inhibitory subunit). The Ca^{2+} signaling process is initiated by calcium binding to a single regulatory site on troponin C and tight binding of TnC and TnI. The process is transmitted to the rest of the thin filament through troponin T and TM. With the exception of TnC, for which there is no evidence for the expression of alternative isoforms in the heart during development, shifts in the isotype population of the thin filaments have been documented and shown to lead to changes in contractile function.[324-326] With cardiac hypertrophy, there is no switch in TnI isoform expression, but for TnT, there is evidence for alternative splicing of the transcripts in animal models and in man which would be expected to lead to changes in the Ca^{2+} sensitivity of the myofilaments.[327]

RELAXATION

In cardiac muscle the sarcoplasmic reticulum (SR) and sarcolemma together play a central role in the contraction-relaxation cycle due to their ability to regulate intracellular free Ca^{2+} concentrations. The coupling of the electrical excitation of the heart to the production of contraction (EC coupling) involves the interaction of a number of cellular proteins which are involved in calcium (Ca^{2+}) homeostasis (Fig. 5.10). Ca^{2+} influx through the sarcolemma initiates release of stored Ca^{2+} from the SR (Ca^{2+}-induced Ca^{2+} release) via the SR Ca^{2+}-release channel (the ryanodine receptor (RYR)). Together both sources of calcium initiate contraction. Relaxation is brought about as calcium is pumped out of the cytoplasm and either into the SR by the SR phospholamban (PLB)-regulated Ca^{2+}-ATPase (SERCA) or extruded from the cell by the Na^+/Ca^{2+} exchanger.[237,241,328-331]

The sarcoplasmic reticulum is an intracellular membrane network that is in close contact with the myofibrils, and it is electrically coupled with the sarcolemma through its close association with the transverse (T) tubules of the sarcolemma. The T tubules in cardiac tissues are relatively few in number and relatively large in diameter when compared to those from skeletal muscle, but in mammalian myocardium, the T tubule system contains the majority of the L-type voltage dependent Ca^{2+} channels (DHP receptors). The sarcoplasmic reticulum can be divided into two components: the longitudinal or light SR and the junctional or heavy SR which are frequently referred to as terminal cisternae. The calcium release channel is associated with the heavy

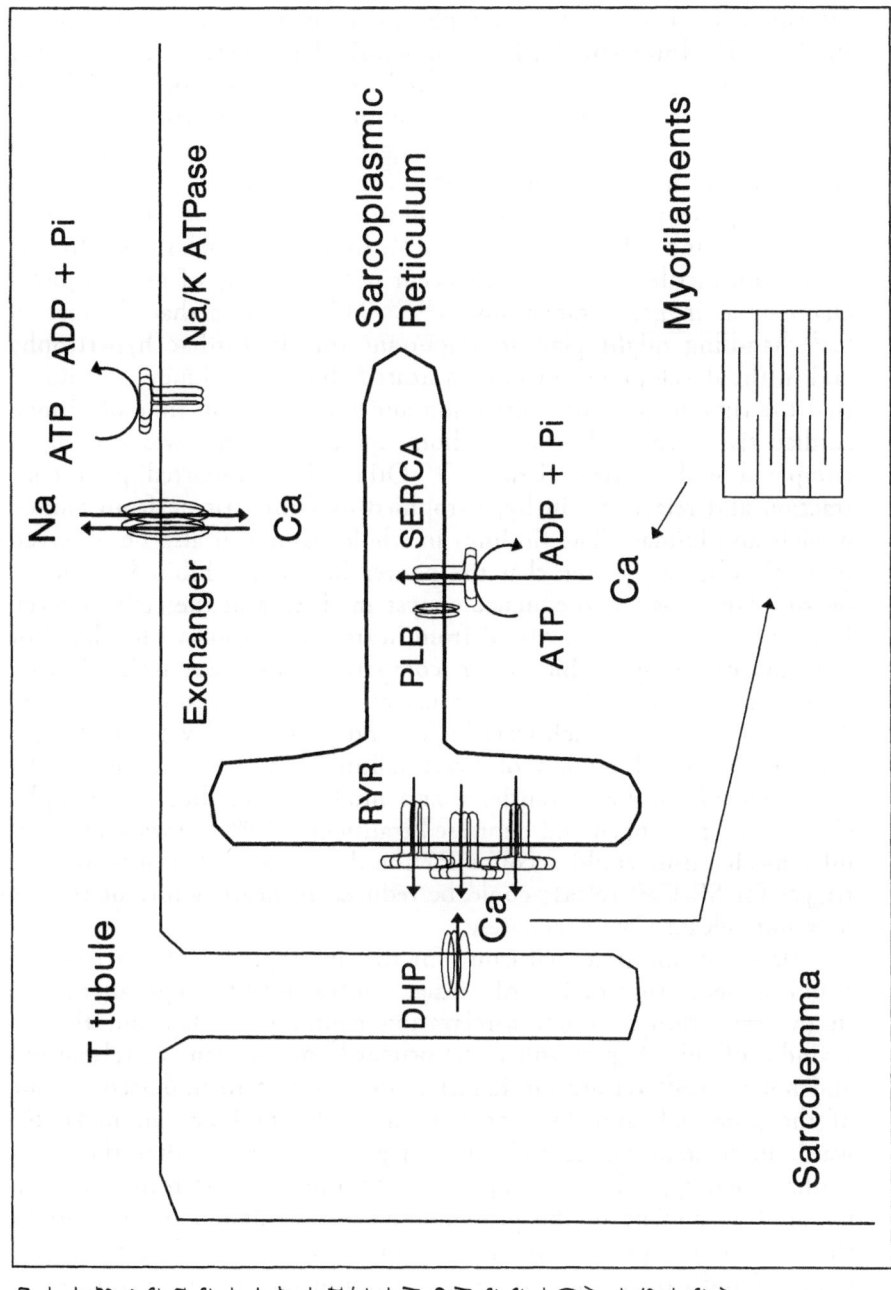

Fig. 5.10. Simplified diagram of excitation-contraction-relaxation coupling mechanisms in heart. Following membrane depolarization, Ca enters the cell via the voltage sensitive L-type Ca channel (DHP) to initiate Ca release from the sarcoplasmic reticulum (SR)-associated ryanodine receptor (RYR). The increase in cytoplasmic Ca and subsequent association with troponin C permits crossbridge formation between actin and myosin. For relaxation to occur, Ca must be removed from the cytoplasm by the combined actions of the phospholamban (PLB)-regulated SR CaATPase (SERCA) and the sarcolemmal Na/Ca exchanger. Na/Ca exchanger activity depends upon electrical and concentration gradients across the sarcolemma established by the Na/K ATPase.

SR whereas the CaATPase and phospholamban are found mostly in the light SR. Interestingly, in avian ventricular myocytes, the majority of terminal cisternae are not associated with T tubules or sarcolemma, and remarkably, in the case of the humming bird ventricle, there are no T tubules. These observations suggest along with other data that E-C coupling in heart results from diffusable signals.[241,328]

In hypertrophy and in heart failure, cardiac tissue displays poor contraction and relaxation,[331-335] and in human ventricles, the changes in the contractile activities must occur independently of myosin heavy chain or actin gene transitions.[63,335,336] This suggests that changes in Ca^{2+} handling might play an important role in cardiac hypertrophy and in the development of human heart failure. Gwathmey et al noted that the time to peak of contraction and the relaxation time of human cardiac tissue isolated from cardiomyopathic patients was prolonged compared with control tissue.[337,338] Others have reported poor contraction and relaxation in hypertrophied or failing tissue from animal models and human. The findings in whole tissue can also be observed in single cells suggesting that the poorer function exhibited by hearts nearing failure is a consequence of systems failing at the cellular level. For example, myocytes isolated from hearts of patients with dilated or ischemic cardiomyopathies when compared with those isolated from healthy donor hearts had Ca^{2+} transients which were slower to relax, diastolic Ca^{2+} levels which were higher and peak Ca^{2+} levels which were smaller.[339] Animal models of heart failure show a reduction in the amplitude of the Ca^{2+} transient, and models of cardiac hypertrophy also display prolonged and depressed transients.[174,340-343] Two major cellular mechanisms could account for the depressed Ca^{2+} transients: the trigger for SR Ca^{2+} release could be reduced in heart failure or the SR may not release Ca^{2+} adequately.

Although one potential cause for the development of heart failure could be depression or loss of cardiac contractility through alterations in the excitation-contraction-relaxation coupling mechanism, the remainder of this chapter will focus primarily on proteins of relaxation, and not on Ca^{2+} release mechanisms, in an effort to underscore some of the potential contributions that molecular biology can make towards understanding how changes in gene expression alter the myocardial phenotype. In so doing, it is important for the reader to keep in mind that many of the proteins regulating calcium movements in the sarcolemma and sarcoplasmic reticulum are characterized by modulation of individual gene expressions without changes in isoforms.[9]

Do changes in the quantity of a gene product alter the function of the myocardium? Molecularly, it is easier to describe how isoform switches, such as those for α- and β-MHC and sarcomeric α-actins, can effect cardiac function than how small changes in gene activities and protein quantities may effect the movements of ions in the hypertrophied myocardium. Why is this the case? First of all, changes in transcriptional activity or mRNA levels do not necessarily imply that there will be concomitant changes in the quantity of a protein or in its activity. This must be determined. Second, experimentally altering the quantity of a gene product in vivo requires that its expression be modified either in a large population of cells, so that sufficient quantities of the protein can be isolated and its function biochemically assayed, or through injection of antibodies to block the activity of an individual protein in conjunction with a very sensitive single cell technique that is capable of monitoring physiological parameters. The latter is conducive to electrophysiological examinations where techniques have been developed to measure ion movements across the sarcolemmal membrane, but these techniques are not always appropriate for measuring ion movements across intracellular membranes. The former requires the use of transgenic animals that over- or under-express a specific gene product or alternative techniques (pharmacological or molecular) that modify the expression of an individual protein. These techniques are limited since the changes produced are generally profound, while those seen with cardiac hypertrophy may be minor. If the long-term detrimental consequences to the heart are, in fact, due to subtle changes in the amounts of a protein, then it is important to determine how alterations in gene expression may contribute to the pathophysiology of the hypertrophied myocardium.

Sarcolemma

One important sarcolemmal protein involved in relaxation is the Na^+/Ca^{2+} exchanger whose primary role is extrusion of calcium as intracellular calcium is increased. To function, this protein uses the energy derived from the electrical potential and the concentration gradient of Na^+ across the sarcolemma to expel Ca^{2+} from the cell. For every three Na^+ ions which enter the cell, one Ca^{2+} ion is expelled.[344] Available evidence suggests that the exchanger competes with the SR CaATPase for Ca^{2+} to bring about relaxation with about 25-30% of relaxation mediated by this route[341,345,346]. The idea discussed by Wier[340] and investigated by Bridge et al,[347] is that the exchanger is mainly a means of extruding the Ca^{2+} which entered the cell via the Ca^{2+} channel (DHP receptor). In addition, at more positive membrane potentials, the exchanger may contribute to the influx of Ca^{2+} which brings about SR Ca^{2+} release.[348-350]

The exchanger has been cloned[351,352] and found to code for a protein of 970 amino acids. Hydropathy analysis shows that the protein has 11 membrane spanning regions and that between the fifth and sixth regions there is a large (519 residue) hydrophilic (cytoplasmic) domain thought to contain sites involved in the regulation of the exchanger. What happens to the expression of the Na^+/Ca^{2+} exchanger in cardiac hypertrophy? At the protein level, a decrease,[353] an increase[354] and the maintenance of exchanger activity have all been reported in compensated hypertrophy. At the RNA level, an increase in the expression of exchanger mRNA was shown after one hour[355] or several weeks of elevated pressure in the rat. It is thus difficult to draw any definitive conclusions from these results, particularly since the data have been analyzed in different species and at different stages of hypertrophy or cardiac insufficiency. However, recent results by Studer et al[356] support the idea that the expression of the Na^+/Ca^{2+} exchanger is increased in heart failure. They report an increase in exchanger mRNA levels in the myocardium of patients with dilated cardiomyopathies (55% increase) and with coronary artery disease (41% increase), trends which were confirmed at the protein level. They furthermore showed that the sarcoplasmic reticulum CaATPase mRNA and protein levels were decreased in these myocardia. The results suggest that increased exchanger function compensates for reduced SR CaATPase function during human heart failure.

An extremely important caveat to Na^+/Ca^{2+} exchanger regulation is its coupling to the Na^+/K^+-ATPase,[357] the protein primarily responsible for maintaining the ionic gradient across the sarcolemmal membrane. The Na^+/K^+-ATPase consists of a catalytic α-subunit and a β subunit.[358] Three α isoforms encoded by three distinct genes have been shown to be hormonally and developmentally regulated. In rat, the α isoforms differ primarily in their affinity for ouabain and other cardiac glycosides. The low affinity α1 isoform is predominant in rat heart throughout development. Relatively abundant levels of the α3 isoform are found in fetal and neonatal rats. The α3 subtype, like the α2 subtype, has a much higher affinity for cardiac glycosides when compared to the α1 isoform. With experimental compensated cardiac hypertrophy,[359] the levels of the predominant α1 isoform and the β1 subunit are unchanged. The α2 isoform is decreased by 35% and up to 64% in mild and severe hypertrophy respectively, and the α3 isoform which is extremely rare in the adult rat myocardium increases by up to 2-fold to a maximum of < 5% of total Na^+/K^+-ATPase α subunits. Unfortunately, it is currently impossible, due to the lack of experimental data, to draw any conclusion about the effects of these isoform switches in cardiac hypertrophy and how changes in levels of expression of these isoforms may alter the function of the Na^+/Ca^{2+} exchanger.

Sarcoplasmic Reticulum

The Ca^{2+}/Mg^{2+}-dependent ATPase of the sarcoplasmic reticulum is a mammalian ion-transport enzyme whose function in heart is to move Ca^{2+} from the cytoplasm to the sarcoplasmic reticulum. In so doing, this enzyme, with the Na^+/Ca^{2+} exchanger, is able to bring about cardiac relaxation on a beat-to-beat basis. It is a member of a multigene family that encodes for five isoforms,[360-368] and determination of its complete primary structure through characterization of full-length cDNAs for this protein has led to predictions about its secondary and tertiary structures. It is composed of a single large polypeptide of about 100 kDa that when expressed in heart accounts for as much as 40% of total SR protein.[328,369] The CaATPase is characterized by 10 transmembrane segments with three extramembranous globular domains that form a headpiece lying on top of a stalk comprised of five helices.[362] The extramembranous domains are probably highly interactive, although the nucleotide binding domain may function relatively independently of the rest of the molecule. The other extramembranous globular domains contain the sites for phosphorylation and transduction. The CaATPase catalyzes the transport of Ca^{2+} from the cytoplasm to the SR lumen in an ATP dependent process through transfer of the terminal phosphate group of ATP to Asp351, formation of an E-P intermediate and subsequent translocation of two Ca^{2+} ions to the lumen of the SR.[362,370,371] Because of its high capacity to remove Ca^{2+} from the cytoplasm, the CaATPase is the primary mechanism responsible for cardiac relaxation. In fact, depending upon the species, the CaATPase removes from 60% to 80% of all Ca^{2+} from the cytoplasm during end-systole and diastole.

CaATPase activity is regulated by phospholamban,[239] a type II noncovalent homopentameric membrane protein complex. It is encoded by a single gene giving rise to a single isoform composed of 52 amino acids.[370,372-374] Amino acid residues 1-32 form the cytosolic, phosphorylatable portion of the protein, while amino acids 33-52 contain a transmembrane domain that is responsible for pentamerization and the formation of a channel.[375-377] On each of the subunits are found three sites at Ser10, Ser16, and Thr17 which can be phosphorylated.[237,378,379] Only the phosphorylation by cAMP-dependent protein kinase and Ca^{2+}/calmodulin-dependent protein kinase of two of these sites (Ser16 and Thr17) seem to be important for PLB to function as a regulatory protein for SERCA.[237] In the non-phosphorylated form, PLB acts as an inhibitor of SERCA; when phosphorylated, PLB stimulates SERCA activity. In addition to phosphorylation, the levels of expression of PLB in relation to SERCA appear to affect the speed of myocardial relaxation.[237,328,380,381]

In severe experimental cardiac hypertrophy, the concentration of the mRNA and protein of the SERCA2a isoform of the CaATPase is decreased in the absence of any isoform switches.[380,382,383] The absolute amount of protein is unchanged, however, suggesting that SERCA2 gene expression is not upregulated with cardiac hypertrophy. In humans, SERCA2a mRNA is also depressed even when MHC mRNA expression appears to be normal compared with controls.[384,385] At the protein level, however, the levels and activity of SERCA2 expression have been reported to be unchanged or depressed, making interpretation of the contribution of the CaATPase to heart failure difficult.[356,386,387] Phospholamban expression is depressed with pressure-overload and may be reduced with heart failure caused by dilated cardiomyopathy, coronary artery disease, and primary pulmonary hypertension.[328,387-389] The decreased expression of PLB has been correlated with an increase in ANF expression. The data at a protein level are however inconsistent, as phospholamban-mediated stimulation of calcium uptake by the sarcoplasmic reticulum is not altered in failing hearts.[390] In an ongoing study with the laboratory of T. Eschenhagen, we have begun looking at PLB and CaATPase expression following administration of β-adrenergic agonists. We have strong data showing that isoproterenol infusions for four days in rats lead to as much as a 50% decrease in SERCA and a 30-35% decrease in PLB mRNA expression. Preinjection with verapamil, while it does not antagonize the effects of hypertrophy, does antagonize the necrosis associated with long-term isoproterenol infusions, and more importantly attenuates the effects on PLB and SERCA down-regulation. Similar results have been seen in treated neonatal cardiocytes. As changes in sympathetic drive are frequently associated with hypertrophic responses, it is likely that this is one mechanism that will play a role in regulating SERCA gene expression.

Characterization of the expression of genes encoding proteins involved in E-C coupling in the heart does not permit one to conclude whether the observed differences reflect causative or responsive phenomena. However, with recent advances in the field of transgenic manipulations, it is possible to isolate and knock out specific genes to study functionally significant changes in an animal's physiology and biochemistry. The importance of PLB expression has been examined in transgenic animals.[381] In mouse ventricles lacking PLB, heart rate and blood pressure were unaltered; however cardiac contraction and relaxation, measured in isolated heart preparations, were both significantly increased. The increased affinity of the SERCA pump for Ca^{2+} in the PLB-deficient mice was to the same extent as the removal of PLB inhibition by β-adrenoceptor stimulation in wild type mice. Furthermore, perfusion of the isolated hearts from the PLB-deficient mice with isoproterenol produced no additional increases in the cardiac parameters. These data reinforce the idea that PLB has an inhibitory role on CaATPase mediated Ca^{2+} uptake in the SR of mammalian hearts. This study emphasizes how transgenics can be used to examine the physiological consequences of over- or under-expression of individual gene products in the cardiac system. What happens to the Ca^{2+} transients in a pressure-overload induced model of cardiac hypertrophy in these PLB-deficient transgenic mice? This is unknown, but one would like to speculate that if a prolonged and depressed Ca^{2+} transient could be induced and measured in hypertrophied hearts of mice, then the changes in the Ca^{2+} transient would be to a lesser extent than those seen in control animals. This would be the hypothesis, since the function of the SR CaATPase would be expected to be maximal in the absence of the inhibitory protein PLB, even with an overall reduction in the quantity of SR CaATPase. The alternative or null hypothesis would state that any changes in the Ca^{2+} transients would be identical in the control and transgenic mice, and if found to be true, would suggest that PLB did not play an important inhibitory role in the hypertrophied myocardium.

CONCLUSIONS

Molecular biological techniques thus have allowed the cloning of many of the cDNAs coding for cardiac proteins involved in the adaptation of the heart to pressure-overload. In conjunction with simpler model systems and the molecular markers of cardiac hypertrophy, these cDNAs have and will continue to serve as tools in the examination of how hypertrophic stimuli effect gene regulation with cardiac hypertrophy. The use of transgenics and other model systems, that permit the over- or under-expression of these proteins, should in future studies contribute greatly to the understanding of the functional role played by these proteins in the normal heart and with cardiac hypertrophy. The data generated will depend on in vitro and in vivo experiments. As shown for the molecular markers of cardiac hypertrophy, it is likely that changes in cardiac protein gene expression will be regulated by specific factors associated with cardiac hypertrophy, and examination of their expression and properties in model systems should enable the cardiac molecular biologist to make important contributions to the elucidation of how these proteins may be regulated and eventually how their (dys)regulation may effect the function of the hypertrophied myocardium.

REFERENCES

1. Nadal-Ginard B, Mahdavi V. Molecular basis of cardiac performance. Plasticity of the myocardium generated through protein isoform switches. J Clin Invest 1989; 84:1693-700.
2. Schneider MD, Roberts R, Parker TG. Modulation of cardiac genes by mechanical stress. Mol Biol Med 1991; 8:167-83.
3. Katz A. Cardiomyopathy of overload: A major determinant of prognosis in congestive heart failure. New Eng J Med 1990; 322:100-10.
4. Swynghedauw B (ed). Cardiac Hypertrophy and Failure. Paris:John Libbey Eurotext Ltd, 1990.
5. Swynghedauw B. Developmental and functional adaptation of contractile proteins in cardiac and skeletal muscles. Physiol Rev 1986; 66:710-71.
6. Lompré AM, Mercadier JJ, Schwartz K. Changes in gene expression during cardiac growth. Int Rev Cytol 1990; 124:137-86.
7. Morgan HE, Baker KM. Cardiac hypertrophy: mechanical, neural, and endocrine dependence. Circulation 1991; 83:13-25.
8. Chien KR, Knowlton KU, Zhu H et al. Regulation of cardiac gene expression during myocardial growth and hypertrophy: molecular studies of an adaptive physiologic response. FASEB J 1991; 5:3037-46.
9. Boheler KR, Schwartz K. Gene expression in cardiac hypertrophy. Trends Cardiovasc Med 1992; 2:176-82.

10. Bishopric NH, Gahlmann R, Wade R, et al. Gene expression during skeletal and cardiac muscle development.In: Fozzard HA (ed): The Heart and Cardiovascular System. 2nd ed. New York:Raven Press Limited, 1992: 1587-1598.

11. Schwartz K, Carrier L, Mercadier J et al. Molecular phenotype of the hypertrophied and failing myocardium. Circulation 1993; 87(suppl VII):5-10.

12. Grossman W. Cardiac hypertrophy: useful adaptation or pathologic process. Am J Med 1980; 69:576-84.

13. Cooper G. Cardiocyte adaptation to chronically altered load. Annu Rev Physiol 1987; 49:501-18.

14. Sonneblick EH, Strobeck JE. Current concepts in cardiology: Derived indexes of ventricular and myocardial function. New Eng J Med 1977; 296:978-82.

15. Levy D, Garrison RJ, Savage DD et al. Prognostic implications of echocardiographically determined left ventricular mass in the Framingham heart study. New Eng J Med 1990; 322:1561-6.

16. Zak (ed) Growth of the Heart in Health and Disease, New York, Raven Press Limited, 1984.

17. Anversa P, Olivetti G, Loud AV. Morphometric study of the early postnatal development in the left and right ventricular myocardium of the rat. I. Hypertrophy, hyperplasia and binucleation of myocytes. Circ Res 1980; 46:495-502.

18. Anversa P, Olivetti G, Loud A. Morphometric studies of left ventricular hypertrophy.In:Tarazi RC, Dunbar JB (eds): Perspectives in Cardiovascular Research. 8th ed. New York:Raven Press, 1983: 27-39.

19. Grove D, Zak R, Nair KG et al. Biochemical correlates of cardiac hypertrophy. IV. Observations on the cellular organization of growth during myocardial hypertrophy in the rat. Circ Res 1969; 25:473-85.

20. Anversa P, Fitzpatrick D, Arganis S et al. Myocyte mitotic division in the aging mammalian rat heart. Circ Res 1991; 69:1159-64.

21. Meerson FZ. The myocardium in hyperfunction hypertrophy and heart failure. Circ Res 1969; 25:1-163.

22. Julian FJ, Morgan DL, Moss RL et al. Myocyte growth without physiological impairment in gradually induced rat cardiac hypertrophy. Circ Res 1981; 49:1300-10.

23. Nair KG, Cutilletta AF, Zak R et al. Biochemical correlates of cardiac hypertrophy. I. Experimental model; changes in heart weight, RNA content and nuclear RNA polymerase activity. Circ Res 1968; 23:451-62.

24. Posner BI, Fanburg BL. Ribonucleic acid synthesis in experimental cardiac hypertrophy in rats. II. Aspects of regulation. Circ Res 1968; 23:137-45.

25. Fanburg BL, Posner BI. Ribonucleic acid synthesis in experimental cardiac hypertrophy in rats. I. Characterization and kinetics of labeling. Circ Res 1968; 23:123-35.

26. Schreiber SS, Oratz M, Evans C et al. Effect of acute overload in cardiac muscle mRNA. Am J Physiol 1968; 215:1250-9.

27. Swynghedauw B, Moalic JM, Bouveret P et al. Messenger RNA content and complexity in normal and overloaded rat heart: A preliminary report. Eur Heart J 1984; 5 (suppl. F):211-7.
28. Schreiber SS, Evans CK, Oratz M et al. Protein synthesis and degradation in cardiac stress. Circ Res 1981; 48:601-11.
29. Ray A, Aumont M, Aussedat J et al. Protein and 28S ribosomal RNA fractional turnover rates in the rat heart after abdominal aortic stenosis. Cardiovasc Res 1987; 21:587-92.
30. Albin R, Dowell RT, Zak R et al. Synthesis and degradation of mito-chondrial components in hypertrophied rat heart. Biochem J 1973; 136:629-737.
31. Boheler KR, Dillmann WH. Cardiac response to pressure overload in the rat: The selective alteration of in vitro directed RNA translation products. Circ Res 1988; 63:448-56.
32. Morkin E, Garrett JC, Fishman AP. Effects of actinomycin D and hypophysectomey on development of myocardial hypertrophy in the rat. Am J Physiol 1968; 214:6-9.
33. Meerson FZ, Alekhina GM, Aleksandrov A et al. Dynamics of nucleic acid and protein synthesis of the myocardium in compensatory hyper-function and hypertrophy of the heart. Am J Cardiol 1684; 22:337-48.
34. Hoh JFY, Mc grath PA, Hale PT. Electrophoretic analysis of multiple forms of rat cardiac myosin: Effects of hypophysectomy and thyroxine replacement. J Mol Cell Cardiol 1978; 10:1053-76.
35. Izumo S, Lompré A, Matsuoka R et al. Myosin heavy chain messenger RNA and protein isoform transitions during cardiac hypertrophy. J Clin Invest 1987; 79:970-7.
36. Schwartz K, Lecarpentier Y, Martin JL et al. Myosin isoenzyme distribu-tion correlates with speed of myocardial contraction. J Mol Cell Cardiol 1981; 13:1071-5.
37. Komuro I, Yazaki Y. Control of cardiac gene expression by mechanical stress. Annu Rev Physiol 1993; 55:55-75.
38. Izumo S, Nadal-Ginard B, Mahdavi V. Protooncogene induction and re-programming of cardiac gene expression produced by pressure overload. Proc Natl Acad Sci USA 1988; 85:339-43.
39. Komuro I, Kaida T, Shibazaki M et al. Stretching cardiac myocytes stimu-lates proto-oncogene expression. J Biol Chem 1990; 265:3591-8.
40. Komuro I, Shibazaki Y, Kurabayashi M et al. Molecular cloning of gene sequences from rat heart rapidly responsive to pressure overload. Circ Res 1990; 66:979-85.
41. Mulvagh SL, Michael LH, Perryman MB et al. A hemodynamic load in vivo induces cardiac expression of the cellular oncogene, c-myc. Biochem Biophys Res Comm 1987; 147:627-36.
42. Komuro I, Kurabayashi M, Takaku F et al. Expression of cellular onco-genes in the myocardium during the developmental stage and pressure-overload hypertrophy of the rat heart. Circ Res 1988; 62:1075-9.
43. Black FM, Packer SE, Parker TG et al. The vascular smooth muscle α-actin gene is reactivated during cardiac hypertrophy provoked by load. J Clin Invest 1991; 88:1581-8.

44. Schwartz K, de la Bastie D, Bouveret P et al. α-Skeletal actin mRNAs accumulate in hypertrophied adult rat hearts. Circ Res 1986; 59:551-5.

45. Delcarpio JB, Nicholas A, Lanson NA et al. Morphological characterization of cardiomyocytes isolated from a transplantable cardiac tumor derived from transgenic mouse atria (AT-1 cells). Circ Res 1991; 69:1591-600.

46. Kimes BW, Brandt BL. Properties of a clonal muscle line from rat heart. Experimental Cell Research 1976; 98:367-81.

47. Sipido KR, Marban E. L-type calcium channels, potassium channels, and novel nonspecific cation channels in a clonal muscle cell line derived from embryonic rat ventricle. Circ Res 1991; 69:1487-99.

48. Hescheler J, Meyer R, Plant S et al. Morphological, biochemical and electrophysiological characterisation of a clonal cell (H9c2) line from rat heart. Circ Res 1991; 69:1476-86.

49. Engelmann GL, Birchenall-Roberts MC, Ruscetti FW et al. Formation of fetal rat cardiac cell clones by retroviral transformation: retention of select myocyte characteristics. J Mol Cell Cardiol 1993; 25:197-213.

50. Steinhelper ME, Lanson NAJ, Dresdner KP et al. Proliferation in vivo and in culture of differentiated adult atrial cardiomyocytes from transgenic mice. Am J Physiol 1990; 259:H1826-34.

51. Mahdavi V, Chambers AP, Nadal Ginard B. Cardiac alpha- and beta-myosin heavy chain genes are organized in tandem. Proc Natl Acad Sci U S A 1984; 81:2626-30

52. Morkin E. Regulation of myosin heavy chain genes in the heart. Circulation 1993; 87:1451-60.

53. Seidel U, Bober E, Winter B et al. Alkali myosin light chains in man are encoded by a multigene family that includes the adult skeletal muscle, the embryonic or atrial, and nonsarcomeric isoforms. Gene 1988; 66:135-46.

54. Reiser PJ, Moss RL, Giulian GG et al. Shortening velocity in single fibers from adult rabbit soleus muscles is correlated with myosin heavy chain composition. J Biol Chem 1985; 260:9077-80.

55. Reiser PJ, Greaser ML, Moss RL. Myosin heavy chain composition of single cells from avian slow skeletal muscle is strongly correlated with velocity of shortening during development. Dev Biol 1988; 129:400-7.

56. Barany M. ATPase activity of myosin correlated with speed of muscle shortening. J Gen Physiol 1967; 50:Suppl:197-218.

57. Mercadier JJ, Bouveret P, Gorza L et al. Myosin isoenzymes in normal hypertrophied human ventricular myocardium. Circ Res 1983; 53:52-62.

58. Hamrell BB, Alpert NR. The mechanical characteristics of hypertrophied rabbit cardiac muscle in the absense of congestive heart failure. Circ Res 1977; 40:20-5.

59. Maughan D, Low E, Litten R, III. et al. Calcium-activated muscle from hypertrophied rabbit hearts: Mechanical and correlated biochemical changes. Circ Res 1979; 44:279-87.

60. Buccino RA, Spann JF, Jr., Pool PE et al. Influence of the thyroid state on the intrinsic contractile properties and energy stores of the myocardium. J Clin Invest 1967; 46:1669-82.

61. Holubarsch C, Goulette RP, Litten RZ et al. The economy of isometric force development, myosin isoenzyme pattern and myofibrillar ATPase activity in normal and hypothyroid rat myocardium. Circ Res 1985; 56:78-86.

62. Alpert NR, Mulieri LA. Increased myothermal economy of isometric force development in compensated cardiac hypertrophy induced by pulmonary artery constriction in the rabbit. Circ Res 1982; 50:491-500.

63. Hasenfuss G, Mulieri LA, Blanchard EM et al. Energetics of isometric force development in control and volume-overload human myocardium. Circ Res 1991; 68:836-46.

64. Lompré A, Nadal-Ginard B, Mahdavi V. Expression of the cardiac ventricular α- and β-myosin heavy chain genes is developmentally and hormonally regulated. J Biol Chem 1984; 259:6437-46.

65. Mahdavi V, Izumo S, Nadal Ginard B. Developmental and hormonal regulation of sarcomeric myosin heavy chain gene family. Circ Res 1987; 60:804-14.

66. Ng WA, Grupp IL, Subramaniam A et al. Cardiac myosin heavy chain mRNA expression and myocardial function in the mouse heart. Circ Res 1991; 68:1742-50.

67. Schwartz K, Lompré AM, Bouveret P et al. Comparisons of rat 'cardiac' myosins at fetal stages in young animals and in hypothyroid adults. J Biol Chem 1981; 257:14412-8.

68. Everett AW, Sinha AM, Umeda PK et al. Regulation of myosin synthesis by thyroid hormone: relative change in the alpha- and beta-myosin heavy chain mRNA levels in rabbit heart. Biochemistry 1984; 23:1596-9.

69. Effron MB, Bhatnagar GM, Spurgeon HA et al. Changes in myosin isoenzymes, ATPase activity, and contraction duration in rat cardiac muscle with aging can be modulated by thyroxine. Circ Res 1987; 60:238-45.

70. Lompré A, Schwartz K, D'Albis A et al. Myosin isoenzyme redistribution in chronic heart overload. Nature 1979; 282:105-7.

71. Mercadier JJ, Lompré A, Wisnewsky C et al. Myosin isoenzyme changes in several models of rat cardiac hypertrophy. Circ Res 1981; 49:525-32.

72. Litten RZ, Low BJ, Alpert NR. Altered myosin isozyme patterns from pressure overloaded and thyrotoxic hypertrophied rabbit hearts. Circ Res 1982; 50:856-64.

73. Tsuchimochi T, Sugi HM, Kuro-o M et al. Isozymic changes in myosin of human atrial myocardium induced by overload: immunohistochemical study using monoclonal antibodies. J Clin Invest 1984; 74:662-5.

74. Gorza L, Mercadier JJ, Schwartz K et al. Myosin types in the human heart. An immunofluorescence study of normal and hypertrophied atrial and ventricular myocardium. Circ Res 1984; 54:694-702.

75. Schiaffino S, Samuel JL, Sassoon D et al. Nonsynchronous accumulation of α-skeletal actin and β-myosin heavy chain mRNAs during early stages

of pressure-overload-induced cardiac hypertrophy demonstrated by *in situ* hybridization. Circ Res 1989; 64:937-48.

76. Mercadier JJ, de la Bastie D, Menasche P et al. Alpha-myosin heavy chain isoform and atrial size in patients with various types of mitral valve dysfunction: A quantitative study. J Am Coll Cardiol 1987; 9:1024-30.

77. Mayer Y, Czosnek H, zeelon PE et al. Expression of the genes coding for the skeletal muscle and cardiac actins in the heart. Nucleic Acids Res 1984; 12:1087-100.

78. Minty AJ, Caravatti M, Robert B et al. Mouse actin messenger RNAs. J Biol Chem 1981; 256:1008-14.

79. Gunning P, Ponte P, Kedes L et al. Chromosomal location of the co-expressed human skeletal and cardiac actin genes. Proc Natl Acad Sci USA 1984; 81:1813-7.

80. Gunning P, Ponte P, Okayama H et al. Isolation and characterization of full-length cDNA clones for human α-, β-, and γ-actin mRNAs: skeletal but not cytoplasmic actins have an amino-terminal cysteine that is subsequently removed. Mol Cell Biol 1983; 3:787-95.

81. Ponte P, Gunning P, Blau H et al. Human actin genes are single copy for α-skeletal and α-cardiac actin but multicopy for β- and γ-cytoskeletal genes: 3' untranslated regions are isotype specific but are conserved in evolution. Mol Cell Biol 1983; 3:1783-91.

82. Vandekerckhove J, Bugaisky G, Buckingham M. Simultaneous expression of skeletal muscle and heart actin proteins in various striated muscle tissues and cells. J Biol Chem 1986; 261:1838-43.

83. Carrier L, Boheler KR, Chassagne C et al. Expression of the sarcomeric actin isogenes in the rat heart with development and senescence. Circ Res 1992; 70:999-1005.

84. Sassoon DA, Garner I, Buckingham M. Transcripts of α-cardiac and α-skeletal actins are markers for myogenesis in the mouse embryo. Development 1988; 104:155-64.

85. Babai F, Musevi-Aghdam J, Schurch W et al. Coexpression of α-sarcomeric actin, α-smooth muscle actin and desmin during myogenesis in rat and mouse embryos. Differentiation 1990; 44:132-42.

86. Winegrad S, Wisnewsky C, Schwartz K. Effect of thyroid hormone on the accumulation of mRNA for skeletal and cardiac α-actin in hearts from normal and hypophysectomized rats. Proc Natl Acad Sci USA 1990; 87:2456-60.

87. Collie ESR, Muscat GEO. The human skeletal α-actin promoter is regulated by thyroid hormone: identification of a thyroid hormone response element. Cell Growth & Differentiation 1992; 3:31-42.

88. Gunning P, Ponte P, Blau H et al. α-Skeletal and α-cardiac actin genes are coexpressed in adult human skeletal muscle and heart. Mol Cell Biol 1983; 3:1985-95.

89. Boheler KR, Carrier L, de la Bastie D et al. Skeletal actin mRNA increases in the human heart during ontogenic development and is the major isoform of control and failing adult hearts. J Clin Invest 1991; 88:323-30.

90. Wikman-Coffelt J, Parmley WW, Mason DT. The cardiac hypertrophy process. Analysis of factors determining pathological vs. physiological development. Circ Res 1979; 45:697-707.

91. Gunning P, Ponte P, Kedes L et al. Expression of human cardiac actin in mouse L cells: a sarcomeric actin associates with a nonmuscle cytoskeleton. Cell 1984; 36:709-15.

92. Harris DE, Warshaw DM. Smooth and skeletal muscle actin are mechanically indistinguishable in the in vivo motility assay. Circ Res 1993; 72:219-24.

93. Hewett TE, Grupp IL, Grupp G et al. α-Skeletal actin is associated with increased contractility in the mouse heart. Circ Res 1994; 74:740-6.

94. Lattion AL, Michel JB, Arnauld E et al. Myocardial recruitment during ANF mRNA increase with volume overload in the rat. Am J Physiol 1986; 251:H890-6.

95. Mercadier J, Samuel J, Michel JL et al. Atrial natriuretic factor gene expression in rat ventricle during experimental hypertension. Am J Physiol 1989; 257:H979-87.

96. Day ML, Schwartz D, Wiegand RC et al. Ventricular atriopeptin. Unmasking of messenger RNA and peptide synthesis by hypertrophy or dexamethasone. Hypertension 1987; 9:485-91.

97. Saito Y, Nakao K, Arai H et al. Augmented expression of atrial natriuretic polypeptide gene in ventricle of human failing heart. J Clin Invest 1989; 83:298-305.

98. Takayanagi R, Imada T, Inagami T. Synthesis and presence of atrial natriuretic factor in rat ventricle. Biochem Biophys Res Commun 1987; 142:483-8.

99. Knowlton KU, Baracchini E, Ross RS et al. Co-regulation of the atrial natriuretic factor and cardiac myosin light chain-2 genes during α-adrenergic stimulation of neonatal rat ventricular cells. J Biol Chem 1991; 266:7759-68.

100. Mercadier JJ, Michel JB. Neurohormonal axis in cardiac hypertension and failure. In:Swynghedauw B (ed): Research in Cardiac Hypertrophy and Failure. Paris:INSERM/John Libbey Eurotext, 1990: 401-403.

101. Zeller R, Bloch KD, Williams BS et al. Localized expression of the atrial natriuretic factor gene during cardiac embryogenesis. Genes Devel 1987; 1:693-8.

102. Wu JP, Deschepper CF, Gardner DG. Perinatal expression of the atrial natriuretic factor gene in rat cardiac tissue. Am J Physiol 1988; 255:E388-96.

103. Arai H, Nakao K, Saito Y et al. Augmented expression of atrial natriuretic polypeptide gene in ventricles of spontaneously hypertensive rats (SHR) and SHR-stroke prone. Circ Res 1988; 62:926-30.

104. Franch HA, Dixon RAF, Blaine EH et al. Ventricular atrial natriuretic factor in the cardiomyopathic hamster model of congestive heart failure. Circ Res 1988; 62:31-6.

105. Edwards BS, Ackermann DM, Lee ME et al. Identification of atrial natriuretic factor within ventricular tissue in hamsters and humans with congestive heart failure. J Clin Invest 1988; 81:82-6.

106. Drexler H, Hanze J, Finckh M et al. Atrial natriuretic peptide in a rat model of cardiac failure. Atrial and ventricular mRNA, atrial content, plasma levels, and effect of volume loading. Circulation 1989; 79:620-33.

107. Gutkowska J, Horky K, Finckh M et al. Atrial natriuretic factor in spontaneously hypertensive rats. Hypertension 1986; 8:I137-40.

108. Simpson P, Savion S. Differentiation of rat myocytes in single cell cultures with and without proliferating nonmyocardial cells. Cross-striations, ultrastructure, and chronotropic response to isoproterenol. Circ Res 1982; 50:101-16.

109. Simpson P. Norepinephrine-stimulated hypertrophy of culture rat myocardial cells is an α1-adrenergic response. J Clin Invest 1983; 72:732-8.

110. Simpson P. Stimulation of hypertrophy of cultured neonatal rat heart cells through an α_1-adrenergic receptor and induction of beating through an α_1- and β_1-adrenergic receptor interaction: evidence for independent regulation of growth and beating. Circ Res 1985; 56:884-94.

111. Nagai R, Pritzi N, Low RB et al. Myosin isozyme synthesis and mRNA levels in pressure-overloaded rabbit hearts. Circ Res 1987; 60:692-9.

112. Engelmann GL, McTiernan C, Gerrity RG et al. Serum-free primary cultures of neonatal rat cardiomyocytes: cellular and molecular applications. Technique 1990; 2:279-91.

113. Ueno H, Perryman MB, Roberts R et al. Differentiation of cardiac myocytes after mitogen withdrawal exhibits three sequential states of the ventricular growth response. J Cell Biol 1988; 107:1911-8.

114. Claycomb WC, Bradshaw HDJ. Acquisition of multiple nuclei and the activity of DNA polymerase alpha and reinitiation of DNA replication in terminally differentiated adult cardiac muscle cells in culture. Dev Biol 1983; 99:331-7.

115. Katzberg AA, Farmer BB, Harris RA. The predominance of binucleation in isolated rat heart myocytes. Am J Anat 1977; 149:489-99.

116. Korecky B, Sweet S, Rakusan K. Number of nuclei in mammalian cardiac myocytes. Can J Physiol Pharmacol 1979; 57:1122-9.

117. Bishopric NH, Kedes L. Adrenergic regulation of the skeletal α-actin gene promoter during myocardial cell hypertrophy. Proc Natl Acad Sci USA 1991; 88:2132-6.

118. Marzluff WF, Huang RCC. Transcription of RNA in isolated nuclei. In:Hames BD, Higgins SJ (eds): Transcription and Translation. Washington:IRL Press, 1984: 89-129.

119. Bishopric NH, Simpson PC, Ordahl CP. Induction of the skeletal α-actin gene in α_1-adrenoceptor-mediated hypertrophy of rat cardiac myocytes. J Clin Invest 1987; 80:1194-9.

120. Boheler KR, Chassagne C, Martin X et al. Cardiac expressions of α- and β-myosin heavy chains and sarcomeric α-actins are regulated through transcriptional mechanisms. J Biol Chem 1992; 267:12979-85.

121. Sittman DB, Graves RA, Marzluff WF. Histone mRNA concentrations are regulated at the level of transcription and mRNA degradation. Proc Natl Acad Sci USA 1983; 80:1849-53.

122. Cox RD, Garner I, Buckingham ME. Transcriptional regulation of actin and myosin genes during differentiation of a mouse muscle cell line. Differentiation 1990; 43:183-91.

123. McCully JD, Liew C. RNA transcription in myocardial-cell nuclei during postnatal development. Biochem J 1988; 256:441-5.

124. Jackowski G, Liew C. Fractionation of rat ventricular nuclei. Biochem J 1980; 188:363-73.

125. Muller FU, Boheler KR, Eschenhagen T et al. Isoprenaline stimulates gene transcription of the inhibitory G protein α-subunit Giα-2 in rat heart. Circ Res 1993; 72:696-700.

126. Knoll R, Arras M, Zimmermann R et al. Changes in gene expression following short coronary occlusions studied in porcine hearts with run-on assays. Cardiovasc Res 1994; 28:1062-9.

127. Lee HR, Henderson SA, Reynolds R et al. Alpha 1-adrenergic stimulation of cardiac gene transcription in neonatal rat myocardial cells. Effects on myosin light chain-2 gene expression. J Biol Chem 1988; 263:7352-8.

128. Long CS, Ordahl CP, Simpson PC. $α_1$-Adrenergic receptor stimulation of sarcomeric actin isogene transcription in hypertrophy of cultured rat heart muscle cells. J Clin Invest 1989; 83:1078-82.

129. Komuro I, Katoh Y, Kaida T et al. Mechanical loading stimulates cell hypertrophy and specific gene expression in cultured rat cardiac myocytes. J Biol Chem 1991; 266:1265-8.

130. Simpson P. Stimulation of hypertrophy of cultured neonatal rat heart cells through an α1-adrenergic receptor and induction of beating through an α1- and β-adrenergic receptor interaction: evidence for independent regulation of growth and beating. Circ Res 1985; 56:884-94.

131. Bishopric NH, Sato B, Webster KA. β-adrenergic regulation of a myocardial actin gene via a cyclic AMP-independent pathway. J Biol Chem 1992; 267:20932-6.

132. Dubus I, Rappaport L, Barrieux A et al. Contractile protein gene expression in serum-free cultured adult rat cardiac myocytes. Eur J Physiol 1993; 423:455-61.

133. Dubus I, Samuel JL, Marotte F et al. β-Adrenergic agonists stimulate the synthesis of non-contractile but not contractile proteins in cultured myocytes isolated from adult rat heart. Circ Res 1990; 66:867-74.

134. Schneider MD, Parker TG. Cardiac growth factors. Prog Growth Factor Res 1991; 3:1-26.

135. Parker TG, Packer SE, Schneider MD. Peptide growth factors can provoke "fetal" contractile protein gene expression in rat cardiac myocytes. J Clin Invest 1990; 85:507-14.

136. Neer EJ, Clapham DE. Signal transduction through G protein in the cardiac myocyte. Trends Cardiovasc Med 1992; 2:6-11.

137. Cummins P. Fibroblast and transforming growth factor expression in the cardiac myocyte. Cardiovasc Res 1993; 27:1150-4.

138. Subramaniam A, Gulick J, Neumann J et al. Transgenic analysis of the thyroid-responsive elements in the α-cardiac myosin heavy chain gene promoter. J Biol Chem 1993; 268:4331-6.

139. Gustafson TA, Markham BE, Bahl JJ et al. Thyroid hormone regulates expression of a transfected alpha-myosin heavy chain fusion gene in fetal heart cells. Proc Natl Acad Sci U S A 1987; 84:3122-6.

140. Izumo S, Mahdavi V. Thyroid hormone receptor alpha isoforms generated by alternative splicing differentially activate myosin HC gene transcription. Nature 1988; 334:539-42.

141. Flink IL, Morkin E. Interaction of thyroid hormone receptors with strong and weak *cis*-acting elements in the human α-myosin heavy chain gene promoter. J Biol Chem 1990; 265:11233-7.

142. Simon MI, Strathmann MP, Gautam N. Diversity of G proteins in signal transduction. Science 1991; 252:802-8.

143. Feldman AM. Experimental issues in assessment of G protein function in cardiac disease. Circulation 1991; 84:1852-61.

144. Port JD, Malbon CC. Integration of transmembrane signaling: Cross-talk among G-protein-linked receptors and other signal transduction pathways. Trends Cardiovasc Med 1993; 3:85-92.

145. Fleming JW, Wisler PL, Watanabe AM. Signal transduction by G proteins in cardiac tissues. Circulation 1992; 85:420-33.

146. Gilman AG. G proteins: transducers of receptor-generated signals. Annu Rev Biochem 1987; 56:615-49.

147. Johnson GL, Dhanasekaran N. The G-protein family and their interaction with receptors. Endocr Rev 1989; 10:317-31.

148. Birnbaumer L, Abramowitz J, Brown AM. Receptor-effector coupling by G proteins. Biochim Biophys Acta 1990; 1031:163-224.

149. Ross EM. Signal sorting and amplification through G protein-coupled receptors. Neuron 1989; 3:141-52.

150. Baldwin JM. The probable arrangement of the helices in G protein-coupled receptors. EMBO J 1993; 12:1693-703.

151. Wang JYJ, McWhirter JR. Tyrosine-kinase-dependent signaling pathways. Trends Cardiovasc Med 1994; 4:270

152. Fantl WJ, Johnson DE, Williams LT. Signalling by receptor tyrosine kinases. Annu Rev Biochem 1993; 62:453-81.

153. Wang JYJ. Nuclear protein tyrosine kinases. Trends Biochem Sciences 1994; 19:373-6.

154. Lefkowitz RJ, Caron MG. Adrenergic receptors. Adv Second Messenger Phosphoprotein Res 1988; 21:1-10.

155. Brown JH, Buxton IL, Brunton LL. Alpha 1-adrenergic and muscarinic cholinergic stimulation of phosphoinositide hydrolysis in adult rat cardiomyocytes. Circ Res 1985; 57:532-7.

156. Scholz J, Schaefer B, Schmitz W et al. Alpha-1 adrenoceptor-mediated positive inotropic effect and inositol trisphosphate increase in mammalian heart. J Pharmacol Exp Ther 1988; 245:327-35.

157. Otani H, Das DK. Alpha 1-adrenoceptor-mediated phosphoinositide breakdown and inotropic response in rat left ventricular papillary muscles. Circ Res 1988; 62:8-17.

158. Kohl C, Schmitz W, Scholz H et al. Evidence for alpha 1-adrenoceptor-mediated increase of inositol trisphosphate in the human heart. J Cardiovasc Pharmacol 1989; 13:324-7.

159. Schmitz W, Scholz H, Scholz J et al. Increase in IP3 precedes alpha-adrenoceptor-induced increase in force of contraction in cardiac muscle. Eur J Pharmacol 1987; 140:109-11.

160. Schmitz W, Kohl C, Neumann J et al. On the mechanism of positive inotropic effects of alpha-adrenoceptor agonists. Basic Res Cardiol 1989; 84 Suppl 1:23-33.

161. Bohm M, Diet F, Feiler G et al. Alpha-adrenoceptors and alpha-adrenoceptor-mediated positive inotropic effects in failing human myocardium. J Cardiovasc Pharmacol 1988; 12:357-64.

162. Bristow MR, Minobe W, Rasmussen R et al. Alpha-1 adrenergic receptors in the nonfailing and failing human heart. J Pharmacol Exp Ther 1988; 247:1039-45.

163. Buxton IL, Brunton LL. Action of the cardiac alpha 1-adrenergic receptor. Activation of cyclic AMP degradation. J Biol Chem 1985; 260:6733-7.

164. Schmitz W, Scholz H, Scholz J et al. Pertussis toxin does not inhibit the alpha 1-adrenoceptor-mediated effect on inositol phosphate production in the heart. Eur J Pharmacol 1987; 134:377-8.

165. Buxton ILO, Brunton LL. Direct analysis of β-adrenergic receptor subtypes on intact adult ventricular myocytes of the rat. Circ Res 1985; 56:126-32.

166. Chevalier B, Mansier P, Callens el Amrani F et al. Beta-adrenergic system is modified in compensatory pressure cardiac overload in rats. physiological and biochemical evidence. J Cardiovasc Pharmacol 1989; 13:412-20.

167. Moalic JM, Bourgeois F, Mansier P et al. β_1 adrenergic receptor and Gαs mRNAs in rat heart as a function of mechanical load and thyroxine intoxication. Cardiovasc Res 1993; 27:231-7.

168. Limas C, Limas CJ. Reduced number of beta-adrenergic receptors in the myocardium of spontaneously hypertensive rats. Biochem Biophys Res Commun 1978; 83:710-4.

169. Limas CJ. Increased number of beta-adrenergic receptors in the hypertrophied myocardium. Biochim Biophys Acta 1979; 588:174-8.

170. Vatner DE, Homcy CJ, Sit SP et al. Effects of pressure overload, left ventricular hypertrophy on beta-adrenergic receptors, and responsiveness to catecholamines. J Clin Invest 1984; 73:1473-82.

171. Swynghedauw B. Changes in membrane proteins in chronic mechanical overload of the heart. Am J Cardiol 1990; 65:30G-3G.

172. Frielle T, Kobilka B, Lefkowitz RJ et al. Human β1- and β2-adrenergic receptors: structurally and functionally related receptors derived from distinct genes. Trends Neurosci 1988; 11:321-4.

173. Bristow MR, Hershberger RE, Port JD et al. β-adrenergic pathways in nonfailing and failing human ventricular myocardium. Circulation 1990; 82:112-25.

174. Barry WH, Bridge JHB. Intracellular calcium homeostasis in cardiac myocytes. Circulation 1993; 87:1806-15.

175. Yanagisawa T, Ishii K, Hashimoto H et al. Differential coupling to positive inotropic responses of cyclic AMP produced by stimulation of β1- and β2-adrenergic receptors. Journal of Cardiovascular Pharmacology 1989; 13:64-75.

176. del Monte F, Kaumann AJ, Poole-Wilson PA et al. Coexistence of functioning β1- and β2-adrenoceptors in single myocytes from human ventricle. Circulation 1993; 88:854-63.

177. Bristow MR, Ginsburg R, Umans V et al. B1-and B2-adrenergic receptor subpopulations in nonfailing and failing human ventricular myocardium: Coupling of both receptor subtypes to muscle contraction and selective B1-receptor down-regulation in heart failure. Circ Res 1986; 59:297-309.

178. Steinfath M, Danielsen W, von der Leyen H et al. Reduced α1- and β2-adrenoceptor-mediated positive inotropic effects in human end-stage heart failure. Br J Pharmacol 1992; 105:463-9.

179. Bristow MR, Hershberger RE, Port JD et al. Beta 1- and beta 2-adrenergic receptor-mediated adenylate cyclase stimulation in nonfailing and failing human ventricular myocardium. Mol Pharmacol 1989; 35:295-303.

180. Ungerer M, Bohm M, Elce JS et al. Altered expression of β-adrenergic receptor kinase and β1-adrenergic receptors in the failing human heart. Circulation 1993; 87:454-63.

181. Waspe LE, Ordahl CP, Simpson PC. The cardiac β-myosin heavy chain isogene is induced selectively in α1-adrenergic receptor-stimulated hypertrophy of cultured rat heart myocytes. J Clin Invest 1990; 85:1206-14.

182. Kass-Eisler A, Falck-Pedersen E, Alvira M et al. Quantitative determination of adenovirus-mediated gene delivery to rat cardiac myocytes in vitro and in vivo. Proc Natl Acad Sci USA 1993; 90:11498-502.

183. Meidell RS, Sen A, Henderson SA et al. Coordinate increases in myofibrillar protein synthesis during α-1 adrenergic stimulation of cultured rat myocardial cells. Am J Physiol 1986; 251:H1076-84.

184. Simpson PC, Long CS, Waspe LE et al. Transcription of early developmental isogenes in cardiac myocyte hypertrophy. J Mol Cell Cardiol 1989; 21 (Suppl V):79-89.

185. Sharp WW, Terracio L, Borg TK et al. Contractile activity modulates actin synthesis and turnover in cultured neonatal rat heart cells. Circ Res 1993; 73:172-83.

186. Engelmann GL, Dionne CA, Jaye MC. Acidic fibroblast growth factor and heart development: Role in myocyte proliferation and capillary angiogenesis. Circ Res 1993; 72:7-19.

187. Engelmann GL, Boehm KD, Birchenall Roberts MC et al. Transforming growth factor-beta 1 in heart development. Mech Develop 1992; 38:85-97.

188. Weiner HL, Swain JL. Acidic fibroblast growth factor mRNA is expressed by cardiac myocytes in culture and the protein is localized to the extracellular matrix. Proc Natl Acad Sci U S A 1989; 86:2683-7.

189. Villareal RJ, Dillmann WH. Cardiac hypertrophy-induced changes in mRNA levels for TGF-β1, fibronectin and collagen. Am J Physiol 1992; 262:H1861-6.

190. Parker TG, Chow KL, Schwartz RJ et al. Differential regulation of skeletal α-actin transcription in cardiac muscle by two fibroblast growth factors. Proc Natl Acad Sci USA 1990; 87:7066-70.

191. Baker KM, Aceto JA. Characterization of avian angiotensin II cardiac receptors: coupling to mechanical activity and phosphoinositide metabolism. J Mol Cell Cardiol 1989; 21:375-82.

192. Ishihata A, Endoh M. Pharmacological characteristics of the positive inotropic effect of angiotensin II in the rabbit ventricular myocardium. Br J Pharmacol 1993; 108:999-1005.

193. Rogers TB, Lokuta AJ. Angiotensin II signal transduction pathways in the cardiovascular system. Trends Cardiovasc Med 1994; 4:110-6.

194. Baker K, Booz G, Dostal D. Cardiac actions of angiotensin II: role of an intracardiac renin-angiotensin system. Annu Rev Physiol 1992; 54:227-41.

195. Sechi LA, Griffin CA, Grady EF et al. Characterization of angiotensin II receptor subtypes in rat heart. Circ Res 1992; 71:1482-9.

196. Timmermans PB, Wong PC, Chiu AT et al. Nonpeptide angiotensin II receptor antagonists. Trends Pharmacol Sci 1991; 12:55-62.

197. Mukoyama M, Nakajima M, Horiuchi M et al. Expression cloning of type 2 angiotensin II receptor reveals a unique class of seven-transmembrane receptors. J Biol Chem 1993; 268:24539-42.

198. Rogg H, Schmid A, de Gasparo M. Identification and characterization of angiotensin II receptor subtypes in rabbit ventricular myocardium. Biochem Biophys Res Commun 1990; 173:416-22.

199. Iwai N, Inagami T. Identification of two subtypes in the rat type I angiotensin II receptor. FEBS Lett 1992; 298:257-60.

200. Sadoshima J, Izumo S. Mechanical stretch rapidly activates multiple signal transduction pathways in cardiac myocytes: potential involvement of an autocrine/paracrine mechanism. EMBO J 1993; 12:1681-92.

201. Dzau VJ. Autocrine and paracrine mechanisms in the pathophysiology of heart failure. Am J Cardiol 1992; 70:4C-11C.

202. Urata H, Hoffmann S, Ganten D. Tissue angiotensin II system in the human heart. Eur Heart J 1994; 15 (suppl D):68-78.

203. Soubrier F, Cambien F. Renin-angiotensin system genes as candidate genes in cardiovascular diseases. Trends Endocrinol Metab 1993; 3:250-8.

204. Schunkert H, Dzau VJ, Tang SS et al. Increased rat cardiac angiotensin converting enzyme activity and mRNA expression in pressure overload left ventricular hypertrophy. J Clin Invest 1990; 86:1913-20.

205. Zierhut W, Zimmer H-G, Gerdes AM. Effect of angiotensin converting enzyme inhibition on pressure-induced left ventricular hypertrophy in rats. Circ Res 1991; 69:609-17.

206. Linz W, Schaper J, Wiemer G et al. Ramipril prevents left ventricular hypertrophy with myocardial fibrosis without blood pressure reduction: on one year study in rats. Br J Pharmacol 1992; 107:970-5.

207. Linz W, Schölkens, Ganten D. Converting enzyme inhibition specifically prevents the development and induces regression of cardiac hypertrophy in rats. Clinical Experimental Hypertrophy—Theory and Practice 1989; A11:1325-50.

208. Miyata S, Haneda T. Hypertophic growth of cultured neonatal rat heart cells mediated by type I angiotensin II receptor. Am J Physiol 1994; 266:H2443-51.

209. Mizuno K, Tani M, Hashimoto S et al. Effects of losartan, a nonpeptide angiotensin II receptor antagonist, on cardiac hypertrophy and the tissue angiotensin II content in spontaneously hypertensive rats. Life Sciences 1992; 51:367-74.

210. Suzuki J, Matsubara H, Urakami M et al. Rat angiotensin II (type 1A) receptor mRNA regulation and subtype expression in myocardial growth and hypertrophy. Circ Res 1993; 73:439-47.

211. Rockman HA, Wachorst SP, Mao L et al. ANG II receptor blockade prevents ventricular hypertrophy and ANF gene expression with pressure overload in mice. Am J Physiol 1994; 266:H2468-75.

212. Cooper G, Kent RL, Uboh CE et al. Hemodynamic versus adrenergic control of cat right ventricular hypertrophy. J Clin Invest 1985; 75:1403-14.

213. Kira Y, Kochel PJ, Gordon EE et al. Aortic perfusion pressure as a determinant of cardiac protein synthesis. Am J Physiol 1984; 246:C247-58.

214. McDermott PJ, Morgan HE. Contraction modulates the capacity for protein synthesis during growth of neonatal heart cells in culture. Circ Res 1989; 64:542-53.

215. Cooper G, Mercer WE, Hoober JK et al. Load regulation of the properties of adult feline cardiocytes. The role of substrate adhesion. Circ Res 1986; 58:692-705.

216. Mann DL, Kent RL, Cooper G. Load regulation of the properties of adult feline cardiocytes: growth induction by cellular deformation. Circ Res 1989; 64:1079-90.

217. von Harsdorf R, Lang RE, Fullerton M et al. Myocardial stretch stimulates phosphatidylinositol turnover. Circ Res 1989; 65:494-501.

218. Sadoshima J, Jahn L, Takahashi T et al. Molecular characterization of the stretch-induced adaptation of cultured cardiac cells. J Biol Chem 1992; 267:10551-60.

219. Komuro I, Yazaki Y. Intracellular signaling pathways in cardiac myocytes induced by mechanical stress. Trends Cardiovasc Med 1994; 4:117-21.

220. Sadoshima J, Izumo S. Signal transduction pathways of angiotensin II-induced *c-fos* gene expression in cardiac myocytes in vitro. Circ Res 1993; 73:424-38.

221. Sadoshima J, Takahashi T, Jahn L et al. Roles of mechano-sensitive ion channels, cytoskeleton, and contractile activity in stretch-induced immediate-early gene expression and hypertrophy of cardiac myocytes. Proc Natl Acad Sci USA 1992; 89:9905-9.

222. Sadoshima J, Izumo S. Molecular characterization of angiotensin II-induced hypertrophy of cardiac myocytes and hyperplasia of cardiac fibroblasts. Circ Res 1993; 73:413-23.

223. Sadoshima J, Xu Y, Slayter HS et al. Autocrine release of angiotensin II mediates stretch-induced hypertrophy of cardiac myocytes in vitro. Cell 1993; 75:977-84.

224. Ito H, Hiroe M, Hirata Y et al. Insulin-like growth factor-I induces hypertrophy with enhanced expression of muscle specific genes in cultured rat cardiomyocytes. Circulation 1993; 87:1715-21.

225. Shubeita HE, McDonough PM, Harris AN et al. Endothelin induction of inositol phospholipid hydrolysis, sarcomere assembly, and cardiac gene expression in ventricular myocytes. A paracrine mechanism for myocardial cell hypertrophy. J Biol Chem 1990; 265:20555-62.

226. Ito H, Hirata Y, Hiroe M et al. Endothelin-1 induces hypertrophy with enhanced expression of muscle-specific genes in cultured neonatal rat cardiomyocytes. Circ Res 1991; 69:209-15.

227. Ito H, Hirata Y, Adachi S et al. Endothelin-1 is an autocrine/paracrine factor in the mechanism of angiotensin II-induced hypertrophy in cultured rat cardiomyocytes. J Clin Invest 1993; 92:398-403.

228. Marx J. Two major signal pathways linked. Science 1993; 262:988-9.

229. Ross EM, Gilman AG. Resolution of some components of adenylate cyclase necessary for catalytic activity. J Biol Chem 1977; 252:6966-9.

230. Pfeuffer T, Helmreich EJM. Activation of pigeon erythrocyte membrane adenylate cyclase by guanyl nucleotide analogs and separation of nucleotide binding protein. J Biol Chem 1975; 250:867-76.

231. Neer EJ, Clapham DE. Roles of G protein subunits in transmembrane signalling. Nature 1988; 333:129-34.

232. Roesler WJ, Vandenbark GR, Hanson RW. Cyclic AMP and the induction of eukaryotic gene transcription. J Biol Chem 1988; 263:9063-6.

233. Lamph WW, Dwarki VJ, Ofir R et al. Negative and positive regulation by transcription factor cAMP response element-binding protein is modulated by phosphorylation. Proc Natl Acad Sci U S A 1990; 87:4320-4.

234. Robishaw JD, Foster KA. Role of G proteins in the regulation of the cardiovascular system. Annu Rev Physiol 1989; 51:229-44.

235. Bristow MR, Port JD. Receptor pharmacology of the human heart. In:Garfein OE (ed): Current concepts in cardiovascular physiology. New York:Academic Press, 1990:

236. Brodde OE. Beta 1- and beta 2-adrenoceptors in the human heart: properties, function, and alterations in chronic heart failure. Pharmacol Rev 1991; 43:203-42.

237. Colyer J. Control of the calcium pump of cardiac sarcoplasmic reticulum. A specific role for the pentameric structure of phospholamban? Cardiovasc Res 1993; 27:1766-71.

238. Simmerman HK, Collins JH, Theibert JL et al. Sequence analysis of phospholamban. Identification of phosphorylation sites and two major structural domains. J Biol Chem 1986; 261:13333-41.

239. Tada M, Katz AM. Phosphorylation of the sarcoplasmic reticulum and sarcolemma. Annu Rev Physiol 1982; 44:401-23.

240. Arai M, Otsu K, MacLennan DH et al. Regulation of sarcoplasmic reticulum gene expression during cardiac and skeletal muscle development. Am J Physiol 1992; 31:C614-20.

241. Lompré AM, Anger M, Levitsky D. Sarco(endo)plasmic reticulum calcium pumps in the cardiovascular system: function and gene expression (Review). J Mol Cell Cardiol 1994; 26:1109-21.

242. Watson PA, Haneda T, Morgan HE. Effect of higher aortic pressure on ribosome formation and cAMP content in rat heart. Am J Physiol 1989; 256:C1257-61.

243. Xenophontos XP, Watson PA, Chua BHL et al. Increased cyclic AMP content accelerates protein synthesis in rat heart. Circ Res 1989; 65:647-56.

244. Stanton HC, Brenner G, Mayfield ED. Studies on isoproterenol-induced cardiomegaly in rats. Am Heart J 1969; 77:72-80.

245. Johnson MD, Grignolo A, Kuhn CM et al. Hypertension and cardiovascular hypertrophy during chronic catecholamine infusion in rats. Life Sci 1983; 33:169-80.

246. Zierhut W, Zimmer HG. Significance of myocardial alpha- and beta-adrenoceptors in catecholamine-induced cardiac hypertrophy. Circ Res 1989; 65:1417-25.

247. Eschenhagen Y, Mende U, Nose M et al. Isoprenaline-induced increase in mRNA levels of inhibitory G-protein α-subunits in rat heart. Naunyn-Schmiedeberg's Arch Pharmacol 1991; 343:609-15.

248. Stiles GL, Caron MG, Lefkowitz RJ. Beta-adrenergic receptors: biochemical mechanisms of physiological regulation. Physiol Rev 1984; 64:661-743.

249. Hausdorff WP, Caron MG, Lefkowitz RJ. Turning off the signal: desensitization of beta-adrenergic receptor function [published erratum appears in FASEB J 1990 Sep; 4(12):3049]. FASEB J 1990; 4:2881-9.

250. Hadcock JR, Port JD, Gelman MS et al. Cross-talk between tyrosine kinase and G-protein-linked receptors. Phosphorylation of beta 2-adrenergic receptors in response to insulin. J Biol Chem 1992; 267:26017-22.

251. Hadcock JR, Ros M, Watkins DC et al. Cross-regulation between G-protein-mediated pathways. Stimulation of adenylyl cyclase increases expression of the inhibitory G-protein, Gi alpha 2. J Biol Chem 1990; 265:14784-90.

252. Bohm M, Gierschik P, Erdmann E. Quantification of Gi alpha-proteins in the failing and nonfailing human myocardium. Basic Res Cardiol 1992; 87 Suppl 1:37-50.

253. Bohm M, Gierschik P, Jakobs KH et al. Increase of Gi alpha in human hearts with dilated but not ischemic cardiomyopathy. Circulation 1990; 82:1249-65.

254. Feldman AM, Cates AE, Bristow MR et al. Altered expression of alpha-subunits of G proteins in failing human hearts. J Mol Cell Cardiol 1989; 21:359-65.

255. Muller FU, Eschenhagen T, Reidemeister A et al. In vivo β-adrenergic stimulation leads to biphasic regulation of Giα-2 gene transcriptional activity in rat heart. J Mol Cell Cardiol 1994; 26:869-75.

256. Hunter T, Karin M. The regulation of transcription of phosphorylation. Cell 1992; 70:375-87.

257. Yamamoto KK, Gonzalez GA, Biggs WH, III. et al. Phosphorylation-induced binding and transcriptional efficacy of nuclear factor CREB. Nature 1988; 334:494-8.

258. Iwaki K, Sukhatme VP, Shubeita HE et al. Alpha- and beta-adrenergic stimulation induces distinct patterns of immediate early gene expression in neonatal rat myocardial cells. J Biol Chem 1990; 265:13809-17.

259. Fain JN, Wallace MA, Wojcikiewicz RJ. Evidence for involvement of guanine nucleotide-binding regulatory proteins in the activation of phospholipases by hormones. FASEB J 1988; 2:2569-74.

260. Henrich CJ, Simpson PC. Differential acute and chronic response of protein kinase C in cultured neonatal rat heart myocytes to alpha 1-adrenergic and phorbol ester stimulation. J Mol Cell Cardiol 1988; 20:1081-5.

261. Kaku T, Lakatta E, Filburn C. Alpha-adrenergic regulation of phosphoinositide metabolism and protein kinase C in isolated cardiac myocytes. Am J Physiol 1991; 260:C635-42.

262. Shubeita HE, Martinson EA, van Bilsen M et al. Transcriptional activation of the cardiac myosin light chain 2 and atrial natriuretic factor genes by protein kinase C in neonatal rat ventricular myocytes. Proc Natl Acad Sci U S A 1992; 89:1305-9.

263. Rhee SG, Suh PG, Ryu SH et al. Studies of inositol phospholipid-specific phospholipase C. Science 1989; 244:546-50.

264. Majerus PW, Ross TS, Cunningham TW et al. Recent insights in phosphatidylinositol signaling. Cell 1990; 63:459-65.

265. Okumura K, Kawai T, Hashimoto H et al. Sustained diacylglycerol formation in norepinephrine-stimulated rat heart is associated with alpha 1-adrenergic receptor. J Cardiovasc Pharmacol 1988; 11:651-6.

266. Berridge MJ, Irvine RF. Inositol phosphates and cell signalling. Nature 1989; 341:197-205.

267. Nishizuka Y. The molecular heterogeneity of protein kinase C and its implications for cellular regulation. Nature 1988; 334:661-5.

268. Parker PJ, Coussens L, Totty N et al. The complete primary structure of protein kinase C—the major phorbol ester receptor. Science 1986; 233:853-9.

269. Coussens L, Parker PJ, Rhee L et al. Multiple, distinct forms of bovine and human protein kinase C suggest diversity in cellular signaling pathways. Science 1986; 233:859-66.

270. Ono Y, Fujii T, Ogita K et al. The structure, expression, and properties of additional members of the protein kinase C family. J Biol Chem 1988; 263:6927-32.

271. Ono Y, Fujii T, Ogita K et al. Protein kinase C zeta subspecies from rat brain: its structure, expression, and properties. Proc Natl Acad Sci U S A 1989; 86:3099-103.

272. Schaap D, Parker PJ, Bristol A et al. Unique substrate specificity and regulatory properties of PKC-epsilon: a rationale for diversity. FEBS Lett 1989; 243:351-7.

273. Bogoyevitch MA, Parker PJ, Sugden PH. Characterization of protein kinase C isotype expression in adult rat heart. Protein kinase C-epsilon is a major isotype present, and it is activated by phorbol esters, epinephrine, and endothelin. Circ Res 1993; 72:757-67.

274. Mochly Rosen D, Henrich CJ, Cheever L et al. A protein kinase C isozyme is translocated to cytoskeletal elements on activation. Cell Regul 1990; 1:693-706.

275. Kariya K, Karns LR, Simpson PC. Expression of a constitutively activated mutant of the β-isozyme of protein kinase C in cardiac myocytes stimulates the promoter of the β-myosin heavy chain isogene. J Biol Chem 1991; 266:10023-6.

276. Kariya K, Farrance IK, Simpson PC. Transcriptional enhancer factor-1 in cardiac myocytes interacts with an alpha 1-adrenergic- and beta-protein kinase C-inducible element in the rat beta-myosin heavy chain promoter. J Biol Chem 1993; 268:26658-62.

277. Kariya K, Karns LR, Simpson PC. An enhancer core element mediates stimulation of the rat β-myosin heavy chain promoter by an α1-adrener-

gic agonist and activated β-protein kinase C in hypertrophy of cardiac myocytes. J Biol Chem 1994; 269:3775-82.

278. Smith MR, DeGudicibus SJ, Stacey DW. Requirement for c-ras proteins during viral oncogene transformation. Nature 1986; 320:540-3.

279. Downward J. Regulatory mechanisms for ras proteins. BioEssays 1992; 14:177-84.

280. Finney RE, Robbins SM, Bishop JM. Association of pRas and pRaf-1 in a complex correlates with activation of a signal transduction pathway. Current Biology 1993; 3:805-12.

281. Thorburn A, Thorburn J, Chen SY et al. HRas-dependent pathways can activate morphological and genetic markers of cardiac muscle cell hypertrophy. J Biol Chem 1993; 268:2244-9.

282. Cook SJ, McCormick F. Inhibition by cAMP of Ras-dependent activation of Raf. Science 1993; 262:1069-72.

283. Green NK, Gammage MD, Franklyn JA et al. Regulation by thyroid status of c-myc, c-fos, and H-ras mRNAs in the rat myocardium. J Endocrinol 1991; 130:239-44.

284. Moodie SA, Willumsen BM, Weber MJ et al. Complexes of Ras.GTP with Raf-1 and mitogen-activated protein kinase kinase. Science 1993; 260:1658-61.

285. Van Aelst L, Barr M, Marcus S et al. Complex formation between RAS and RAF and other protein kinases. Proc Natl Acad Sci U S A 1993; 90:6213-7.

286. Schaap D, van der Wal J, Howe LR et al. A dominant negative mutant of ras blocks mitogen-activated protein kinase activation by growth factors and oncogenic p21ras. J Biol Chem 1993; 20232-6.

287. Davis RJ. The mitogen-activated protein kinase signal transduction pathway. J Biol Chem 1993; 268:14553-6.

288. Troppmair J, Bruder JT, App H et al. Ras controls coupling of growth factor receptors and protein kinase C in the membrane to Raf-1 and B-Raf protein serine kinases in the cytosol. Oncogene 1992; 7:1867-73.

289. Wood KW, Sarnecki C, Roberts TM et al. Ras mediates nerve growth factor receptor modulation of three signal-transducing protein kinases: MAP kinase, Raf-1, and RSK. Cell 1992; 68:1041-50.

290. Kolch W, Heidecker G, Lloyd P et al. Raf-1 protein kinase is required for growth of induced NIH/3T3 cells. Nature 1991; 349:426-8.

291. Bogoyevitch MA, Glennon PE, Sugden PH. Endothelin-1, phorbol esters and phenylephrine stimulate MAP kinase activities in ventricular cardiomyocytes. FEBS Lett 1993; 317:271-5.

292. Pulverer BJ, Kyriakis JM, Avruch J et al. Phosphorylation of c-jun mediated by MAP kinases. Nature 1991; 353:670-4.

293. Kerppola TK, Curran T. Fos-jun heterodimers and jun homodimers bend DNA in opposite orientations: implications for transcription factor cooperativity. Cell 1991; 66:317-26.

294. Webster KA, Discher DJ, Bishopric NH. Regulation of fos and jun immediate-early genes by redox or metabolic stress in cardiac myocytes. Circ Res 1994; 74:679-86.

295. Webster KA, Discher DJ, Bishopric NH. Induction and nuclear accumulation of Fos and Jun proto-oncogenes in hypoxic cardiac myocytes. J Biol Chem 1993; 268:16852-8.

296. Cohen DR, Ferreira PC, Gentz R et al. The product of a fos-related gene, fra-1, binds cooperatively to the AP-1 site with Jun: transcription factor AP-1 is comprised of multiple protein complexes. Genes Dev 1989; 3:173-84.

297. Nishina H, Sato H, Suzuki T et al. Isolation and characterization of fra-2, an additional member of the fos gene family. Proc Natl Acad Sci U S A 1990; 87:3619-23.

298. Zerial M, Toschi L, Ryseck RP et al. The product of a novel growth factor activated gene, fos B, interacts with JUN proteins enhancing their DNA binding activity. EMBO J 1989; 8:805-13.

299. Hirai SI, Ryseck RP, Mechta F et al. Characterization of junD: a new member of the jun proto-oncogene family. EMBO J 1989; 8:1433-9.

300. Prywes R, Dutta A, Cromlish JA et al. Phosphorylation of serum response factor, a factor that binds to the serum response element of the c-FOS enhancer. Proc Natl Acad Sci U S A 1988; 85:7206-10.

301. Bishopric NH, Jayasena V, Webster KA. Positive regulation of the skeletal α-actin gene by Fos and Jun in cardiac myocytes. J Biol Chem 1992; 267:25535-40.

302. Gupta MP, Gupta M, Zak R et al. Egr-1, a serum inducible zinc finger protein regulates transcription of the rat cardiac α-myosin heavy chain gene. J Biol Chem 1991; 266:12813-6.

303. Jackson T, Allard MF, Sreenan CM et al. The *c-myc* proto-oncogene regulates cardiac development in transgenic mice. Mol Cell Biol 1990; 7:3709-16.

304. Robbins R, Swain JL. C-*myc* protooncogene modulates cardiac hypertrophic growth in transgenic mice. Am J Physiol 1992; 262:H590-7.

305. Karns LR, Kariya K, Simpson PC. M-CAT, CArG, and Sp1 elements are required for α₁-adrenergic induction of the skeletal α-actin promoter during cardiac myocyte hypertrophy. J Biol Chem 1995; 270:410-7.

306. Rayment I, Rypniewski WR, Schmidt-Bäse K et al. Three-dimensional structure of myosin subfragment-1: a molecular motor. Science 1993; 261:50-6.

307. Rayment I, Holden H, Whittaker M et al. Structure of the actin-myosin complex and its implications for muscle contraction. Science 1993; 261:58-65.

308. Gulick J, Subramaniam A, Neumann J et al. Isolation and characterization of the mouse cardiac myosin heavy chain genes. J Biol Chem 1991; 266:9180-5.

309. McNally EM, Kraft R, Bravo Zehnder M et al. Full-length rat alpha and beta cardiac myosin heavy chain sequences. Comparisons suggest a molecular basis for functional differences. J Mol Biol 1989; 210:665-71.

310. Jaenicke T, Diederich KW, Haas W et al. The complete sequence of the human beta-myosin heavy chain gene and a comparative analysis of its product. Genomics 1990; 8:194-206.

311. Liew CC, Sole MJ, Yamauchi-Takihara K et al. Complete sequence and organization of the human cardiac beta-myosin heavy chain gene. Nuc Acids Res 1990; 18:3647-51.

312. Eldin P, Cathiard AM, Leger J et al. Probing function regions in cardiac isomyosins with monoclonal antibodies. Biochemistry 1993; 32:2542-7.

313. Kabsch W, Mannherz HG, Suck D et al. Atomic structure of the actin: DNAse I complex. Nature 1990; 347:37-44.

314. Holmes KC, Popp D, Gebhard W et al. Atomic model of the actin filament. Nature 1990; 347:44-9.

315. Ponte P, Ng S, Engel J et al. Evolutionary conservation in the untranslated regions of actin mRNAs: DNA sequence of a human beta-actin cDNA. Nuc Acids Res 1984; 12:1687-96.

316. Sutoh K. Identification of myosin-binding sites on the actin sequence. Biochemistry 1982; 21:3654-61.

317. Drummond DR, Peckham M, Sparrow JC et al. Alteration in crossbridge kinetics caused by mutations in actin. Nature 1990; 348:440-2.

318. Garner I, Minty AJ, Alonso A et al. A 5' duplication of the α-cardiac actin gene in BALB/c mice is associated with abnormal levels of α-cardiac and α-skeletal actin mRNAs in adult cardiac tissue. EMBO J 1986; 5:2559-67.

319. Alonso S, Garner I, Vandekerckhove J et al. Genetic analysis of the interaction between cardiac and skeletal actin gene expression in striated muscle of the mouse. J Mol Biol 1990; 211:727-38.

320. Gordon DJ, Boyer JL, Korn ED. Comparative biochemistry of non-muscle actins. J Biol Chem 1977; 252:8300-9.

321. Sobieszek A, Small JV. Regulation of the actin-myosin interaction in vertebrate smooth muscle: activation via a myosin light-chain kinase and the effect of tropomyosin. J Mol Biol 1977; 112:559-76.

322. Nadal-Ginard B, Mahdavi V. Molecular basis of cardiac performance. J Clin Invest 1989; 84:1693-700.

323. Humphreys JE, Cummins P. Regulatory proteins of the myocardium. Atrial and ventricular tropomyosin and troponin-I in the developing and adult bovine and human heart. J Mol Cell Cardiol 1984; 16:643-57.

324. McAuliffe JJ, Gao L, Solaro RJ. Changes in myofibrillar activation and troponin C Ca^{++} binding associated with troponin T isoform switching developing rabbit heart. Circ Res 1990; 66:1204-16.

325. Martin AF, Ball K, Gao LZ et al. Identification and functional significance of troponin I isoforms in neonatal rat heart myofibrils. Circ Res 1991; 69:1244-52.

326. Dieckman LJ, Solaro RJ. Effect of thyroid status on thin-filament Ca^{2+} regulation and expression of troponin I in perinatal and adult rat hearts. Circ Res 1990; 67:344-51.

327. Anderson PAW, Malouf NN, Oakeley AE et al. Troponin T isoform expression in humans: a comparison among normal and failing heart, fetal heart, and adult and fetal skeletal muscle. Circ Res 1991; 69:1226-33.

328. Arai M, Alpert NR, MacLennan DH et al. Alterations in sarcoplasmic reticulum gene expression in human heart failure: A possible mechanism

for alterations in systolic and diastolic properties of the failing myocardium. Circ Res 1993; 72:463-9.

329. Grossman W. Diastolic dysfunction in congestive heart failure. New Eng J Med 1991; 325:1557-64.

330. Moss RL. Ca²⁺ regulation of mechanical properties of striated muscle. Circ Res 1992; 70:865-84.

331. Levy AJ, Brooksby P, Hancox JC. One hump or two? The triggering of calcium release from the sarcoplasmic reticulum and the voltage dependence of contraction in mammalian cardiac muscle. Cardiovasc Res 1993; 27:1743-57.

332. Perreault CL, Meuse AJ, Bentivegna LA et al. Abnormal intracellular calcium handling in acute and chronic heart failure: role in systolic and diastolic dysfunction. Eur Heart J 1990; 11 (suppl C):8-21.

333. Morgan JP, Erny RE, Allen PD et al. Abnormal intracellular calcium handling: a major cause of systolic and diastolic dysfunction in ventricular myocardium from patients with heart failure. Circulation 1990; 81 (suppl III):21-32.

334. D'Agnolo A, Luciani GB, Mazzucco A et al. Contractile properties and Ca²⁺ Release activity of the sarcoplasmic reticulum in dilated cardiomyopathy. Circulation 1992; 85:518-25.

335. Hasenfuss G, Reinecke H, Studer R et al. Relation between myocardial function and expression of sarcoplasmic reticulum Ca²⁺-ATPase in failing and nonfailing human myocardium. Circ Res 1994; 75:434-42.

336. Hasenfuss G, Mulieri LA, Leavitt B et al. Alteration of the contractile function and excitation-contraction coupling in dilated cardiomyopathy. Circ Res 1992; 70:1225-32.

337. Gwathmey JK, Copelas L, MacKinnon R et al. Abnormal intracellular calcium handling in myocardium from patients with end-stage heart failure. Circ Res 1987; 61:70-6.

338. Gwathmey JK, Slawsky MT, Hajjar RJ et al. Role of intracellular calcium handling in force-interval relationships of human ventricular myocardium. J Clin Invest 1990; 85:1599-613.

339. Beuckelmann DJ, Nabauer M, Erdmann E. Intracellular calcium handling in isolated ventricular myocytes from patients with terminal heart failure. Circulation 1992; 85:1046-55.

340. Wier WG. Sodium-calcium exchange in intact cardiac cells: Exchange currents and intracellular calcium transients. Ann N Y Acad Sci 1991; 639:366-74.

341. Bers DM, Bridge JH. Relaxation of rabbit ventricular muscle by Na-Ca exchange and sarcoplasmic reticulum calcium pump. Circ Res 1989; 65:334-42.

342. Naqvi RU, del Monte F, O'Gara P et al. Characteristics of myocytes isolated from hearts of renovascular hypertensive guinea pigs. Am J Physiol (Heart Circ Physiol) 1994; 266 (35):H1886-95.

343. Naqvi RU, MacLeod KT. Effect of hypertrophy on mechanisms of relaxation in isolated cardiac myocytes from guinea pig. Am J Physiol (Heart Circ Physiol) 1994; 267 (36):H1851-61.

344. Schulze D, Kofuji P, Hadley R et al. Sodium/calcium exchanger in heart muscle: molecular biology, cellular function, and its special role in excitation-contraction coupling. Cardiovasc Res 1993; 27:1726-34.

345. Hryshko LV, Stiffel V, Bers DM. Rapid cooling contractures as an index of sarcoplasmic reticulum calcium content in rabbit ventricular myocytes. Am J Physiol 1989; 257:H1369-77.

346. Banijamali HS, Gao WD, MacIntosh BR et al. Force-interval relations of twitches and cold contractures in rat cardiac trabeculae. Effect of ryanodine. Circ Res 1991; 69:937-48.

347. Bridge JH, Smolley JR, Spitzer KW. The relationship between charge movements associated with ICa and INa-Ca in cardiac myocytes [see comments]. Science 1990; 248:376-8.

348. Kohmoto O, Levi AJ, Bridge JHB. Relation between reverse sodium-calcium exchange and sarcoplasmic reticulum calcium release in guinea pig ventricular cells. Circ Res 1994; 74:550-4.

349. Levesque PC, Leblanc N, Hume JR. Release of calcium from guinea pig cardiac sarcoplasmic reticulum induced by sodium-calcium exchange. Cardiovasc Res 1994; 28:370-8.

350. Leblanc N, Hume JR. Sodium current-induced release of calcium from cardiac sarcoplasmic reticulum [see comments]. Science 1990; 248:372-6.

351. Komuro I, Wenninger KE, Philipson KD et al. Molecular cloning and characterization of the human cardiac Na$^+$/Ca^{2+} exchanger cDNA. Proc Natl Acad Sci USA 1992; 89:4769-73.

352. Nicoll DA, Longoni S, Philipson KD. Molecular cloning and functional expression of the cardiac sarcolemmal Na-Ca exchanger. Science 1990; 250:562-5.

353. Hanf R, Drubaix I, Marotte F et al. Rat cardiac hypertrophy: Altered sodium-calcium exchange activity in sarcolemmal vesicles. FEBS Lett 1988; 236:145-9.

354. Nakanishi H, Makino N, Hata T et al. Sarcolemmal Ca2+ transport activities in cardiac hypertrophy caused by pressure overload. Am J Physiol 1989; 257:H349-56.

355. Kent RL, Vozich JD, McCollam PL et al. Rapid expression of the Na-Ca exchanger in response to cardiac pressure overload. Am J Physiol 1993; 265:H1024-9.

356. Studer R, Reinecke H, Rilger I et al. Gene expression of the cardiac NA+-CA2+ exchanger in end-stage human heart failure. Circ Res 1994; 75:443-53.

357. Moore EDW, Etter EF, Philipson KD et al. Coupling of the Na$^+$/Ca^{2+} exchanger, Na$^+$/K$^+$ pump and sarcoplasmic reticulum in smooth muscle. Nature 1993; 365:657-60.

358. Charlemagne D, Mayoux E, Poyard M et al. Identification of two isoforms of the catalytic subunit of Na,K-ATPase in myocytes from adult rat heart. J Biol Chem 1987; 262:8941-3.

359. Charlemagne D, Orlowski J, Oliviero P et al. Alteration of Na,K-ATPase subunit mRNA and protein levels in hypertrophied rat heart. J Biol Chem 1994; 269:1541-7.

360. Brandl CJ, Green NM, Korczak B et al. Two Ca2+ ATPase genes: Homologies and mechanistic implications of deduced amino acid sequences. Cell 1986; 44:597-607.

361. Korczak B, Zarain Herzberg A, Brandl CJ et al. Structure of the rabbit fast-twitch skeletal muscle Ca2+-ATPase gene. J Biol Chem 1988; 263:4813-9.

362. MacLennan DH, Brandl CH, Korczak B et al. Amino-acid sequence of a $Ca^{2+}Mg^{2+}$-dependent ATPase from rabbit muscle sarcoplasmic reticulum, deduced from its complementary DNA sequence. Nature 1985; 316: 696-700.

363. Lompré A, de la Bastie D, Boheler KR et al. Characterization and expression of the rat heart sarcoplasmic reticulum Ca^{2+}-ATPase mRNA. FEBS Lett 1989; 249:35-41.

364. Zarain-Herzberg A, MacLennan DH, Periasamy M. Characterization of rabbit cardiac sarco(endo)plasmic reticulum Ca^{2++}-ATPase gene. J Biol Chem 1990; 265:4670-7.

365. Lytton J, MacLennan DH. Molecular cloning of cDNA from human kidney coding for two alternatively spliced products of the cardiac Ca^{2+}-ATPase gene. J Biol Chem 1988; 263:15024-31.

366. Brandl CJ, deLeon S, Martin DR et al. Adult forms of Ca^{2+}-ATPase of sarcoplasmic reticulum: expression in developing skeletal muscle. J Biol Chem 1987; 262:3768-74.

367. Burk SE, Lytton J, MacLennan DH et al. cDNA cloning, functional expression, and mRNA tissue distribution of a third organellar Ca^{2+} pump. J Biol Chem 1989; 264:18561-8.

368. Gunteski-Hamblin AM, Greb J, Shull G. A novel Ca^{2+} pump expressed in brain,kidney, and stomach is encoded by an alternative transcript of the slow-twitch muscle sarcoplasmic reticulum Ca-ATPase gene. J Biol Chem 1988; 263:15032-40.

369. MacLennan DH, Reithmeier RAF. In:Fleischer S, Tonomura Y (eds): Structure and function of sarcoplasmic reticulum. Orlando:Academic Press, 1985: 91-100.

370. Fujii J, Zarain-Herzberg A, Willard HF et al. Structure of the rabbit phospholamban gene, cloning of the human cDNA, and assignment of the gene to human chromosome 6. J Biol Chem 1991; 266:11669-75.

371. Champeil P, Guillain F, Venien C et al. Interaction of magnesium and inorganic phosphate with calcium-deprived sarcoplasmic reticulum adenosinetriphosphatase as reflected by organic solvent induced perturbation. Biochemistry 1985; 24:69-81.

372. Fujii J, Weno A, Kitano K et al. Complete complementary DNA-derived amino acid sequence of canine cardiac phospholamban. J Clin Invest 1987; 79:301-4.

373. Ganim JR, Luo W, Ponniah S et al. Mouse phospholamban gene expression during development in vivo and in vivo. Circ Res 1992; 71:1021-30.

374. Moorman AFM, Vermeulen JLM, Koban MU et al. Patterns of expression of sarcoplasmic reticulum Ca^{2+}ATPase and phospholamban mRNAs during rat heart development. Circ Res 1995; 76:616-25.

375. Kovacs RJ, Nelson MT, Simmerman HKB et al. Phospholamban forms Ca²⁺-selective channels in lipid bilayers. J Biol Chem 1988; 263:18364-8.

376. Voss J, Jones LR, Thomas DD. The physical mechanisms of calcium pump regulation in the heart. Biophys J 1994; 67:190-6.

377. Arkin IT, Adams PD, MacKenzie KR et al. Structural organization of the pentameric transmembrane α-helicies of phospholamban, a cardiac ion channel. EMBO J 1994; 13:4757-64.

378. Wegener AD, Simmerman HKB, Lindemann JP et al. Phospholamban phosphorylation in intact ventricles. J Biol Chem 1989; 264:11468-74.

379. Movsesian MA, Nishikawa M, Adelstein RS. Phosphorylation of phospholamban by calcium-activated, phospholipid-dependent protein kinase. J Biol Chem 1984; 259:8029-32.

380. Nagai R, Zarain-Herzberg A, Brandly J et al. Regulation of myocardial Ca²⁺-ATPase and phospholamban mRNA expression in response to pressure overload and thyroid hormone. Proc Natl Acad Sci USA 1989; 86:2966-70.

381. Luo W, Grupp IL, Harrer J et al. Targeted ablation of the phospholamban gene is associated with markedly enhanced myocardial contractility and loss of β-agonist stimulation. Circ Res 1994; 75:401-9.

382. de la Bastie D, Levitsky D, Rappaport L et al. Function of the sarcoplasmic reticulum and expression of its Ca²⁺-ATPase gene in pressure overload-induced cardiac hypertrophy in the rat. Circ Res 1990; 66:554-64.

383. Feldman AM, Weinberg EO, Ray PE et al. Selective changes in cardiac gene expression during compensated hypertrophy and the transition to cardiac decompensation in rats with chronic aortic banding. Circ Res 1993; 73:184-92.

384. Mercadier JJ, Lompre AM, Duc P et al. Altered sarcoplasmic reticulum Ca²⁺-ATPase gene expression in the human ventricle during end-stage heart failure. J Clin Invest 1990; 85:305-9.

385. Arai M, Matsui H, Periasamy M. Sarcoplasmic reticulum gene expression in cardiac hypertrophy and heart failure. Circ Res 1994; 74:555-64.

386. Hassenfuss G, Reinecke H, Studer R et al. Relationship between myocardial function and expression of sarcoplasmiic reticulum Calcium-ATPase in failing and nonfailing human myocardium. Circ Res 1994; 75:434-42

387. Movsesian MA, Karimi M, Green K et al. Ca²⁺-transporting ATPase, phospholamban, and calsequestrin levels in nonfailing and failing human myocardium. Circulation 1994; 90:653-7.

388. Baertschi AJ. Antisense oligonucleotide strategies in physiology. Mol Cell Endo 1994; 101:R15-24.

389. Feldman AM, Ray PE, Silan CM et al. Selective gene expression in failing human heart. Circulation 1991; 83:1866-72.

390. Schwartz K, Boheler KR, de la Bastie D et al. Switches in cardiac muscle gene expression as a result of pressure and volume overload. Am J Physiol 1992; 31:R364-9.

CHAPTER 6

MOLECULAR BASIS OF GENETIC DISORDERS OF THE HEART

In recent years, advances in the methodology of molecular biology have provided an extraordinarily sophisticated set of tools with which to determine the molecular basis of inherited genetic disorders.[1,2] In many cases this has led to the development of improved diagnostic techniques and the design of experiments that hold promise for the eventual treatment of genetic diseases. The application of reverse genetics (positional cloning) and "candidate gene" deduction has led to the identification of the disease loci or, in some cases, the affected genes, including: cystic fibrosis (CF), Duchenne muscular dystrophy (DMD), Huntington's disease and some inherited cardiac disorders such as familial hypertrophic cardiomyopathy (FHC), Marfan syndrome (MFS) and DiGeorge syndrome.

What do we mean by reverse genetics (positional cloning)? In brief, it is the use of DNA markers to find an allele(s) at a defined chromosomal location that co-segregates with the disease, followed by the use of molecular biological techniques to systematically examine the DNA in the vicinity of the allele(s) until the gene responsible for the disease is identified. Although theoretically straightforward, it can be a technically demanding process and the magnitude of such a problem can be envisaged as follows. In the human haploid genome there are approximately 3.0×10^9 base pairs and the average chromosome contains about 120×10^6 base pairs. Each gene contains, on average, from $1\text{-}200 \times 10^3$ base pairs and there are probably about $1\text{-}2 \times 10^5$ genes per genome.[3] However, the minimum number of mutations necessary to cause a disease is, in fact, one base pair. The technical aspect of identifying such point mutations in order to understand the genetic basis of the disease state is thus a monumental task.

POSITIONAL CLONING AND CANDIDATE GENES

Identification of the affected gene in a particular disorder can be achieved using one of two approaches, both of which rely heavily on

Molecular Biology of Cardiac Development and Growth, by Paul J.R. Barton, Kenneth R. Boheler, Nigel J. Brand, Penny S. Thomas. © 1995 R.G. Landes Company.

molecular biology techniques. In the first, called positional cloning, and where the biology of the defect is often completely unknown, the DNA of affected patients and their families is examined by linkage analysis in order to define a particular region of a chromosome in which the affected gene is thought to lie.[4] If two loci on the same chromosome lie very close to one another, there is a high likelihood that they will co-segregate during the crossing-over of chromosomes and exchange of DNA (recombination) that occurs at meiosis. However, if they are separated by a considerable distances on the chromosome, it is highly likely that recombination will separate the loci. If a disease locus can be predicted to lie close to a recognized DNA marker, such as a region of DNA sequence polymorphism (see below) whose precise chromosomal location is already established, the pattern of inheritance of the disease locus can be traced in a patient and relatives from (hopefully) several generations to see whether or not it always co-segregates with the marker. If it appears statistically likely that the disease locus co-segregates with the marker, then it is reasonable to hypothesize that the affected gene underlying the disorder lies close to the marker. Linkage analysis often permits mapping of a gene to a region or locus containing several million base pairs of DNA. Pinpointing the actual gene and determining its DNA sequence, exonic structure and the amino acid sequence of the protein it encodes (Fig. 6.1) can still be an immensely time-consuming and often frustrating task, as experienced with Huntington's disease.[5] Positional cloning was used to identify the previously unknown Cystic Fibrosis Transmembrane Regulator (CFTR) gene as the site of mutation in CF (see ref. 6 for review and references). Analysis of the CFTR gene and of its protein product, which spans the cell membrane of bronchial epithelial cells and functions as a channel for chloride ions, has, in a relatively short period of time, led to a large body of work directed at identifying and understanding the gene defect at the molecular level. It has led to the development of novel gene therapy strategies for treating CF in the clinical environment. For example, the use of disabled adenoviral vectors to direct the expression of the CFTR cDNA to epithelial cells in the lungs of CF patients, or the use of topically applied liposome aerosols to deliver recombinant CFTR cDNA or protein to the lungs underlie two forms of gene transfer currently undergoing clinical trials.[6]

A second approach to identifying gene defects is the candidate gene approach. This involves hypothesizing that a particular disorder is due to a defect in a certain type of gene and then looking for that gene within a region of DNA identified through linkage analysis of affected patients and their families. In this way fibrillin, a constituent of several connective tissues affected in Marfan syndrome (MFS) was proposed, and later shown, to be the site of the Marfan phenotype.[7-10] As more genes are identified and their DNA sequences and chromosomal locations entered into international databases as part of the effort to

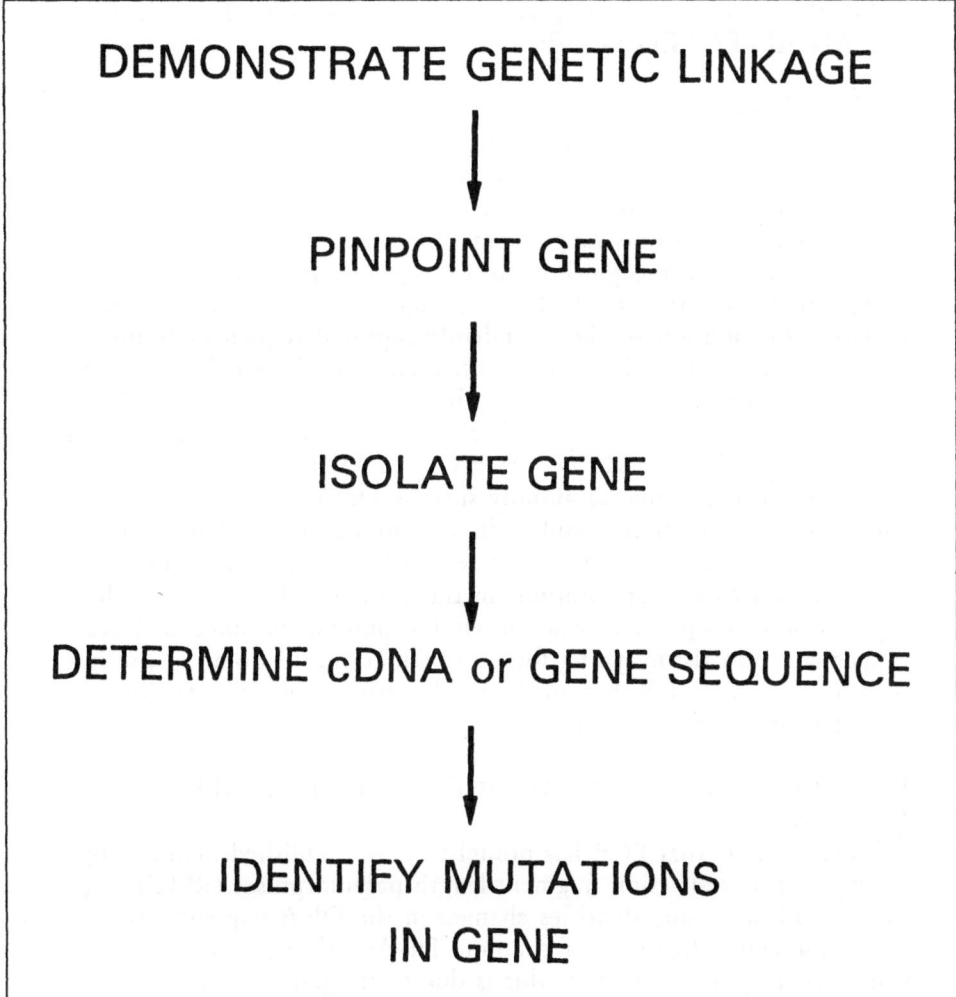

DEMONSTRATE GENETIC LINKAGE

↓

PINPOINT GENE

↓

ISOLATE GENE

↓

DETERMINE cDNA or GENE SEQUENCE

↓

**IDENTIFY MUTATIONS
IN GENE**

Fig. 6.1. Steps involved in positional cloning.

sequence the entire human genome organized by the Human Genome Organization (HUGO), it will facilitate the assignment of candidate genes relative to a particular disease locus. Consequently, it has been predicted that the two methodologies of positional cloning and the candidate gene approach will converge to a great extent, giving a field of genetic research which has been called the "positional candidate strategy".[11] First, linkage analysis will allow the localization of a disease locus to a sub-chromosomal site. DNA sequence databases can then be screened for candidate genes located within that region, which can subsequently be examined by PCR analyses for mutations associated with the disease.

MOLECULAR TECHNIQUES USED IN IDENTIFICATION OF AN AFFECTED GENE

PCR in Diagnosis

Human DNA is highly polymorphic, with more than 5000 different loci, of which about a third have been mapped to part of a chromosome or a specific locus showing allelic variation within the population.[1] These variations range from single base changes, which often create or destroy a recognition site for a restriction endonuclease, allowing their detection by RFLP mapping (see below), to inheritable variation in numbers of short tandemly repeated sequences (mini- or micro-satellites). The ability to identify and amplify single-copy gene sequences using PCR has proved invaluable in the development of new techniques for detecting genetic variation or in the improvement of existing ones. The speed and sensitivity of PCR (see chapter 1) has meant that it is possible to amplify specific DNA sequences from very small amounts of DNA template in a short period of time, allowing rapid screening of mutations associated with some inherited disorders. PCR has many other applications in the pathology laboratory, including screening biopsy material for viral genomes, chromosomal rearrangements and tissue typing for transplantation.[12-15] Some of the improvements that PCR has made to detecting genetic variation are described in the following sections.

Restriction Fragment Length Polymorphism (RFLP) Mapping

The benefits that PCR has brought to an established methodology are typified by restriction fragment length polymorphism (RFLP) mapping. RFLP mapping identifies changes in the DNA sequences of two alleles following digestion of genomic DNA with a particular restriction enzyme.[4] At its simplest, this is due to single-base changes in the DNA sequence which either create or abolish a recognition site for a particular restriction enzyme (Fig. 6.2). In traditional RFLP mapping, genomic DNA is isolated from patients and digested with a restriction enzyme which detects a polymorphism. In the example shown in Figure 6.2 the gene in question is flanked by sites for the restriction endonuclease *Eco*RI in allele A, yielding a large restriction fragment after digestion. The other allele (B) contains an additional *Eco*RI site which has arisen through a single base change. Digestion with *Eco*RI produces a shorter DNA fragment that can be identified by electrophoresis through an agarose gel. (As total genomic DNA contains many sites for *Eco*RI, the DNA fragments are first separated by electrophoresis, then transferred and immobilized onto a nylon or nitrocellulose membrane (Southern blot) and the fragment(s) specific for the gene detected by hybridization with a gene-specific DNA probe). In this way, allelic variation can be determined: Figure 6.3 shows inheritability of a point polymorphism which maps close to the CF gene. PCR has

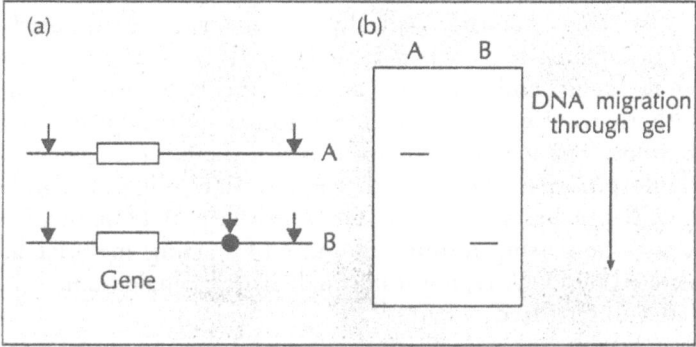

Fig. 6.2. RFLP mapping. (a), two different alleles, A and B, of a gene (open rectangle) differ in the distribution of recognition sites for the enzyme EcoRI (vertical arrows). (b) Polymorphism for EcoRI sites is detected by electrophoresis and by probing a Southern blot of the resulting gel with a probe specific to the gene. (From Santis and Barton, Annual of Cardiac Surgery, 5th edition; 1992:41-6)

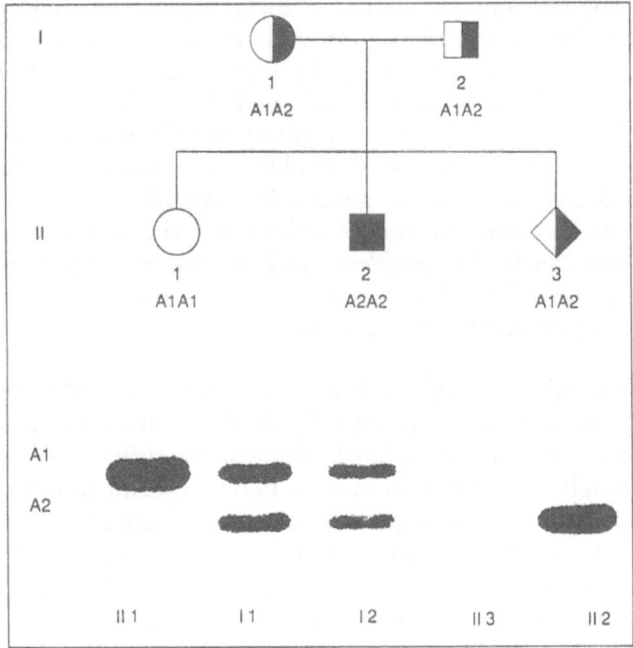

Fig. 6.3. Two-generation pedigree for a family carrying an inheritable point polymorphism (A1/A2) which maps close to the cystic fibrosis gene. Both parents (I 1 and I 2) are heterozygous for the polymorphism, whereas child II 2 is homozygous. The fetus II 3 is also heterozygous, although the bands are extremely faint. (From Weatherall, New Genetics and Clinical Practice;1991:103-37, with permission of Oxford University Press)

brought many improvements to RFLP mapping. It requires much less DNA as starting material, typically 10 ng or so as opposed to the several micrograms required in traditional RFLP mapping techniques, is faster to set up and carry out and, by and large, obviates the need for Southern transfer and detection through hybridization to a gene-specific probe. The major requirement is that the DNA sequence flanking the polymorphism is known so that appropriate oligonucleotide primers for PCR can be synthesized. The sensitivity of PCR in identifying genetic variation using minute amounts of starting material has been demonstrated by DNA typing from single hairs,[16] single human sperm[17,18] or isolated unfertilized oocytes.[19]

Many polymorphisms do not create a change in a restriction site. Some polymorphisms involve expansion of short repeated sequences called Variable Number of Tandem Repeats or VNTRs. When flanked by recognition sites for a particular enzyme these hypervariable regions can be detected by electrophoresis as restriction fragments of varying sizes. PCR has improved these techniques by enabling the identification of such polymorphisms, requiring the DNA sequence at two positions flanking a polymorphic region but not the presence of convenient flanking restriction sites. This is useful when estimating the length of VNTR regions where the repeat is often just a few dozen base pairs in length. The analysis of VNTRs by PCR is used routinely in forensic medicine to identify individuals[20] and the technique has received widespread media coverage recently during the OJ Simpson trial. However, only a small percentage of VNTRs are amenable to PCR-based detection. Many VNTRs are genetically unstable and some are too large for efficient amplification. Furthermore, the repetitive nature of the sequences can lead to problems such as primer slippage during the annealing phase of the PCR cycle, leading to the accumulation of a heterologous population of fragments.[13]

SINGLE STRAND CONFORMATION POLYMORPHISM (SSCP)

Point mutations are detectable by PCR and form the basis of approaches for screening individuals thought to suffer from DMD or CF. The Single-Strand Conformation Polymorphism (SSCP) assay allows the identification of single-base changes by the amplification of a given locus in the presence of a labeled dNTP or primer.[21] The resulting labeled amplification products (either radioactively or non-isotopically tagged) are denatured, then separated by polyacrylamide gel electrophoresis. Single-base changes in the amplified region are revealed by their altered electrophoretic mobility with respect to a wild-type sequence amplified with the same primers. With no need to depend on convenient restriction enzyme sites, PCR primers can be designed to flank any polymorphism, as long as the surrounding gene sequence is known. Mutant sequence identified by SSCP can be eluted from the gel and re-amplified for DNA sequence determination.[21]

ALLELE-SPECIFIC OLIGONUCLEOTIDE HYBRIDIZATION (ASO)

Allele-specific oligonucleotide hybridization or ASO allows the detection of point mutations on the basis that an oligonucleotide containing a single mismatch hybridizes to an imperfect match more weakly than to a perfect one. Under carefully controlled washing conditions it is possible to distinguish a point mismatch by changing the temperature of the washing buffer by a few degrees. PCR-amplified sequences spanning polymorphic sites are immobilized on nylon or nitrocellulose membranes and then probed with an oligonucleotide of wild-type sequence. This requires a separate hybridization for each polymorphism under scrutiny. Alternatively, oligonucleotides specific for different polymorphisms at a specific locus are immobilized on a filter and hybridized to a PCR-generated DNA fragment spanning the locus under investigation (Fig. 6.4). This approach has been used to screen variation within the β-globin gene, associated with β-thalassemia, and at the HLA-DQA locus.[22] In these examples, sequence-specific oligonucleotides were immobilized to nylon membranes via long poly-dT tails attached to the 3' end of each primer. These were subsequently hybridized to a PCR-amplified 242 bp fragment from a compact hypervariable region of the DQA locus or with the entire β-globin gene. Single base-pair mismatches were detected by washing just below the empirically determined melting temperature (Tm) for a perfect match between the labeled oligonucleotide and wild-type target.[23] This technique can be done with a radiolabeled DNA probe or, as

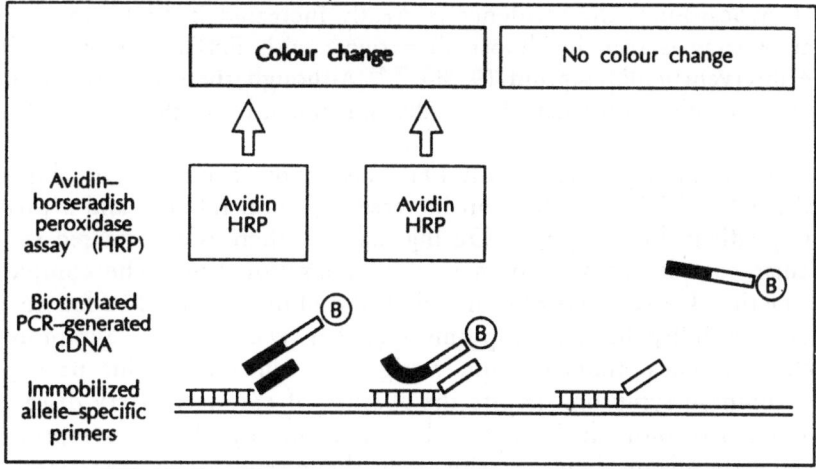

Fig. 6.4. Allele-specific oligonucleotide hybridization (ASO). Oligonucleotide primers specific for different polymorphic regions of the same allele (shown by black and gray rectangles) or an unrelated allele (white rectangle) are immobilized on a membrane (bottom). The membrane is hybridized to a PCR-generated biotinylated (B) probe specific for the black and gray polymorphisms. After washing to remove non-specific binding, biotinylated DNA-primer complexes are detected using a non-radioactive avidin-horseradish peroxidase colorimetric test. (From Brand, Annual of Cardiac Surgery, 4th edition; 1991:21-8)

shown in Figure 6.4, a non-isotopic labeling strategy.[24] In the latter case, the PCR-amplified DNA fragment is biotinylated, allowing the bound probe to be detected by a simple colorimetric assay using the horseradish peroxidase (HRP) enzyme system (Fig. 6.4).

To illustrate the key steps in detecting the causes of inherited disorders and to show how molecular biology and the techniques described above are used to study the incidence and causes of cardiac pathologies, we describe in this chapter how the gene defects underlying FHC and Marfan syndrome have been identified.

FAMILIAL HYPERTROPHIC CARDIOMYOPATHIES

Cardiomyopathies are by definition diseases of cardiac muscle of unknown etiology. The prevalence of hypertrophic cardiomyopathies (FHC) in the United States is about 19.7 cases per 100,000 persons;[25,26] the proportion of familial cases is, however, uncertain. Although variable, FHC is characterized by an increase in cardiac mass predominantly of the left ventricle without dilatation of the ventricular cavities involving predominantly the ventricular septum.[25-27] The increase in mass is due to an increase in the thickness of the ventricular wall, and the left ventricular chamber is generally of normal or of only a slightly diminished size. Histologically the myocytes are characterized by myofibril disarray and the presence of a significant degree of fibrosis. The myofibrils are frequently oriented obliquely or longitudinally in the cell and certain myofilaments apparently are inserted in adjacent Z lines of other myofilaments. Although this may occur in normal myocardium, the incidence is greatly increased in FHC. Myocardial disarray is seen in almost all patients with FHC, particularly in the interventricular septum (95%).[28-33] Although these features occur in other cardiomyopathies, the incidence is much lower than that linked with FHC.

Functionally, patients with FHC also show a number of abnormalities.[30,31,33] These include an increased ejection fraction and a pressure gradient during systole. During diastole, there is a decreased ventricular compliance and an increase in relaxation time. The complex interaction between systolic and diastolic abnormalities has led to a great variability in the symptoms seen between patients. Symptoms include dyspnea, thoracic pain, vertigo and syncope. Angina pectoris is a common symptom despite the absence of coronary artery disease. A number of individuals do not show signs of any of these symptoms making the diagnosis more difficult in the absence of previous family history. One serious consequence associated with FHC is the increased incidence of sudden death, particularly in individuals aged 10 to 35 years, that occurs at an annual rate of 2% to 4%.[34,35] Sudden death is only correlated with the symptom syncope, and not with the degree of ventricular hypertrophy, which is generally considered as one of the best diagnostic tools for predicting morbidity and mortality (see chapter 5).[36] The majority (60%) of sudden death cases occur during sedentary activity or with moderate activities, but in the rest of the cases the patients die during or just after intense physical exercise, particularly in young athletes.

Patients with FHC show a high degree of clinical variability. Some remain clinically stable and relatively asymptomatic over long periods of time, some die suddenly and others show severe symptoms and die from progressive cardiac insufficiency. Left ventricular cardiac hypertrophy might be detectable from birth, but can develop progressively between infancy and adolescence. This presents a number of problems diagnostically, as cardiac hypertrophy may not develop until after adolescent growth is complete. Diagnosis has until recently been based on two-dimensional echocardiography. Clinically, medical treatments can diminish the symptoms associated with FHC, but do not seem to offer any protection from the risk of sudden death.[30] In certain severe cases surgery can be indicated.

From genetic studies, FHC has been shown to be inherited as an autosomal dominant trait. A significant proportion of FHC cases are sporadic, that is they develop in the absence of any known family history.[37] FHC thus shows a large variability in its morphologic expression, clinical symptoms and natural history and this variability is both inter- and intra-familial. The fact that this disease is inherited in an autosomally dominant manner has permitted the use of reverse genetic techniques to identify the genes responsible.

The first gene found to co-segregate with FHC was the β-MHC gene.[38-41] Identification of this affected allele was achieved primarily through the use of RFLP mapping of a large afflicted French-Canadian family (Fig. 6.5). Initially, neither candidate genes nor known cytogenic abnormalities were associated with the disease locus. A genetic linkage map spanning the human genome was constructed and, through the use of 41 polymorphic DNA probes, one was identified that was tightly linked to the locus for FHC. This clone was derived from chromosome 14 and provided the basis for identifying the gene that caused the disorder in this family.

Once the genetic locus was identified, it was still possible that the marker and disease allele were separated by thousands of base pairs. In the case of the marker CRI-L436, the investigators were fortunate, as the disease locus mapped very close (within 5 centimorgans) to the well-known cardiac α- and β-myosin heavy chain genes on chromosome 14q11.[38,39,42] As these genes are expressed in cardiac muscle, the data suggested that cardiac α- or β-MHC might be candidates for the disease gene in FHC and that mutations in them would be found in FHC patients. Detailed analyses subsequently identified a point mutation in exon 13 of the β-MHC gene that was present in all individuals affected with FHC in the large French-Canadian kindred.[40] This mutation led to the substitution of a glutamine for an arginine at position 403 (Arg403Gln) in a highly conserved region of the protein (Fig. 6.6), suggesting that a critical function of the myosin molecule might be

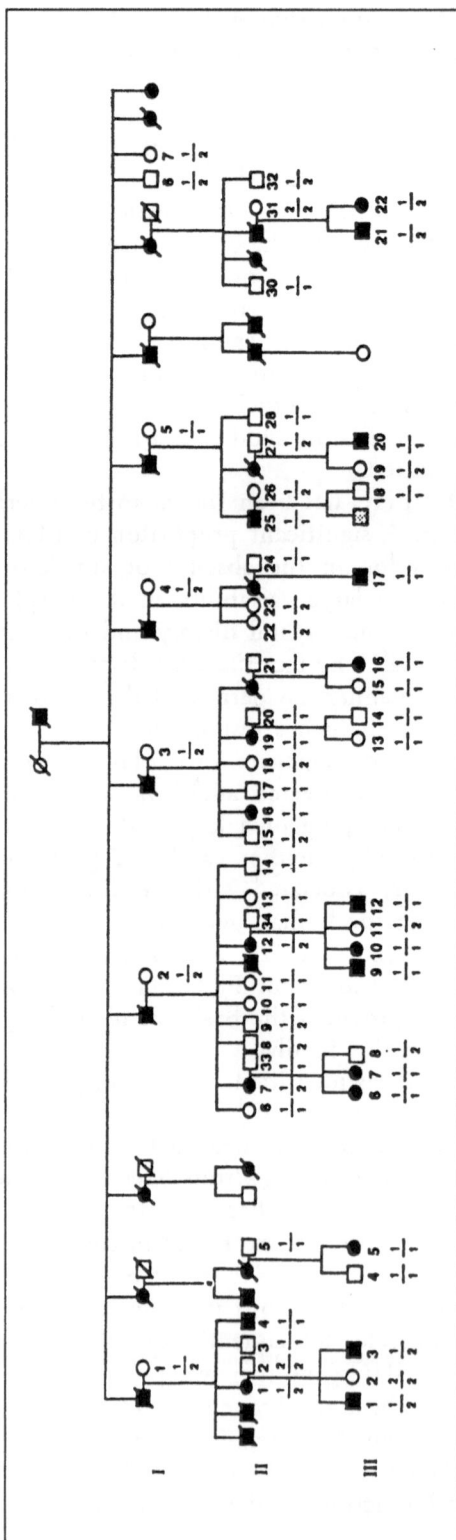

Fig. 6.5. Pedigree of inheritance of familial hypertrophic cardiomyopathies in a large Canadian kindred. It was by examination of this family through the use of positional cloning techniques that the first point mutations for FHC were identified. Darkened symbols denote affected individuals, while open symbols denote unaffected. Stippled symbols denote individuals who were not examined. The genotype of the individuals is shown where allele 1 designates a 3.5-kb fragment detected by probe pSC14, and allele 2 designates a 1.3-kb fragment detected by probe pSC14. Linkage is to allele 1. (From Solomon et al, Am J Hum Genet 1990; 47:389-394, by permission of the University of Chicago Press).

```
Unaffected      luAlaAspLysSerAlaTyrLeuMetGlyLeuAsnSerAlaAspLeuLeuLysGlyLeuCys
           TCCAGAGGCTGACAAGTCTGCCTACCTCATGGGGCTGAACTCAGCCGACCTGCTCAAGGGGCTGTGC  10154
Affected   ...................................................................

Unaffected HisProArgValLysValGlyAsnGluTyrValThrLysGlyGlnAsnValGlnGln
           CACCCTCGGGTGAAAGTGGGCAATGAGTACGTCACCAAGGGGCAGAATGTCCAGCAGGTGGGTCCAT  10221
Affected   .......A...........................................................
                 Gln
```

Fig. 6.6. DNA sequence showing the difference between unaffected and affected individuals with a point mutation in exon 13 of the β-myosin heavy chain gene. The result of this mutation is the creation of a missense mutation and the predicted change in amino acid sequence from Arg to Gln at position 403. (From Giesterfer-Lowrance et al, Cell 1990; 62:999-1006)

disrupted. At least 29 mutations in myosin heavy chains have been published from different affected families and as many as 38 have been identified to date.[27] The data are therefore highly suggestive of a role for the mutations of β-MHC in the development of FHC. The most conclusive evidence that mutations in myosin heavy chain genes cause FHC comes from the sporadic cases where de novo mutations have been defined in individuals whose parents are genetically and clinically unaffected.[37]

Myosin Mutations and Function

The myosin molecule is composed of a number of components including heavy and light chains (see chapter 3). Previous knowledge of the head region of the heavy chain had been obtained from tryptic digestions from which three major regions had been described: a 25 kD NH_2-terminal nucleotide binding region, a central 50 kD segment and a 20 kD COOH-terminal segment with the last two regions capable of binding actin.[43] The general structure of the S1 fragment of myosin has also been determined by X-ray crystallography from which it is now possible to ascribe various consequences to the identified mutations in the heavy chains. The "head" region (amino terminus) of the molecule contains an actin binding domain composed of three regions (amino acids 626 to 647, 529 to 553, and 403 to 416), and a nucleotide substrate pocket located on the opposite side of the actin binding sites (Trp131, Ser181, Ser243, and Ser324). The mutation Arg403Gln seen in the first identified point mutation for β-MHC is thus located in the important actin binding domain of the molecule. Several point mutations (Arg403 to Gln, Leu or Trp; Lys615Asn; Val606Met) in the vicinity of the actin binding site have been described as well as several in the nucleotide binding site (Thr124Ile; Asn187Lys; Asn232Ser; Phe244Leu). Notably, all the mutations identified with FHC have been shown to be present in heavy meromyosin and, to date, none with light meromyosin.

A number of the identified point mutations in the myosin heavy chains have been studied in experimental systems to elucidate the underlying molecular mechanisms responsible for FHC. These include a number of molecular techniques and the in vitro motility assay. Seven mutations have been analyzed (Arg403Gln, Tyr162Cys, Val606Met, Leu908Val, Try162Cys, Thr124Ile, Gly256Glu) with the motility assay and, relative to non-mutated forms, each of the mutations leads to a reduction in force and velocity.[44] In COS cells transfected with mutated forms of β-MHC, it appears that the stability of the proteins is not affected, but their ability to form filaments does appear to be impaired.[45] These data together with those generated from the genetic analyses strongly suggest that β-MHC mutations are indeed responsible for FHC (Fig. 6.7). How these mutations lead to the histological abnormalities normally seen with FHC is still unclear.

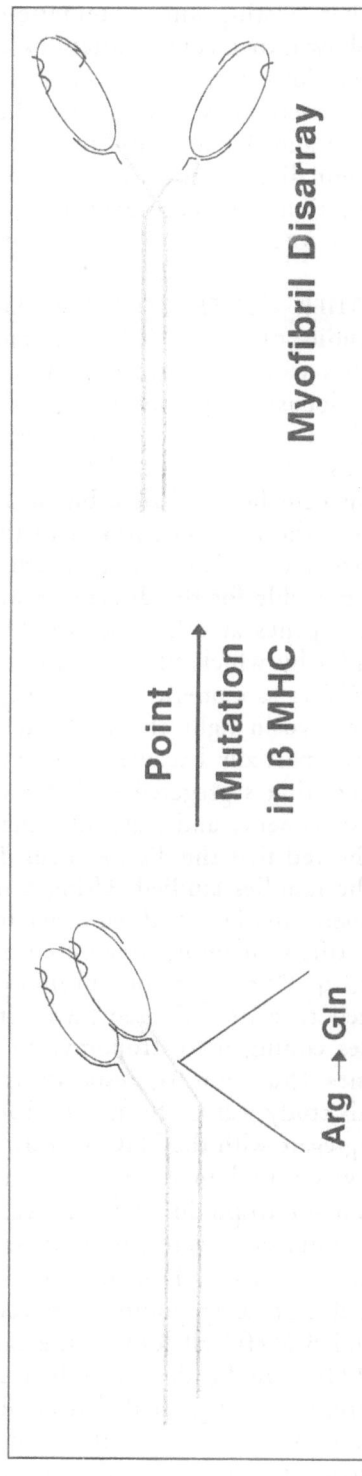

Fig. 6.7. Cartoon showing the potential for how a point mutation could lead to some of the symptoms associated with familial hypertrophic cardiomyopathy, which include sudden death and myofibril disarray. (From Boheler, Annual of Cardiac Surgery, 7th edition, 1994:41-47)

Retrospective studies relating known mutations with the clinical prognosis have also shown that some mutations in the heavy chain molecules can be more serious than others. For example, the Arg403Gln is malignant and fully penetrant; whereas, two other mutations in the same codon (Arg403Trp and Arg403Leu) are associated with different penetrances and different disease severities.[27,40,43,46-50] Clinically, this is important, once a mutation has been identified, as it would allow a more accurate prognosis to be given to the patients.

FHC LINKED TO OTHER CONTRACTILE PROTEIN GENES

After the initial studies on MHC, it became apparent that a number of unrelated families with FHC did not show linkage to β-MHC (designated the FHC-1 locus), suggesting that the disease causing FHC could be caused by defects at more than one locus.[39,51] In fact, less than 30% of all FHC cases localize to the FHC-1 locus. FHC is therefore not just a disease of myosin heavy chains, but is a genetically heterogeneous disease. Because the first gene linked to the disease was associated with the sarcomere, a candidate gene approach was used to identify other loci or genes responsible for the disease. It was hoped that identification of the disease genes at other loci would help elucidate the underlying mechanism(s) by which FHC-causing mutations produced the FHC phenotype. Other sarcomeric-associated genes were therefore assessed, including the myosin light chain 1 gene[52] and cardiac α-actin.[51] These were chosen partially because of the availability of convenient molecular markers that segregated with these loci. However, in these cases, genetic experiments and statistical analyses (logarithm of odds (LOD) scores) showed that the disease locus did not co-segregate with these genes in the families studied. Using a more defined set of highly informative genetic markers,[53] three other loci linked with the disease FHC were identified, including one on chromosome 1q3,[52] one on chromosome 11p13-q13[54] and one on chromosome 15q2.[55] A fifth locus has been reported to exist.[56] Subsequently, it was reported that mutations in the genes coding for α-tropomyosin and cardiac troponin T on chromosomes 15q2 and 1q respectively were also responsible for FHC.[55,57] This study was performed on four families that had FHC which did not segregate with the FHC-1 locus. For α-tropomyosin, a missense mutation was found in exon 5, which encodes part of a putative binding region for troponin T. In the case of cardiac troponin T, two missense mutations were found in exons 8 and 9, and another in a splice site (see chapter 1) in intron 15 that was predicted to produce a truncated cardiac troponin T. Because α-tropomyosin, cardiac troponin T and β-MHC all lead to the same disease state, it was concluded that FHC can be described best as a disease of the sarcomere. Furthermore, it was suggested that an abnormal stoichiometry of sarcomeric proteins in these patients might be the underlying cause of the cardiac hypertrophy seen in this disease.

MARFAN SYNDROME (MFS)

MFS IS A DISEASE OF CONNECTIVE TISSUE

Marfan syndrome (MFS) is an autosomal dominant disorder of connective tissue with an incidence of about 1 in 20,000 which, if left untreated, can be fatal through severe cardiovascular complications. The disease presents with one or more of several characteristic phenotypes which fall into three main categories: skeletal disorders, including elongation of the tubular bones (dolichostenomelia), extreme flexibility of the joints, and arachnodactyly or "spider fingers"; the occular defect ectopia lentis, in which the lens may easily become detached due to weakening of the supporting suspensory filaments (ciliary zonule), and cardiovascular complications, including progressive aortic root dilatation and mitral valve prolapse. Whilst the skeletal and occular phenotypes are fairly easy to detect from an early age (the lens may detach during play or sport; parents or siblings may be unusually tall or, in rare instances, MFS manifests on one side of the body only[58]), cardiovascular symptoms are often difficult to detect, particularly if the other phenotypes are absent. Indeed, it is not unknown for otherwise fit individuals with no obvious characteristics of MFS to die due to sudden rupture of the weakened aortic wall.[59]

The search for a common link between the seemingly disparate manifestations of MFS led many research groups to examine the major components of connective tissues as candidates for the site of the Marfan defect and these studies have been reviewed recently.[58,60] The possible linkage of the genes for collagens and elastin to MFS was investigated by several groups, as these genes were plausible candidates for the genetic lesion associated with MFS. This was because early findings showed a notable disarray of elastin fibers and altered collagen fibers in aorta wall or fibroblast cell lines cultured from skin of Marfan patients (reviewed in ref. 58). Changes in the levels of type I collagens present in aortal tissue and skin of some patients were also noted, but no clear common element could be found between different families, ruling out defects in these collagens as the major cause of MFS.[58] A few patients with MFS have been shown to have reduced levels of decorin, a proteoglycan which binds to type I and II collagens and affects the rate of formation of collagen fibrils in affected tissues.[61] However, a common link between the majority of affected patients remained unclear until recently. The breakthrough in determining a candidate gene for the site of the Marfan defect came about through a combination of linkage analysis, which localized the gene to chromosome 15 in several geographically dispersed families,[9,62,63] and biochemical characterization of fibrillin, a novel 350 kDa glycoprotein.[64] When viewed under the electron microscope, fibrillin has an extended rod-like structure, with pronounced "beads" along its length. It is a major component of microfibrils present in elastic and connec-

tive tissues in the aorta, skin, lung and the lens suspensory filaments.[58,60,64-66] The fibrillin protein, as deduced from translating the
DNA sequence derived from cDNAs spanning its 10 kb mRNA, contains 1974 amino acids. Fourteen percent of these are cysteines, the
majority of which are grouped into two types of repeating unit. At
least 34 units of 42-44 amino acids in length, each containing six
cysteine residues at conserved positions, are dispersed throughout the
length of the protein (Fig. 6.8). These show significant homology to
repeats found in the epidermal growth factor (EGF) precursor protein
and some may bind calcium in order to stabilize secondary struc-
ture.[7,8,58,67-69] Several larger repeats, each containing 8 cysteines, are similar
to repeated sequences found in transforming growth factor β1 (TGF-
β1) binding protein. The presence of repeats containing a high proportion of conserved cysteines suggests that they may participate in
forming inter-molecular disulphide bridges in order to stabilize the
arrangement of fibrillin molecules and stiffen the microfibrils. Significantly, fibrillin was found to be deficient in tissues from MFS patients and this deficiency co-segregated with the disease in several independently studied families.[58]

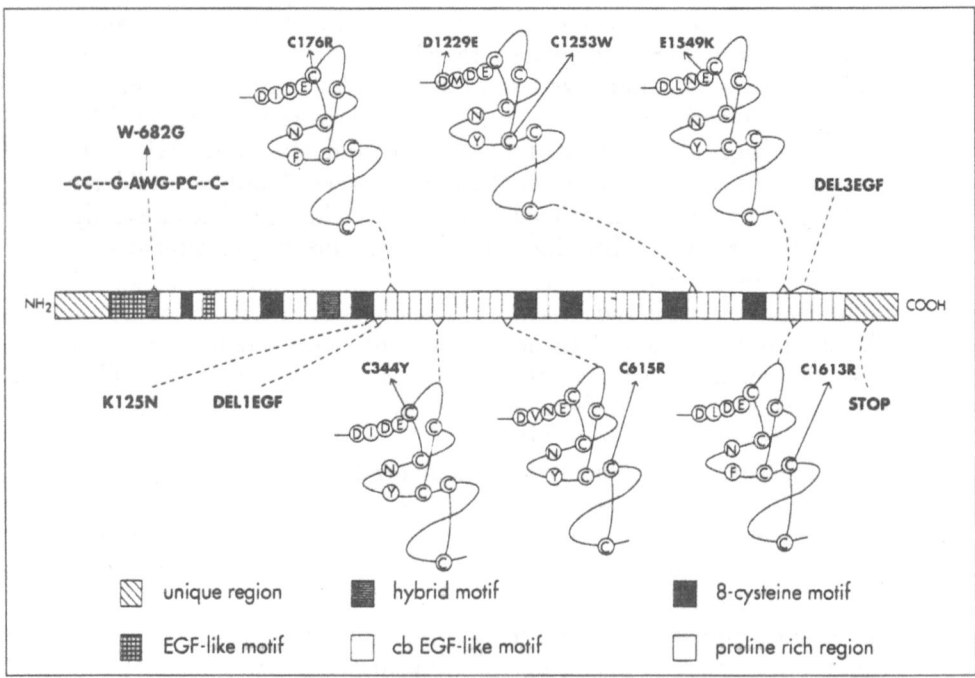

Fig. 6.8. Schematic representation of the different motifs that make up the fibrillin polypeptide. Also
shown are the position and nature of several point mutations associated with MFS, using the single letter
amino acid code. DEL1EGF, mutation in which one cb-EGF motif is deleted, DEL3EGF, three motifs
deleted. cb, calcium-binding. (From Kainulainen et al, Nature Genet 1994; 6:64-9)

LINKAGE OF THE FIBRILLIN GENE TO MFS

These various lines of research culminated with the firm linkage of MFS to the fibrillin gene (designated FBN1).[7,8,10] cDNA clones for fibrillin were isolated[8] and the first identification made of mutations within the fibrillin gene of MFS patients.[10] Linkage analysis indicated that the disease locus was located on chromosome 15 with odds of over 1025 to 1 and a combined LOD score of 7.56.[9,62,63] Additionally, the genes for many connective tissue proteins such as collagen and fibronectin had been cloned, providing probes which were used in Southern analyses to exclude those genes as candidates for the Marfan gene.[58] In 1991, the localization of the fibrillin gene to 15q21, close to the best MFS marker, was reported.[7,70] This was made possible because purified peptide fragments of fibrillin had been partially sequenced, allowing degenerate oligonucleotide sets to be synthesized based on the amino acid sequence. These primers were used in PCR cloning experiments to amplify partial cDNA fragments for fibrillin which were labeled and used as probes for in situ hybridization to human metaphase chromosome spreads. The crucial evidence was the discovery of mutations within the Marfan gene in the DNA of affected patients, but not in normal controls. The first mutations, described by Dietz et al,[10] identified the same single base mutation in the fibrillin gene from two unrelated Marfan patients who both presented with a severe clinical phenotype marked by pronounced skeletal and cardiovascular symptoms. ASO hybridization showed that the mutation was not present in genomic DNA from over 200 unrelated individuals with no symptoms of MFS. The mutation, in the triplet coding for an arginine at position 239 in the fibrillin protein, converts a G on the coding strand to a C, changing the arginine-encoding triplet CGC to CCC, which codes for proline.[10]

How can a single-base mutation account for what manifests to the clinician as a severe Marfan phenotype? Proline is well recognized for its ability to disrupt protein secondary structure (particularly within α-helices) because, unlike other amino acids, its side chain is covalently bonded to the α-carbon, generating a rigid ring. Proteins that have an elongated rod-like structure such as fibrillin are often made up of repeated amino acid motifs. Replacement of Arg239 by proline significantly alters the structure of the fibrillin protein and this probably disrupts the formation of inter-molecular disulphide bridges when fibrillin molecules pack together, leading ultimately to microfibrillar disarray. Subsequently, other mutations in the fibrillin gene have been determined in MFS patients which result in the replacement of a conserved amino acid by an inappropriate one. As shown in Figure 6.8 these often fall within the EGF-like domains and may affect calcium binding and intramolecular stability.[69,71] Many mutations result in the replacement of conserved cysteines which participate in disulphide bridges formation or affect other conserved positions.[67] In addition, mutations other than single-base changes resulting in amino acid replacement have been detected. Mutations which lead to the creation of premature termination signals in the fibrillin coding sequence result in the production of truncated proteins[71] and, in some cases, in-frame deletions remove a substantial part of the protein.[67,72] In all, over 20 different mutations have been reported to date, dispersed throughout the gene sequence.[73]

DETECTION OF MUTATIONS AND LINKAGE TO DISEASE SEVERITY

As described earlier, PCR has greatly aided the detection of polymorphisms, allowing detection from a few nanograms of genomic DNA (perhaps taken from a blood spot or an oral wash), and can be used to identify genetic variation in utero from samples of amniotic fluid.[14,74] However, the size of the fibrillin gene and the nature of the fibrillin protein, containing over 30 EGF-like repeats, makes screening by PCR technically demanding. Nonetheless, there is evidence to suggest that heterogeneity amongst patients with respect to severity of symptoms might be associated with a particular type of mutation. For example, the mutation Glu1549Lys (see Fig. 6.8: E1549K) in one British four-generation MFS family segregates with dominantly inherited ectopia lentis.[67] If the replacement of Arg209 with proline, a severe phenotype mutation, introduces a non-flexible kink into the fibrillin, this effect may be exacerbated as proteins encoded by both wild-type and mutant fibrillin alleles attempt to assemble into the same microfibril (Fig. 6.9). In contrast, the effect of a mutation at one allele which results in the expression of a truncated protein may be less deleterious to microfibril assembly if truncated and wild-type fibrillin proteins can polymerize to form a more normal microfibril, giving a less severe phenotype.[69,71]

Some forms of MFS might be detectable through pre-natal diagnosis due to clustering of mutations into a "hot-spot". Mutations associated with severe or mild MFS phenotypes tend to cluster to particular regions of the protein, rather than be dispersed randomly throughout its length.[67] It has been reported that in at least three cases of the lethal severe neonatal form of the disease, in which enlarged heart and valve malformations can often be detected in utero through the use of echocardiography, the mutations in FBN1 cluster to the beginning of the longest stretch of EGF-like repeats (see Fig. 6.8; mutations K125N and C176R (point mutations) and a deletion of one EGF-like motif, DEL1EGF). It has been proposed that the positioning of these mutations at the beginning of repeat regions may interfere with the way in which adjacent fibrillin molecules pack together in an orderly array and may have greater bearing on the severity of the disease than the actual nature of the mutation itself. Analysis of the *Drosophila* gene *Notch*, whose gene product also contains repeated EGF-like motifs, indicates that particular phentotypes are associated with mutations at specific positions in the *Notch* gene.[67]

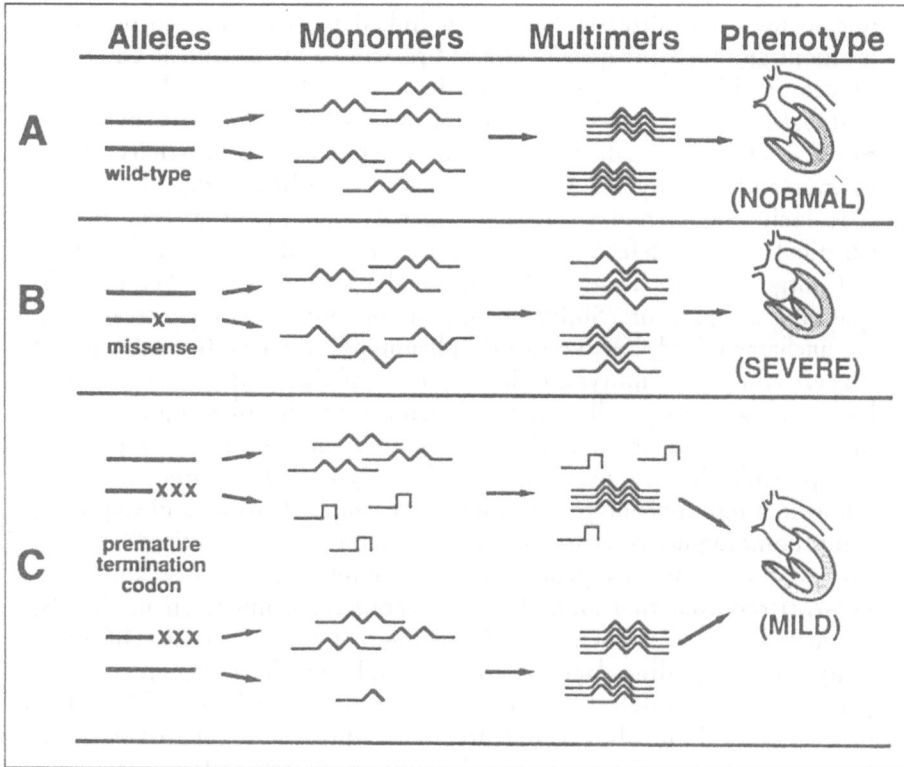

Fig. 6.9. Possible effect of different types of mutation upon severity of MFS. (A) proteins produced from the two alleles of the wild-type fibrillin gene polymerize to form normal microfibrils. (B) a missense mutation in one allele leads to the production of proteins with altered amino acid sequence: these polymerize with normal molecules encoded by the wild-type allele and interfere with the orderly assembly of fibrillin into microfibrils, producing a severe MFS phenotype (depicted as dilation of aortic root). (C) a mutation which creates a premature termination codon leads to the production of a truncated fibrillin molecule. The incorporation of these into microfibrils may result in a mild MFS phenotype. (From Dietz and Pyeritz, reproduced, with permission, from Annual Review of Physiology 56:763-96, copyright 1994, by Annual Reviews Inc).

Alternative strategies to DNA screening include analyzing changes in microfibril synthesis and deposition in extracellular matrix using primary fibroblasts cultures grown from skin or aortic wall biopsies.[67] Electron microscopic examination of fibrillin preparations could also aid identification by revealing abnormal microfibrillar structures that correlate with a particular MFS phenotype.

DIGEORGE SYNDROME

An increasing number of disease loci implicated in a variety of cardiac disorders are being identified[69] although, as described above for FHC, it is unlikely that all diseases are linked to defects at just one locus. Some diseases have been linked to particular chromsomal locations although the gene(s) affected remain to be identified. DiGeorge syndrome, like Marfan syndrome, is a multiphenotypic disorder (in that patients often present with one or more of a variety of symptoms), including facial dysmorphism, absence or hypoplasticity of the thymus and parathyroid glands and various cardiovascular abnormalities, including tetralogy of Fallot, persistent truncus arteriosus and ventricular septal defect.[75] Molecular genetics studies suggest that most DiGeorge patients have a chromosomal deletion at 22q11, possibly spanning a region of 750,000 base pairs or more.[76] The genes affected are uncharacterized, but the various phenotypes resulting from this defect suggest abnormal embryonic development of the neural crest, a population of migratory cells which contribute to the pharyngeal arches, including thymus and parathyroid regions, and the developing heart. It is possible that the perturbation of one gene in DiGeorge syndrome results in a multiphenotypic disorder. For example, Chisaka and Capecchi, using homologous recombination (see chapter 1) to inactivate expression of the homeobox gene Hoxa-3 in mouse (previously known as *hox-1.5*) reported that animals homozygous for a mutation in this locus have a particular phenotype displaying many of the classic DiGeorge symptoms, including lack of thymus and parathyroids, craniofacial abnormalities and defects of the heart and great arteries.[77] As described in chapter 2, homeobox genes are transcriptional regulators and are thought to be key determinants of many developmental processes. Although there are significant similarities in the phenotypes of DiGeorge patients and the *Hoxa-3⁻/Hoxa-3⁻* homozygous knock-out mice, absence of the human homologue HOXA3 is not the genetic lesion resulting in DiGeorge, as HOXA3 displays an autosomal dominant pattern of inheritance and is located on chromosome 7, not 22. However, the similarity of the two phenotypes suggests that proteins encoded by the two loci may contribute to parts of the same developmental pathway and that the *Hoxa-3⁻* mouse is a useful research tool for understanding DiGeorge syndrome.

CONCLUSIONS AND FUTURE DIRECTIONS

The growth in research at the interface between the basic and clinical sciences offers real promise that some cardiovascular disorders may be amenable to gene therapy strategies sooner rather than later. These may include: direct introduction of DNA into the myocardium; the use of antisense oligonucleotides to knock-out selective gene expression; delivery of recombinant DNA by viral vectors or liposomes, and the grafting of skeletal muscle, cardiac or other cells onto the myocardium, either to assist contractile properties or as a platform for the local or systemic delivery of pharmacologic agents such as angiogenic or anti-clotting agents. Some examples of these are given in the following sections.

ANTISENSE OLIGONUCLEOTIDES

Antisense oligonucleotide technology is a growing research area with important therapeutic implications. If cells in culture are transfected with antisense oligonucleotides specific for a particular mRNA, they may complex with the mRNA, leading to its destruction, probably through endogenous cellular ribonucleases and/or modifying enzymes.[78] The selective introduction of oligonucleotides into particular tissues or groups of cells therefore has great potential for altering the expression of a specific gene. The difficulty will be introducing oligonucleotides into every cell of an organ or, conversely, into a discrete subset of cells. An early attempt in vivo was the apparently successful, short-term use of antisense oligonucleotides to the proto-oncogene *c-myb* to inhibit intimal vascular smooth muscle cell proliferation in rat. The intimal lining of the carotid arteries was damaged by using a catheter and the *c-myb* antisense oligonucleotides introduced within a gel into the affected area.[79] The result of this was, as predicted, inhibition of both *c-myb* expression and intimal cell growth.

VIRAL VECTORS—ADENOVIRUS

The use of disabled adenovirus vectors to deliver recombinant genes to a particular tissue is being pursued in many centers (see refs. 80 and 81 for reviews). Disabled viral vectors are unable to replicate (that is, duplicate the genetic material they carry and package it inside a viral protein coat) because they lack certain essential genes encoding, for example, viral coat proteins. The virus can only be propagated in cell-lines which have been engineered to supply the missing viral factors necessary for correct propagation and packaging. Examples of potential adenoviral-mediated gene therapies include their use in the treatment and study of atherosclerosis[81,82] and CF.[6] This technique is not without its limitations. In the case of the CF clinical trials, recombinant CFTR, whether delivered in a disabled adenoviral vector or by an alternative method (as a complex with synthetic liposomes), is administered to the patient as an aerosol spray. This means that the upper airways are the predominant site of uptake of the recombinant CFTR. Disabled adenoviral vectors present added potential complications, including immunological reaction leading to destruction of the adenovirus vectors over short periods of time or recombination with a wild-type adenovirus strain infecting the patient, perhaps as the result of a cold. Due to these limitations, other delivery systems are being explored for their use in eventual gene therapy, including the use of cationic liposomes as a means of integrating DNA into the membranes of target cells easily.[6,81,83]

SKELETAL MUSCLE CELL TRANSPLANTATION

The mouse skeletal muscle cell line C2 has been well-characterized at the cellular and biochemical levels.[84] Recently, myoblasts of the derived line C2C12 have been grafted directly into the ventricular myocardium of syngeneic mice by Field and co-workers to assess their potential to differentiate in situ and form a permanent graft.[85] All grafts survived and the myoblasts differentiated and expressed muscle-specific proteins. Immunological analysis revealed no chronic graft rejection. All animals exhibited normal sinus rhythm and there was no sign of arrythmia in response to grafting. In addition, in some animals, myofibers formed by cell transplantation organized themselves directionally, suggesting that careful grafting of cells could have some potential for increasing cardiac output. Similar studies have been undertaken in dog. Undifferentiated skeletal myoblasts (satellite cells) were isolated, transfected with a plasmid containing the bacterial *lacZ* (β-galactosidase) gene and subsequently grafted into the myocardium of the same animal.[86] Up to eight weeks after transplantation, cells that stained for β-galactosidase activity could be detected around the site of the graft, confirming the survival of the engineered cells. Furthermore, there was evidence in two out of five animals of differentiation of the satellite cells, marked by the appearance of intercalated disc-like structures.[86] These experiments suggest that syngeneic grafting might be a way to introduce recombinant proteins to the myocardium by gene transfer into the cells to be grafted. For example, cells could be used as a platform for the delivery of angiogenic proteins to promote revascularization following an infarct episode. Human skeletal muscle satellite cells (primary myoblasts), which retain the potential to differentiate in response to muscle injury, might be suitable cells for similar grafts in man as has been indicated by Phase II clinical trials of gene therapy for DMD. Normal human myoblasts expressing the endogenous dystrophin gene (the site of the DMD defect) were injected into multiple sites of lower limb muscles in a group of 21 boys suffering from DMD in order to strengthen limb muscles and to assess the efficacy and safety of myoblast transplantation.[87] In animal systems, mouse primary myoblasts containing a stable expression construct for human growth hormone (hGH) have been injected into muscle, resulting in systemic release of hGH.[88,89]

TRANSPLANTATION OF FETAL CARDIAC MYOCYTES

A recent report describes the transplantation of fetal cardiac myocytes into the myocardium of young adult mice. Cardiac myocytes were prepared from fetal mice transgenic for *lacZ* under the control of the α-MHC promoter. These were injected into the myocardium of syngeneic non-transgenic animals.[90] The cells contributed to the ventricular myocardium, forming stable grafts which were stained for β-galactosidase activity. Grafting was not deleterious to normal cardiac function as determined by assaying for macrophage and lymphocyte infiltration and by ECG analysis of grafted and sham-operated animals. Desmosomes, detected by electron microscopy, had formed between graft and host tissues, indicating the successful grafting of host and transplanted cells. Furthermore, some structures tentatively identified as gap junctions were visible between host and donor cells, suggesting that functional electrical connections had been made, an important consideration if transplanted cardiac myocytes are to be of use in assisting the pumping ability of the heart.[90,91] Despite the ethical problems concerning the use of human fetal tissue, cardiac grafts could represent a way of assisting the failing heart or provide a stable platform for the local delivery of therapeutic agents to the heart.

These examples and others may represent the beginning of gene therapy in the cardiovascular system. Basic molecular biology research will continue to provide tools which may be of use in targeting therapeutic agents to the myocardium. As described in chapter 4, by understanding the molecular mechanisms behind tissue-specific expression, it is possible to identify enhancers within genes which bind nuclear transcription factors and are responsible for directing cardiac-specific expression. This raises the possibility of creating defined cardiac enhancer-promoter constructs for the expression of recombinant cDNAs only in cardiac myocytes. Such constructs could be delivered by viral vectors, regional injection or by cell transplantation. Studies on the processes of DNA synthesis and cell division will undoubtedly eventually identify why terminally differentiated cardiac myocytes cannot re-enter the cell cycle. These may, in turn, lead to the development of cardiac cell lines capable of proliferation and of use in cell transplantation. It may even allow the injection of recombinant DNAs encoding cell cycle proteins into differentiated cardiac myocytes close to an infarct in order to induce those cells to re-enter the cell cycle and proliferate. It is too early to say whether these types of therapy (or others yet to be developed) will be appropriate in the treatment of cardiovascular disorders, but the convergence of the viral, molecular, cellular and physiological techniques developed during the last 15 years, and described in some detail in this book, will allow us to test the true potential of gene therapy in the near future.

REFERENCES

1. McKusick VA. Current trends in mapping human genes. FASEB J 1991; 5:12-20.
2. Rossiter BJF, Caskey CT. Molecular studies of human genetic disease. FASEB J 1991; 5:21-7.
3. Menon AG, Klanke CA, Su YR. Identification of disease genes by positional cloning. Trends Cardiovasc Med 1994; 4:97-102.
4. Weatherall DJ. The New Genetics and Clinical Practice, Oxford: Oxford University Press, 1991:1-376.
5. The Huntington's disease collaborative research group. A novel gene containing a trinucleotide repeat that is expanded and unstable on Huntington's disease chromosomes. Cell 1993; 72:971-83.

6. Colledge WH. Cystic fibrosis gene therapy. Curr Op Genet Develop 1994; 4:466-71.
7. Lee B, Godfrey M, Vitale E et al. Linkage of Marfan syndrome and a phenotypically related disorder to two different fibrillin genes. Nature 1991; 352:330-4.
8. Maslen CL, Corson GM, Maddox BK et al. Partial sequence of a candidate gene for the Marfan syndrome. Nature 1991; 352:334-7.
9. Dietz HC, Pyeritz RE, Hall BD et al. The Marfan syndrome locus: confirmation of assignment to chromosome 15 and identification of tightly linked markers at 15q15-q21.3. Genomics 1991; 9:355-61.
10. Dietz HC, Cutting GR, Pyeritz RE et al. Marfan syndrome caused by a recurrent *de novo* missense mutation in the fibrillin gene. Nature 1991; 352:337-9.
11. Collins FS. Positional cloning moves from perditional to traditional. Nature Genet 1995; 9:347-50.
12. Williamson R. Molecular genetics and the transformation of clinical chemistry. Clin Chem 1989; 35:2165-8.
13. Erlich HA, Bugawan TL. HLA DNA typing.In:Innis MA, Gelfand DH, Sninsky JJ, White TJ (eds): PCR Protocols. San Diego:Academic Press, 1990: 261-271.
14. Erlich H, Arnheim N. Genetic analysis using the polymerase chain reaction. Annu Rev Genet 1992; 26:479-506.
15. Brand NJ. Principles and applications of the polymerase chain reaction. In:Latchman DS (ed): PCR Applications in Pathology. Oxford: Oxford University Press, 1995: 1-16.
16. Higuchi R, von Beroldingen CH, Sensabaugh SA et al. DNA typing from single hairs. Nature 1988; 332:543-6.
17. Li H, Gyllensten UB, Cui X et al. Amplification and analysis of DNA sequences in single human sperm and diploid cells. Nature 1988; 335:414-7.
18. Arnheim N, Li H, Cui X. PCR analysis of DNA sequences in single cells: single sperm gene mapping and genetic disease diagnosis. Genomics 1990; 8:415-9.
19. Coutelle C, Williams C, Handyside A et al. Genetic analysis of DNA from single human oocytes: a model for preimplantation diagnosis of cystic fibrosis. British Medical Journal 1989; 299:22-4.
20. Jeffreys AJ, Wilson V, Thein SL. Hypervariable 'minisatellite' regions in human DNA. Nature 1985; 314:67-73.
21. Hayashi K. PCR-SSCP: a simple and sensitive method for the detection of mutations in the genomic DNA. PCR Meths Applications 1991; 1:34-8.
22. Saiki RK, Walsh PS, Levenson CH et al. Genetic analysis of amplified DNA with immobilized sequence-specific oligonucleotide probes. Proc Natl Acad Sci USA 1989; 86:6230-4.
23. van Mansfeld ADM, Bos JL. PCR-based approaches for detection of mutated *ras* genes. PCR Meths Applications 1992; 1:211-6.

24. Bugawan TL, Saiki RK, Levenson CH et al. The use of non-radioactive oligonucleotide probes to analyze enzymatically amplified DNA for prenatal diagnosis and forensic HLA typing. Biotechnology 1988; 6:943-7.
25. WHO/IFSC Task Force. Report: Definition and classification of cardiomyopathies. Br Heart J 1980; 44:672-3.
26. Codd MB, Sugrue DD, Gersh BJ et al. Epidemiology of idiopathic dilated and hypertrophic cardiomyopathy. Circulation 1989; 80:564-72.
27. Schwartz K, Carrier L, Guicheney P et al. Molecular basis of familial cardiomyopathies. Circulation 1995; 91:532-40.
28. Davies MJ. The current status of myocardial disarray in hypertrophic cardiomyopathy. Br Heart J 1984; 51:361-3.
29. Ferrans VJ, Morrow AG, Roberts WC. Myocardial ultrastructure in idiopathic hypertrophic subaortic stenosis. Circulation 1972; XLV:769-92.
30. Maron BJ, Bonow RO, Cannon RO et al. Hypertrophic cardiomyopathy. Interrelations of clinical manifestations, pathophysiology, and therapy (2). N Engl J Med 1987; 316:844-52.
31. Maron BJ, Gottdiener JS, Epstein SE. Patterns and significance of distribution of left ventricular hypertrophy in hypertrophic cardiomyopathy. A wide angle, two dimensional echocardiographic study of 125 patients. Am J Cardiol 1981; 48:418-28.
32. McKenna WJ, Stewart JT, Nihoyannopoulos P et al. Hypertrophic cardiomyopathy without hypertrophy: two families with myocardial disarray in the absence of increased myocardial mass. Br Heart J 1990; 63:287-90.
33. Wigle ED, Sasson Z, Henderson MA et al. Hypertrophic cardiomyopathy: the importance of the site and extent of hypertrophy. A review. Prog Cardiovasc Dis 1985; 28:1-83.
34. McKenna WJ, Camm AJ. Sudden death in hypertrophic cardiomyopathy: assessment of patients at high risk. Circulation 1989; 80:1489-92.
35. Hecht GM, Klues HG, Roberts WC et al. Coexistence of sudden cardiac death and end-stage heart failure in familial hypertrophic cardiomyopathy. J Am Coll Cardiol 1993; 22:489-97.
36. Levy D, Garrison RJ, Savage DD et al. Prognostic implications of echocardiographically determined left ventricular mass in the Framingham heart study. New Eng J Med 1990; 322:1561-6.
37. Watkins H, Thierfelder L, Hwang DS et al. Sporadic hypertrophic cardiomyopathy due to de novo myosin mutations. J Clin Invest 1992; 90:1666-71.
38. Jarcho JA, McKenna W, Pare JA et al. Mapping a gene for familial hypertrophic cardiomyopathy to chromosome 14q1. N Engl J Med 1989; 321:1372-8.
39. Solomon SD, Geisterfer-Lowrance AA, Vosberg HP et al. A locus for familial hypertrophic cardiomyopathy is closely linked to the cardiac myosin heavy chain genes, CRI-L436, and CRI-L329 on chromosome 14 at q11-q12. Am J Hum Genet 1990; 47:389-94.

40. Geisterfer-Lowrance AA, Kass S, Tanigawa G et al. A molecular basis for familial hypertrophic cardiomyopathy: a beta cardiac myosin heavy chain gene missense mutation. Cell 1990; 62:999-1006.

41. Tanigawa G, Jarcho JA, Kass S et al. A molecular basis for familial hypertrophic cardiomyopathy: an alpha/beta cardiac myosin heavy chain hybrid gene. Cell 1990; 62:991-8.

42. Jaenicke T, Diederich KW, Haas W et al. The complete sequence of the human beta-myosin heavy chain gene and a comparative analysis of its product. Genomics 1990; 8:194-206.

43. Warrick HM, Spudich JA. Myosin structure and function in cell motility. Annu Rev Cell Biol 1987; 3:379-421.

44. Cuda G, Sellers J, Epstein ND, et al. In vitro motility activity of β cardiac myosin heavy chain mutation in hypertrophic cardiomyopathy. Circulation 1993; 88:I-343

45. Straceski AJ, Geisterfer-Lowrance A, Seidman CE et al. Functional analysis of myosin missense mutations in familial hypertrophic cardiomyopathy. Proc Natl Acad Sci USA 1994; 91:589-93.

46. Anan R, Greve G, Thierfelder L et al. Prognostic implications of novel beta cardiac myosin heavy chain gene mutations that cause familial hypertrophic cardiomyopathy. J Clin Invest 1994; 93:280-5.

47. Epstein ND, Cohn GM, Cyran F et al. Differences in clinical expression of hypertrophic cardiomyopathy associated with two distinct mutations in the beta-myosin heavy chain gene. A 908Leu→Val mutation and a 403Arg→Gln mutation. Circulation 1992; 86:345-52.

48. Dausse E, Komajada M, Dubourg O et al. Familial hypertrophic cardiomyopathy: microsatellite haplotyping and identification of a hotspot for mutations in the β-myosin heavy chain gene. J Clin Invest 1993; 92:2807-13.

49. Fananapazir L, Dalakas MD, Cyran F et al. Missense mutations in the β myosin heavy chain gene cause central core disease in hypertrophic cardiomyopathy. Proc Natl Acad Sci U S A 1993; 90:3993-7.

50. Fananapazir L, Epstein ND. Genotype-phenotype correlations in hypertrophic cardiomyopathy: insights provided by comparisons of kindreds with distinct and identical β-myosin heavy chain mutations. Circulation 1994; 89:22-32.

51. Schwartz K, Beckmann J, Dufour C et al. Exclusion of myosin heavy chain and cardiac actin gene involvement in hypertrophic cardiomyopathy of several French families. Circ Res 1992; 71:3-8.

52. Watkins H, MacRae C, Thierfelder L et al. A disease locus for familial hypertrophic cardiomyopathy maps to chromosome 1q3. Nat Gen 1993; 3:333-57.

53. Weissenbach J, Gyapay G, Dib C et al. A second-generation linkage map of the human genome. Nature 1992; 359:794-801.

54. Carrier L, Hengstenberg C, Beckmann JS et al. Mapping of a novel gene for familial hypertrophic cardiomyopathy to chromosome 11. Nat Gen 1993; 4:311-3.

55. Thierfelder L, MacRae C, Watkins H et al. A familial hypertrophic cardiomyopathy locus maps to chromosome 15q2. Proc Natl Acad Sci U S A 1993; 90:6270-4.

56. Hengstenberg C, Schwartz K. Molecular genetics of familial hypertrophic cardiomyopathy. J Mol Cell Cardiol 1994; 26:3-10.

57. Thierfelder L, Watkins H, MacRae C et al. Alpha-tropomyosin and cardiac troponin T mutations cause familial hypertrophic cardiomyopathy: a disease of the sarcomere. Cell 1994; 77:701-12.

58. Kainulainen K, Peltonen L. Marfan syndrome: molecular pathogenesis. Adv Genome Biol 1993; 2:113-33.

59. McKusick VA. The defect in Marfan syndrome. Nature 1991; 352:279-81.

60. Francke U, Furthmayr H. Genes and gene products involved in Marfan Syndrome. Semin Thorac Cardiovasc Surgery 1993; 5:3-10.

61. Pulkkinen L, Kainulainen K, Krusius T et al. Deficient expression of the gene coding for decorin in a lethal form of Marfan syndrome. J Biol Chem 1990; 265:17780-5.

62. Kainulainen K, Pulkkinen L, Savolainen A et al. Location on chromosome 15 of the gene defect causing Marfan syndrome. New Eng J Med 1990; 323:935-9.

63. Tsipouras P, Sarfarazi M, Devi A et al. Marfan syndrome is closely linked to a marker on chromosome 15q1.5→q2.1. Proc Natl Acad Sci USA 1991; 88:4486-8.

64. Sakai LY, Keene DR, Engvall E. Fibrillin, a new 350-kDa glycoprotein, is a component of extracellular microfibrils. J Cell Biol 1986; 103: 2499-509.

65. Hollister DW, Godfrey M, Sakai LY et al. Immunohistologic abnormalities of the microfibrillar-fiber system in the Marfan syndrome. New Eng J Med 1990; 323:152-9.

66. Sakai LY, Keene DR, Glanville RW et al. Purification and partial characterization of fibrillin, a cysteine-rich structural component of connective tissue microfibrils. J Biol Chem 1991; 266:14763-70.

67. Kainulainen K, Karttunen L, Puhakka L et al. Mutations in the fibrillin gene responsible for dominant *ectopia lentis* and neonatal Marfan syndrome. Nature Genet 1994; 6:64-9.

68. Corson GM, Chalberg SC, Dietz HC. Fibrillin binds calcium and is coded by cDNAs that reveal a multidomain structure and alternatively spliced exons at the 5' end. Genomics 1993; 17:476-84.

69. Dietz HC III, Pyeritz RE. Molecular genetic approaches to the study of human cardiovascular disease. Annu Rev Physiol 1994; 56:763-96.

70. Magenis RE, Maslen CL, Smith L et al. Localization of the fibrillin gene to chromosome 15, band 15q21.1. Genomics 1991; 11:346-51.

71. Dietz HC, McIntosh I, Sakai LY. Four novel FBN1 mutations: significance for mutant transcript level and EGF-like domain calcium binding in the pathogenesis of Marfan Syndrome. Genomics 1993; 17:468-75.

72. Kainulainen K, Sakai L, Child A et al. Two unique mutations in Marfan syndrome resulting in truncated fibrillin polypeptides. Proc Natl Acad Sci USA 1992; 89:5917-21.

73. Kainulainen K, Peltonen L. Marfan Syndrome: Molecular Pathogenesis. Advances in Genomic Biology 1993; 2:113-133.

74. Arnheim N, Erlich H. Polymerase chain reaction strategy. Annu Rev Biochem 1992; 61:131-56.

75. Scambler P. A genetic aetiology for DiGeorge syndrome, velo-cardio-facial syndrome and familial congenital heart defect.In:Yacoub M, Pepper J (eds): Annual of Cardiac Surgery, 6th edition. London:Current Science Limited, 1993:5-12.

76. Driscoll DA, Budarf ML, Emanuel B. A genetic etiology for DiGeorge syndrome: consistent deletions and microdeletions of 22q11. Am J Hum Genet 1992; 50:924-33.

77. Chisaka O, Capecchi MR. Regionally restricted developmental defects resulting from targeted disruption of the mouse homeobox gene *hox-1.5*. Nature 1991; 350:473-9.

78. Bass BL, Weintraub H. An unwinding activity that covalently modifies its double-stranded RNA substrate. Cell 1988; 55:1089-98.

79. Simons M, Edelman ER, DeKeyser J et al. Antisense c-myb oligonucleotides inhibit intimal arterial smooth muscle cell accumulation in vivo. Nature 1992; 359:67-70.

80. Gerard RD, Meidell RS. Adenovirus-mediated gene transfer. Trends Cardiovasc Med 1993; 3:171-7.

81. Barr E, Leiden JM. Somatic gene therapy for cardiovascular disease. Trends Cardiovasc Med 1994; 4:57-63.

82. Chang MW, Barr E, Seltzer J et al. Cytostatic gene therapy for vascular proliferative disorders with a constitutively active form of the retinoblastoma gene product. Science 1995; 267:518-22.

83. Zhu N, Liggitt D, Liu Y et al. Systemic gene expression after intravenous DNA delivery into adult mice. Science 1993; 261:209-11.

84. Yaffe D, Saxel O. Serial passaging and differentiation of myogenic cells isolated from dystrophic mouse muscle. Nature 1977; 270:725-7.

85. Koh GY, Klug MG, Soonpaa MH et al. Differentiation and long-term survival of C2C12 myoblast grafts in heart. J Clin Invest 1993; 92:1548-54.

86. Zibaitis A, Greentree D, Ma F et al. Myocardial regeneration with satellite cell implantation. Transp Proc 1994; 26:3294

87. Law PK, Goodwin TG, Fang Q et al. Feasibility, safety, and efficacy of myoblast transfer therapy on Duchenne muscular dystrophy boys. Cell Transplant 1992; 1:235-44.

88. Barr E, Leiden JM. Systemic delivery of recombinant proteins by genetically modified myoblasts. Science 1991; 254:1507-9.

89. Dhawan J, Pan LC, Pavlath GK et al. Systemic delivery of human growth hormone by injection of genetically engineered myoblasts. Science 1991; 254:1509-12.

90. Soonpaa MH, Koh GY, Klug MG et al. Formation of nascent intercalated disks between grafted fetal cardiomyocytes and host myocardium. Science 1994; 264:98-101.

91. Nowak R. New cell transplants may mend a broken heart. Science 1994; 264:31.

GLOSSARY

ALLELE. One of several alternative forms of a gene occupying a particular locus on a chromosome.

ALTERNATIVE SPLICING. Molecular 'editing' of RNA transcripts by alternate pathways such that different exon-derived sequences are joined together. This results in different mRNAs being produced from the same gene (see also splicing).

BAND-SHIFT ASSAY. Technique for identifying protein-DNA interactions. The assay detects the binding of protein to a radiolabeled DNA fragment as this retards the migration of the DNA in standard gel electrophoresis. The result is a band which appears shifted in position (retarded in migration). Also known as electrophoretic mobility shift assay (EMSA), gel retardation assay and gel shift assay. Supershift assay refers to detecting bound protein with specific antibodies as this results in an additional shift in the position of the band if the antibody binds to the initial complex.

BASE. The purine (adenine-A, and guanine-G) and pyrimidine (thymine-T, and cytosine-C) constituents of DNA (in RNA thymine is replaced by uracil-U). Bases are linked to ribose sugar and phosphate to form the corresponding nucleotides; adenosine, guanosine, thymidine and cytosine. AMP; adenosine monophosphate, ADP, adenosine diphosphate, ATP adenosine triphosphate etc.

BASE PAIR. Association of bases of one strand of a double helix with those on the other strand. Base pairing occurs following the strict rule that A pairs with T and G pairs with C. bp: abbreviation of base-pairs used to define distances in DNA (e.g., 330 bp means 330 base pairs of a DNA double helix) kb: kilobase pairs.

CAAT BOX. A conserved DNA sequence motif located in the promoter region of many genes.

CANDIDATE GENE. A gene proposed to be involved in genetic disease by virtue of its close genetic linkage (chromosomal location) to a mapped site of genetic mutation.

"CASTing" (Cyclic amplification and selection of targets). Polymerase chain reaction-based technique for the selection of high affinity binding sites for a transcription factor from a pool of degenerate double-stranded-sequences.

cDNA. Complementary DNA: A DNA molecule that is a copy of an RNA.

CHROMOSOME. Discrete physical unit of the genome consisting of a single DNA double helix and associated proteins.

***CIS*-ACTING ELEMENT.** Promoter sequence which acts on the activity of the adjacent gene.

CODON. Triplet of bases that specify the position and identity of corresponding amino acids in the protein product of a given gene.

CODING STRAND. DNA strand which has the same sequence as the RNA encoded by that gene.

CODON. Triplet of nucleotides that is 'read' as encoding a particular amino-acid during mRNA translation.

COSMIDS. Cloning vectors in which the phage lambda *cos* sites have been incorporated and which can, as a consequence, be packaged in phage particles.

DNaseI FOOTPRINTING. Technique for the identification of regions of DNA to which protein is bound. The method relies on the fact that bound protein protects that regions against digestion with endonuclease DNaseI.

EMSA See band-shift assay.

ENDONUCLEASE. Enzymes which cleave bonds within nucleic acid molecules (see also restriction enzyme).

ENHANCER. Short region of DNA sequence responsible for augmenting the level of transcription from a particular gene.

EXON. Segment of a gene that is represented in mature mRNA (c.f. intron).

EXONUCLEASE. Enzymes which catalyze the digestion of nucleic acids from one or both extremities (c.f. endonuclease).

EXPRESSION VECTOR. Cloning vector designed such that the cloned insert DNA will be expressed.

FRAMESHIFT. Mutations which disrupt the triplet reading frame (arise by insertion or deletion of numbers of nucleotides which are not multiples of 3).

GEL RETARDATION. See band-shift assay.

GENE. Segment of DNA transcribed into RNA. Genes generally consist of several exons (containing sequences present in the mature mRNA) separated by introns (intervening sequences which separate the exons).

GENE FAMILY. Group of genes with related sequences.

GENETIC CODE. The correspondence between nucleotide triplets and amino acids.

GENOTYPE. The genetic makeup of a cell or organism.

GENOME. The collective genetic constitution of an individual cell or organism.

HETEROZYGOTE. Individual with different alleles of a particular gene.

HOMOZYGOTE. Individual with identical alleles of a particular gene.

HYBRIDIZATION. Formation of a DNA:DNA, DNA:RNA or RNA:RNA molecule by base pairing (used in specific detection of DNA or RNA molecules by hybridization with a labeled, complementary, 'probe' molecule), see also *in situ* hybridization.

HOMEOBOX. Describes a conserved sequence element originally described in the *Drosophila* fruit fly homeotic genes. It codes for a DNA-binding domain (the homeodomain) present in many transcription factors.

IN SITU HYBRIDIZATION. Technique for the detection of nucleic acids within cells. Commonly used for detecting the distribution of specific RNA in sections of tissue or in whole embryos (whole-mount *in situ* hybridization).

INTERVENING SEQUENCE. Intron sequence.

INTRON. Segment of non-coding DNA which separates two exons of a gene and which is removed during RNA splicing.

LIBRARY. Set of cloned DNA fragments derived from genomic DNA (genomic library) or from mRNA (cDNA library).

LINKAGE. Tendency of two genes to cosegregate in a family pedigree; quantified in terms of recombination frequency.

LOCUS. Chromosomal position of a particular gene for a particular trait.

MESSENGER RNA (mRNA). The RNA which carries genetic information from the nucleus of a cell to the cytoplasm where it directs the synthesis of a polypeptide chain.

METHYLATION INTERFERENCE. Technique for the identification of regions of DNA to which protein is bound. The method relies on the fact that bound protein interferes with the chemical methylation of DNA residues.

NORTHERN BLOT HYBRIDIZATION. The technique for detecting specific RNA species: RNA is run on an agarose gel to separate molecules according to size, transferred to a membrane filter (by 'blotting') and RNA molecules detected by hybridization (see also Southern blot hybridization).

NUCLEOTIDE Combination of covalently linked base and ribose sugar (see also 'BASE'). Often abbreviated to 'nt' when describing linear size of single stranded RNA or DNA molecules.

OLIGONUCLEOTIDE. Short, single-stranded DNA molecule, typically 17-24 nucleotides long, which are synthesized chemically.

ONCOGENE. Genes with the ability to transform cells such that they proliferate in a manner similar to tumor cells. Oncogenes derived from transforming viruses are given the suffix v- (e.g., v-*myc*).

PCR (POLYMERASE CHAIN REACTION). The technique of amplifying the quantity of a specific DNA molecule by successive rounds of denaturation (to separate the two strands of the DNA double helix), annealing with oligonucleotide primers and synthesis with DNA polymerase.

PHAGE. Bacteriophage, a bacterial virus.

PHENOTYPE. Observed characteristics of a cell, organ or organism.

PLASMID. A self-replicating autonomous circular DNA molecule.

POSITIONAL CLONING. See 'Reverse Genetics'.

POLYMORPHISM. Refers to the occurrence of different alleles of a gene (producing different phenotypes or resulting in subtle changes in DNA sequence manifest, for example, as RFLPs).

PRIMARY TRANSCRIPT. The initial RNA molecule derived by transcription of a complete gene.

PROMOTER. The region of DNA located next to the gene which is responsible for directing its transcription.

PROTO-ONCOGENE. Cellular counterparts to the viral oncogenes (see oncogene). Proto-oncogenes are given the prefix c- to denote cellular gene (e.g. *c-myc*).

READING FRAME. One of the three possible ways in which DNA can be 'read' in terms of triplet codons.

RECOMBINATION. Exchange of DNA by crossing-over between homologous chromosomes.

REPORTER GENE. A gene whose product can be easily assayed, e.g. the LacZ (ß-galactosidase) gene and CAT (chloramphenicol acetyl transferase) gene.

RESTRICTION ENZYME (RESTRICTION ENDONUCLEASE). An enzyme which catalyses the cleavage of DNA at a specific recognition sequence (e.g. the restriction enzyme EcoR1 cuts DNA at the sequence GAATTC).

RESTRICTION FRAGMENT. Fragment of DNA produced by cleavage with a restriction enzyme.

RETROVIRUS. An RNA virus which propagates through its conversion to DNA by reverse transcription.

REVERSE GENETICS (positional cloning). The identification and cloning of (new) genes involved in genetic disease, using genetic information to identify the chromosomal region involved and subsequently identifying (and cloning) genes from that region.

REVERSE TRANSCRIPTION. The synthesis of DNA from an RNA molecule. Achieved using the viral enzyme reverse transcriptase.

RFLP. Restriction fragment length polymorphism resulting from polymorphism in DNA sequence at a restriction enzyme cleavage site.

SOUTHERN BLOT HYBRIDIZATION. Technique for detecting specific DNA molecules: DNA is run on an agarose gel to separate molecules according to size, transferred to a membrane filter (by 'blotting') and specific DNA molecules detected by hybridization. Named after the inventor Professor E.M. Southern.

SPLICING. The removal of intronic sequences from an RNA primary transcript and the joining of exonic sequences to produce a mRNA.

TATA BOX. A conserved DNA sequence motif located in the promoter region of many genes.

TRANSCRIPTION. The synthesis of RNA from a gene.

TRANSCRIPTION FACTORS. Protein molecules present in the cell nucleus which interact with DNA to regulate transcription.

TRANSLATION. The synthesis of a polypeptide from mRNA template.

VECTOR. A self-replicating DNA used to 'carry' inserted foreign DNA.

INDEX

Page numbers in italics denote figures (f) or tables (t).

MEDICAL INTELLIGENCE UNIT
AVAILABLE AND UPCOMING TITLES

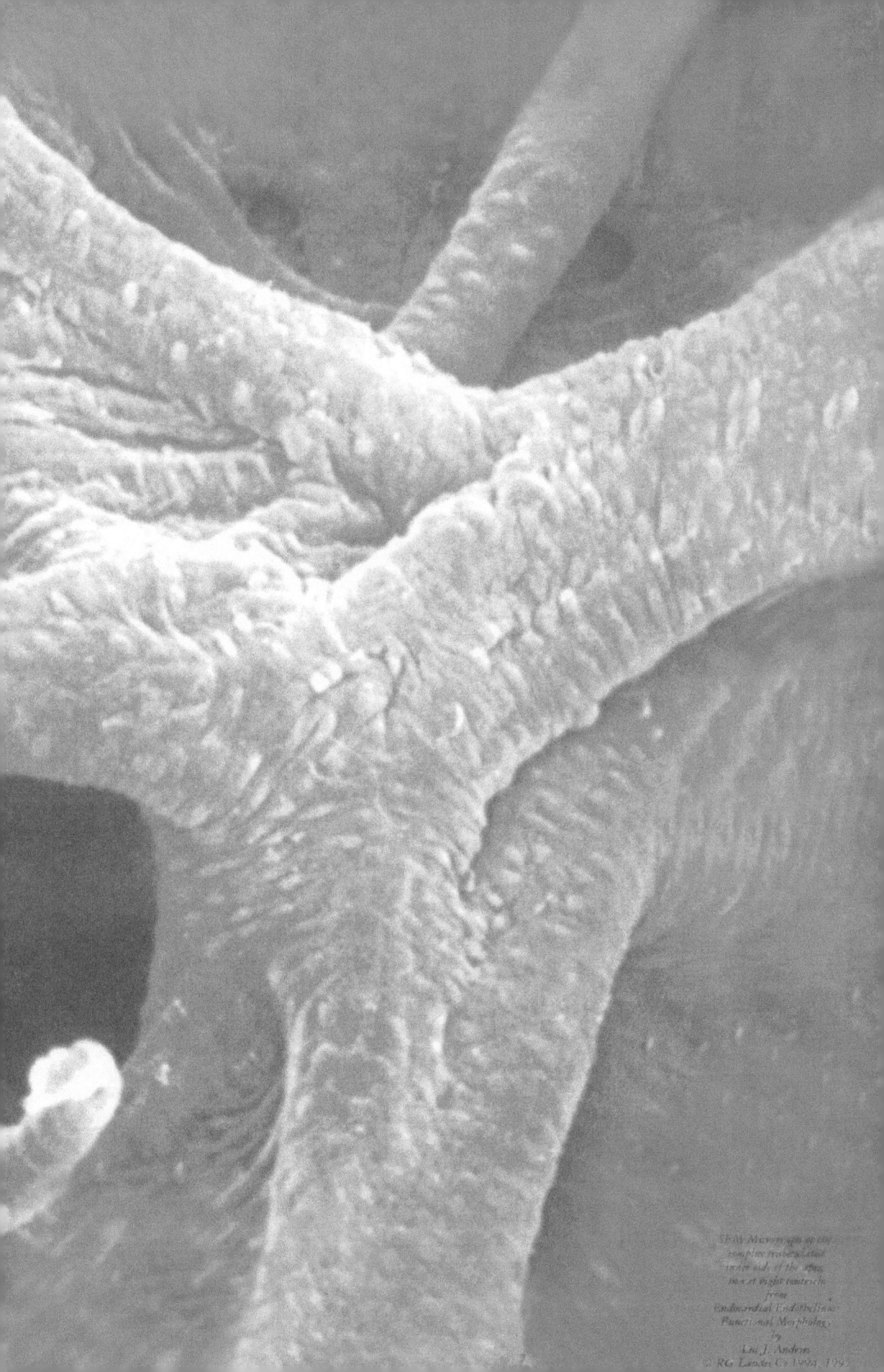

SEM Micrograph of the complex trabeculated inner side of the atria, left et right ventricle from *Endocardial Endothelium Functional Morphology* by Lei J. Andries © R.G. Landes Co 1994-199?